Mechatronics and Machine Vision 2002: Current Practice

ROBOTICS AND MECHATRONICS SERIES

Series Editor: **Professor J. Billingsley**
University of Southern Queensland, Australia

2. Visual Control of Robots: High performance visual servoing
   Peter I. Corke

3. Mechatronics and Machine Vision
   Edited by **John Billingsley**

4. Mechatronics and Machine Vision 2002: Current Practice
   Edited by **Robin Bradbeer** *and* **John Billingsley**

Mechatronics and Machine Vision 2002: Current Practice

Edited by

Robin Bradbeer

Associate Professor
City University of Hong Kong

and

John Billingsley

Professor of Engineering
University of Southern Queensland

RESEARCH STUDIES PRESS LTD.
Baldock, Hertfordshire, England

RESEARCH STUDIES PRESS LTD.
16 Coach House Cloisters, 10 Hitchin Street, Baldock, Hertfordshire, England, SG7 6AE
and
325 Chestnut Street, Philadelphia, PA 19106, USA

Copyright © 2002, by Research Studies Press Ltd.

All rights reserved.
No part of this book may be reproduced by any means, nor transmitted, nor translated into
a machine language without the written permission of the publisher.

Marketing:
Research Studies Press Ltd.
16 Coach House Cloisters, 10 Hitchin Street, Baldock, Hertfordshire, England, SG7 6AE

Distribution:
NORTH AMERICA
Taylor & Francis Inc.
International Thompson Publishing, Discovery Distribution Center, Receiving Dept.,
2360 Progress Drive Hebron, KY. 41048

ASIA-PACIFIC
Taylor & Francis Asia Pacitic
240 Macpherson Road #01-01 Pines Industrial Building, Singapore 348574

AUSTRALIA & NEW ZEALAND
DA Information Services PTY Ltd., 648 Whitehorse Road, Mitcham, Victoria, 3132 Australia

UK & EUROPE
ATP Ltd.
27/29 Knowl Piece, Wilbury Way, Hitchin, Hertfordshire, England, SG4 0SX

Library of Congress Cataloguing-in-Publication Data

Mechatronics and machine vision 2002 : current practice / edited by
Robin Bradbeer and John Billingsley.
    p. cm. -- (Robotics and mechatronics series ; 4)
Includes bibliographical references and index.
 ISBN 0-86380-278-8
1. Mechatronics--Congresses. 2. Computer vision--Congresses. I.
Bradbeer, Robin. II. Billingsley, J. (John) III. Series.
 TJ163.12 M4325 2002
 621--dc21
                        2002012094

British Library Cataloguing in Publication Data
A catalogue record for this book is available from the British Library.

ISBN 0 86380 278 8

Printed in Great Britain by SRP Ltd., Exeter

Acknowledgement

The Editors would like to thank the team of reviewers for their sterling efforts in refereeing the papers in this book. All are internationally respected experts in their field and their efforts allowed us to maintain the very high standard of academic excellence that we needed.

Their task was an arduous one; first they had to select a number of papers from the list of those submitted. Then they reviewed the extended abstracts of their choice, making critical and positive comments on how the full papers could be improved. Finally they reviewed the full papers, again making constructive recommendations for the authors. Of course, during this process a large number of papers were rejected - around 50% of the original submissions. The results of their efforts are in this book.

They are:

Dr Manual Armada, Spain
Prof. Yvan Baudoin, Belgium
Prof. Karsten Berns, Germany
Prof. Peter Brett, UK
Dr Sheng Chen, UK
Prof. Volker Graefe, Germany
Dr Rolf Johansson, Sweden
Dr Sherman Lang, Canada
Prof. C S George Lee, USA
Prof. T H Lee, Singapore
Dr Robin Li, Hong Kong
Prof. Shui-Shong Lu, Taiwan
Dr Bing Luk, Hong Kong
Prof. James Mills, Canada
Prof. Giovanni Muscato, Italy
Dr J Alison Noble, UK
Dr Peter Orban, Canada
Prof. D T Pham, UK
Prof. Ewald von Puttkamer, Germany
Dr Joerg Raczkowsky, Germany
Prof. Imre Rudas, Hungary
Prof. Lakmal Seneviratne, UK
Prof. Gerhard Schweitzer, Switzerland

Dr Dong Sun, Hong Kong
Dr Peter Tsang, Hong Kong
Prof. S K Tso, Hong Kong
Dr Ljobo Vlacic, Australia
Prof. F Wahl, Germany
Dr Juergen Wahrburg, Germany
Prof. Michael Wang, Hong Kong
Prof. Richard Weston, UK
Dr Peter Xu, New Zealand
Prof. Jianwei Zhang, Germany

Contents List

INTRODUCTION xiii

Vision: 1

1. Self-Calibration of an Active Vision System
for 3D Robot Vision 3
Li, Lu and Chen

2. Silicon Retina Sensing Guided by Omni-directional Vision 13
Bečanovi ć, Indiveri, Kobialka, Plöger and Stocker

3. Vision Guidance for a Climbing Cleaning Robot 23
Zhu, Sun, Tso and Mills

4. Measuring Flank Tool Wear on Cutting Tools
with Machine Vision - A Case Solution 33
Pfeifer, Sack, Orth, Stemmer and Roloff

5. A Floating Point Genetic Algorithm for Affine Invariant Matching
of Object Shapes 41
Tsang and Tsang

6. Programmable Focussing Mechanism for the Vision System
of a Gold Wire Bonding Machine 55
Gainekar, Widdowson, Gaungneng, Min and Hui

7. Using a Panoramic Camera for 3D Head Tracking
in an AR Environment 63
Giesler, Salb and Dillmann

Industrial Applications: 75

1. Fabric Defect Classification using Wavelet Frames
and Minimum Classification Error-based Neural Network 77
Pang, Yang and Yung

2. Multi-sensor Fusion in a Flexible Workcell Environment 87
Garg and Kumar

3. Active Control of Internal Turning Operations
Using a Boring Bar 97
Pettersson, Håkansson, Claesson and Olsson

4. Automatic Foundry Brake Disk Inspection
through Different Computer Vision Techniques,
Using a New 3d Calibration Approach 105
Lerones, Fernández, García-Bermejo and Casanova

5. Force-guided Compliant Motion in Robotic Assembly:
Notch-locked Assembly Task 115
Chin, Ratnam and Mandava

6. A PVDF-based Micro-Newton Force Sensing System
for Automated Micro-Manipulation 125
Fung, Li and Xi

Design of Manipulators and MEMS Assembly: 133

1. Resolving the Tasks of the Dynamics for the Control
of a Single-Planimetric Multimobile Manipulator 135
Sholanov

2. Design and Control of a Parallel Robot Based
on the Design for Control Approach 145
Li and Wu

3. ROCON – A Virtual Construction Kit, Visualization Tool
and Remote Control System for Mechatronic Devices 153
Kaiser and Fries

4. Development of a Novel Multi-module Manipulator System:
Dynamic Model and Prototype Design 161
De Silva, Wong and Modl

5. On-line Evolution of a Robot Arm Control Program
Using a Memoized Function 169
Suwannik and Chongstitvatana

6. Design of a Gravity Compensation System
for Flexible Structure-mounted Manipulators 177
Wongratanaphisan, Chew and Fongsamootr

7. Relay-based Friction Modeling Technique
for Servo-mechanical Systems 187
Tan and Jiang

8 Automated Micro-assembly of MEMS by Centrifugal Force 197
Lai and Li

9. An Efficient Distributive Tactile Sensor
for Recognising Contacting Objects 205
Tongpadungrod and Brett

Robots and Machine Vision in Agriculture and Food Processing: **211**

1. A Prototype Mechatronic System for Inspection of Date Fruits 213
Al-Janobi

2. Visual Counting of Macadamia Nuts 221
Billingsley

3. Machine Vision Application to Grading of White Pepper Berries 229
Ratnam, Khor and Lim

4. Autonomous Agricultural Robot 237
Phythian

5. Fuzzy Multivariable Control of Meat Chiller 245
Xu and Cowie

6. Control of the Sugar Cane Harvester Topper 253
McCarthy, Billingsley and Harris

Mobile Robots and Navigation: **261**

1. A new Beacon Navigation System 263
Casanova, Quijada, García-Bermejo and González

2. Space and Time Sensor Fusion and Multi-sensor Integration for Indoor Mobile Robot Navigation 271
Jin, Ko and Lee

3. Multi-purpose Autonomous Robust Carrier for Hospitals (MARCH) 287
Sooraksa, Luk, Tso and Chen

4. Communication with an Underwater ROV Using Ultrasonic Transmission 295
Law, Bradbeer, Yeung, Bin and Zhongguo

5. Reactive Agent Architecture for Underwater Robotics Vehicles 305
Ho, Seet, Lau and Low

Robot-human Interaction: **313**

1. Development of a Chinese Calligraphy Robot 315
Yao, Shao, Takaue and Tamaki

2. Automated People Counting Using Model Template Matching and Head Search 323
Pang and Ng

3. Automatic Facial Expression Recognition for Human-Robot Interaction 333
Kouzani

4. Development of an Expressive Social Robot 341
Quijada, Casanova, García-Bermejo and González

5. Gesture Recognition for Commanding Robots
with the Aid of Mechatronic Data-glove and Hidden Markov Model   349
Liu, Tso and Luk

6. A Novel Mechatronic System for Non-Invasive Lumpectomy
of the Breast Tissue   359
Chauhan

AUTHOR INDEX   **367**

SUBJECT INDEX   **371**

Introduction

The world of mechatronics has evolved rapidly in the two years since the last volume in this series. The introduction of more powerful processors and single board PCs, allowing real-time evaluation and processing of sensor data, for example, has meant that even more data can be processed. This is especially true in the field of machine vision, where it is now possible to manipulate data from high resolution raw video images instead of having to sample via an image grabber, as previously. At the same time, the revolution in micro- and nano- engineering means that mechatronics now encompasses devices that can only be seen through a microscope.

The concurrent evolution of the web has also begun to make its mark on mechatronics and machine vision. Low-cost cameras, originally designed for web communication, are now finding their way into ever more interesting applications, and the adoption of internet protocols and wireless LAN technology is helping to lower the cost of developing mobile robots, as well as remote sensing and control.

These rapid developments are finding their way into manufacturing; one section of this book considers the recent development of integrating intelligent sensors and actuators into the production process itself. Another section looks at similar developments in manipulator design.

Probably the most exciting area of this evolution is the emergence of mechatronic and vision technologies into agriculture and food processing. Although these are early days, the coming few years will see an explosion of applications 'down on the farm'. If just some of these developments meet their promised potential then 21st century agriculture will look nothing like that of the past century.

All of these changes, and more, are reflected in the papers collected in this book; the editors are convinced that the words 'Current Practice' in the title are an accurate description of the contents. The book is divided into six sections, roughly reflecting the developments mentioned above. Inevitably, it has been impossible to bundle all the papers into neat groupings, as subject matter in this area is very broad. Consequently, the section headings are for guidance and convenience only. We hope that you find the papers stimulating, as well as interesting

Robin Bradbeer
John Billingsley

Machine Vision

The applications of machine vision feature prominently throughout this book. Several papers are grouped together in this section for no better reason than that they seem to fit better here than elsewhere.

In general, 3D vision systems either use multiple cameras to achieve binocular sensing or else use the simpler method of structured light. The first paper addresses the need to calibrate a structured light system in terms of both the camera and pattern-projector properties.

Industrial applications provide challenges for machine vision, but they are easy in comparison to some robot games! The second paper describes a vision system destined to guide a goalkeeper in a robot soccer game. The sensor includes the sophistication of on-chip processing to enable optical flow to be analysed and output in the form of two analogue signals. With gyroscopes and CAN-bus connection, this football player threatens to command a high transfer fee!

The third paper is strongly application based. The application is window cleaning and the task of the vision system is to determine what needs cleaning next. Two laser diodes and a CCD camera are involved and once again calibration forms an important of the work.

The next vision system is dedicated to measuring tool wear from observations of the tool flank. Feature extraction is followed by a neural network, trained by back-propagation.

The catchy title of the fifth paper conceals the difficult task of visual identification. An object has three degrees of freedom of rotation and three more of translation, so that the variety of images to be matched presents a very serious problem indeed.

The sixth paper starts from the need to guide the bonding tool that 'stitches' connections together in the fabrication of an integrated circuit. Modern devices may have dies stacked one upon the other, requiring bonding to be performed at different heights. In turn, this requires a focusing system that can change rapidly as the bonds are made. A target repeatability of one micron is quoted for the lens movement.

Finally the worlds of machine vision and artificial reality are combined. To make the projected image consistent with the direction of the gaze of the viewer, it is necessary to measure that direction rapidly and accurately. The work of this paper describes the simple expedient of mounting a video camera on the headset, so that its orientation can be deduced from the perception of target marks in the field of view. Indeed the method goes further and allows the viewer to move through an environment of artificial landmarks.

Self-Calibration of an Active Vision System for 3D Robot Vision

Y. F. Li, R. S. Lu and S. Y. Chen
Department of Manufacturing Engineering and Engineering Management, City University of Hong Kong, Kowloon, Hong Kong

Abstract
In this paper, we describe a calibration method for a 3D vision system using pattern projection. This calibration consists of two phases: off-line calibration of the parameters of the pattern projector by means of the point-to-point method, and on-line calibration of the varying intrinsic and extrinsic parameters of the camera using the line-to-point method or plane-to-point method. During the on-line recalibration, we only need to calibrate those of the two or more arbitrary light planes. The other light stripe planes' homographies, relative to the camera image plane, can then be recovered. By the method we can easily implement recalibration of the 3D vision system with its pattern projection when the intrinsic and extrinsic parameters of the camera are changed.

Keywords: 3D vision, self-calibration, pattern projection, homography.

1. INTRODUCTION

Traditional calibration of an active vision system involves two separate stages: camera calibration and projector calibration. This is normally carried out off-line by using a calibration pattern whose 3D point coordinates of the features must be precisely known [1]. This calibration method is called the *point-to-point* method. This method suffers from the correspondence problem and is suitable only for applications where the system parameters are kept unchanged during the entire measurement process. In some applications, the size or distance of the scene changes sharply, therefore the parameters of the projector and cameras need to be changed as well. In this process, the camera or projector must be recalibrated on-line. In such cases, on-line calibration using the *point-to-point* method would become unfeasible.

The *line-to-point* method was introduced to calibrate a laser-based range finder mounted on the wrist of a robot arm [2]. This method requires at least 6 known world lines to solve the image-to-world transformation matrix. Later, this was extended to *plane-to-point* calibration for a laser range finder [3]. Huynh *et al.* [4] developed an approach based on cross ratio invariant and *point-to-point* method, using four known non-coplanar sets of 3 collinear world points to produce four points on the light stripe plane to be calibrated and their correspondences on the image plane.

The *line-to-point* and *plane-to-point* methods only need the known equations of lines or planes of objects. They are thus suitable for on-line calibration of a robot vision system. However, how to use them to calibrate a vision system with multiple light planes [5] remains to be explored. In this paper, we will introduce our work by taking the advantages of these methods and develop an on-line calibration method for a pattern projection 3D vision system. Our method will implicitly model the 3D vision system based on the homography principle rather than the explicit parameters model of a camera. We will calibrate the 3D vision system in two steps: off-line projector calibration with *point-to-point* method and on-line camera recalibration with *line-to-point* or *plane-to-point* method. During the on-line recalibration, we do not need to directly calibrate the homographic transformations of all the light stripe planes relative to the camera. Rather, we only need to calibrate those of two or more arbitrary light planes. Such a calibration method is suitable for recalibrating a 3D robot vision system when the intrinsic and extrinsic parameters of the camera are changed.

2. THE PATTERN PROJECTION SYSTEM

Fig. 1 illustrates the structure of our vision system. The projector illuminates the scene with a pattern of n light plane stripes. Each of the stripes intersecting with the scene produces a deformation pattern in 3D space F_w.

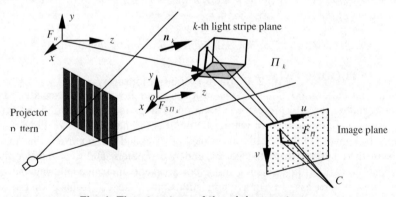

Fig. 1 The structure of the vision system

We model the system via the homographic transformation of each light stripe plane relative to the camera. As for the camera, we will model it as projective one using a pinhole model. As shown in Fig. 1, the optical center of the camera is used as the projective center C. As illustrated in Fig. 2, we define point C as the projective center of the camera with respect to Fig. 1, plane Π as the camera image plane and plane Π_k as the k-th light stripe plane of the projector. Given four or more distinct non-collinear points $\tilde{\mathbf{m}}_i$, where $i = 1, 2, \cdots n$ for $n > 4$, on a plane Π and another four or more distinct non-collinear points $\tilde{\mathbf{m}}_{ki}$ on the light stripe plane Π_k, there is one and only one perspectivity under a projective center C which projects points from the image plane to the corresponding points on the light stripe plane [6]. There exists a unique homography $\mathbf{H}_k$, defined up to a scale, such that

$$\mathbf{H}_k \tilde{\mathbf{m}}_i = \lambda_i \tilde{\mathbf{m}}_{ki} \tag{1}$$

where λ_i are scales, $H_k = \begin{bmatrix} h_{11} & h_{12} & h_{13} \\ h_{21} & h_{22} & h_{23} \\ h_{31} & h_{32} & h_{33} \end{bmatrix}$ is a 3×3 matrix

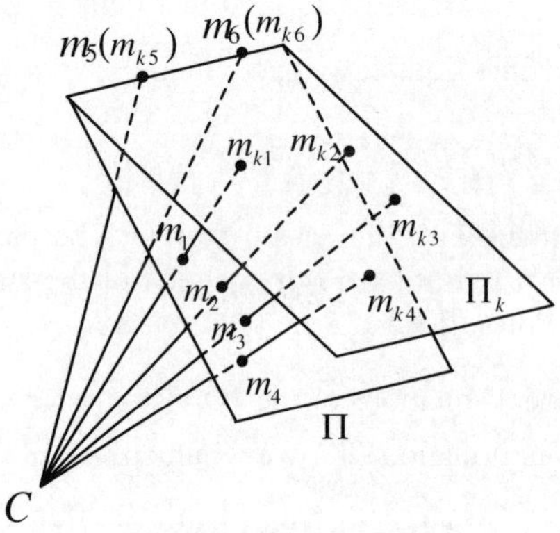

Fig. 2 Plane-to-plane homography

Thus, in order to obtain the coordinates of the points in the *k-th* light stripe plane Π_k with respect to the world coordinate system, we need to know the transformation between them. We define the 3D world coordinate frame as F_w. The *k-th* stripe light plane Π_k has a coordinate frame $F_{3\Pi_k}$ which is obtained by translating the original world coordinate system to the known point *o* on the stripe light plane and then rotating it around a vector perpendicular to its *z*-axis and the normal vector of the light stripe plane until its *xy*-plane aligns with the light stripe plane. Assuming that the original point coordinates of the frame $F_{3\Pi_k}$ is $\overline{\mathbf{M}}_k = (\overline{M}_{k1}, \overline{M}_{k2}, \overline{M}_{k3})^{\mathrm{T}}$, the transformation between the frames F_w and $F_{3\Pi_k}$ can be represented by

$$T_k = \begin{bmatrix} R & -R\overline{M}_k \\ \mathbf{0}^T & 1 \end{bmatrix} \tag{2}$$

If an arbitrary point on the stripe light plane is defined as $\mathbf{M}_{ki} = (M_{k1}, M_{k2}, M_{k3})^T$ and its homogeneous coordinate is $\tilde{\mathbf{M}}_{ki} = (M_{k1}, M_{k2}, M_{k3}, 1)^T$, then the matrix $\mathbf{T}_k$ would transform the world point $\tilde{\mathbf{M}}_{ki}$ in frame F_w into the form in frame $F_{3\Pi_k}$

$$\mathbf{T}_k \tilde{\mathbf{M}}_{ki} = (m_{k1}, m_{k2}, 0, 1)^T \tag{3}$$

Thus the coordinate relation between the world coordinate system F_w and the coordinate system of light stripe plane is as follows

$$\mathbf{T}_k \tilde{\mathbf{M}}_{ki} = \mathbf{S}^T \tilde{\mathbf{m}}_{ki} \tag{4}$$

where $\tilde{\mathbf{m}}_{ki} = (m_{k1}, m_{k2}, 1)$ is the corresponding homogenous coordinates of $\mathbf{M}_k$ in two dimensional frame $F_{2\Pi_k}$.

Substituting Eqn. 1 into Eqn. 4, we can obtain

$$\tilde{\mathbf{M}}_{ki} = \rho_i \mathbf{T}_k^{-1} \mathbf{S}^T \mathbf{H}_k \tilde{\mathbf{m}}_i \tag{5}$$

The formula can be expanded as

$$M_{ki} = \overline{M}_{ki} + \frac{R_k^T Q H_k \tilde{m}_i}{e^T H_k \tilde{m}_i} \tag{6}$$

Eqn. 5 and 6 are the model equations of the 3D vision system.

To obtain the transformation $\mathbf{T}_k$, we assume that the normal vector of the k-th light stripe plane Π_k is $\mathbf{n}_k = (n_{k1}, n_{k2}, n_{k3})^T$ in frame F_w. The angle θ between $\mathbf{n}_k$ and the z-axis of the frame F_w, $\mathbf{z} = (0, 0, 1)^T$, is the inner product of the two vectors:

$$\theta = cos^{-1}(\mathbf{n}^T \mathbf{z}) = cos^{-1}(n_{k3}) \tag{7}$$

The unit vector, $\mathbf{r} = (r_1, r_2, r_3)$, perpendicular to both the z-axis of the frame F_w and the normal vector $\mathbf{n}_k$ is

$$\mathbf{r} = (\mathbf{n}_k \times \mathbf{z}) / \|\mathbf{n}_k \times \mathbf{z}\| \tag{8}$$

Then the rotation matrix can be given by *Rodrigues formula*

$$\mathbf{R} = \cos\theta \cdot \mathbf{I} + (1 - \cos\theta)\mathbf{rr}^T + \sin\theta \cdot \mathbf{r}_\wedge \tag{9}$$

where $r_\wedge$ is the skew-symmetric matrix. If $\mathbf{n}_k$ and $\mathbf{r}$ are parallel, then $\theta = 0$, and $\mathbf{R} = \mathbf{I}$.

3. OFF-LINE CALIBRATION OF THE PROJECTOR

Define
$$T_{wk} = T_k^{-1} S^T H_k \qquad (10)$$

as the k-th image to world (2D-3D) transformation between a 3D point on k-th stripe light plane relative to the frame F_w and its correspondence on the image plane of CCD camera. Here T_{wk} is a 4×3 matrix with 12 unknown parameters, eleven of which are independent. For each stripe light plane and its T_{wk}, a point on the plane and its correspondence on the image plane will give three independent equations. To obtain the transformation T_{wk}, at least four points 3D coordinates on the plane and their correspondences on the image plane are needed.

Here, we use a pattern plane with multiple circles on it as the calibration target to produce many known points in the world coordinate system with a precision single axis mechanical translation stage. When the calibration pattern plane is moved to a different position, the light stripe plane will intersect with the lines, L_i, on the pattern plane, and produce points M_i. The correspondence of the point m_i, will be on the image plane. As mentioned above, any four or more points in the general positions on the light stripe plane and their corresponding image projections define a unique homography H_k. To obtain H_k, the calibration pattern plane will be laid in at least two positions.

To implement this calibration, we to compute the 3D coordinates of the points, M_i ($i \geq 4$), on the k-th stripe light plane relative to the world coordinate frame F_w. Next, we need to compute the unit normal vector $\mathbf{n}_k$ and correct the point M_i coordinates. Finally, we can establish the light stripe plane coordinate system $F_{2\Pi_k}$ and $F_{3\Pi_k}$, and compute the world to light stripe plane transformation T_k. Repeating the above procedures, other light stripe planes' transformations can be obtained.

4. ON-LINE CALIBRATION OF THE CAMERA

In practical applications, the camera parameters including the relative orientation and position between the projector and camera, and the camera's focus, zoom and aperture, may need to be changed to achieve the desired imaging effect. In such a case, the changed parameters need to be recalibrated on-line. Here we assume that the intrinsic parameters of the projector, and its position and orientation with respect to the world coordinate frame, will remain constant, or they can be obtained by other means of sensing which is the case for robotic applications. Thus, we know the normal vector $\mathbf{n}_k$ of the stripe light plane, the origin $\overline{M}_k$ of the frames $F_{2\Pi_k}$ and $F_{3\Pi_k}$, and the world to light stripe plane transformation T_k. Only the parameters of the camera will change. When the normal vector $\mathbf{n}_k$, the origin point $\overline{\mathbf{M}}_k$ and the transformation T_k are calibrated off-line and kept constant, we only need to recalibrate the homographic relations of any two or more stripe light planes relative to the camera image plane. All of the other light stripe planes relative to the camera image plane can be obtained by means of the following relations.

In projective space Ω, if two projective planes Π_1 and Π_2 have homographies $\mathbf{H}_1$ and $\mathbf{H}_2$ with respect to a common projective plane Π under a projective center C, then the two projective homographies will follow

$$\mathbf{H}_2 \mathbf{H}_1^{-1} \mathbf{S} \mathbf{T}_1 \mathbf{P}_i = \lambda \mathbf{S} \mathbf{T}_2 \mathbf{P}_i \tag{11}$$

$$\mathbf{H}_2 \mathbf{H}_1^{-1} \mathbf{Q} \mathbf{R}_1 \mathbf{n}_1 \times \mathbf{n}_2 = \lambda \mathbf{Q} \mathbf{R}_2 \mathbf{n}_1 \times \mathbf{n}_2 \tag{12}$$

where P_i is an arbitrary point on the cross-line between the light planes *1* and *2*, λ is a scale factor.

It can be seen that the scale factor λ_i is unique. That is to say, for all points on the common line of the two planes, the factors λ_i are equal. If we know the homography H_1, $\mathbf{T}_1$, $\mathbf{T}_2$ and point $\mathbf{P}_i$, Eqn. 11 will have nine independent unknowns including the factor λ. If we select two points in an intersection line of the two stripes planes, Eqn. 11 will provide five independent equations for each point. During the on-line calibration, only two arbitrary light stripes planes' homographies relative to the camera need to be recalibrated, thus the other homographies can be derived. The on-line calibration procedure consists of the following major steps:

Step 1. Based on the *line-to-point* method or *plane-to-point* method, on-line calibrate the homographies of any two or more light stripe planes relative to the camera image plane.

Step 2. Using Eqn. 11 or 12, compute all of the other light stripe planes' homographies with respect to the camera image plane.

5. IMPLEMENTATION

The above calibration method was implemented on our 3D vision system consisting of a camera and a line projector which can produce a time-space encoded pattern. This projector can project up to 320 light stripe planes. Using the method described above, we can compute the light stripe planes' normal vectors $\mathbf{n}_k$, the origin points $\overline{\mathbf{M}}_k$ of frames $F_{3\Pi_k}$, the world to light stripe plane transformations $\mathbf{T}_k$, and the homography $\mathbf{H}_k$.

We selected four light stripe planes for the test. If defining the projective center as $\mathbf{C}$, the plane $\mathbf{n}_1$ as the camera image plane, we can compute the homographies of the other three stripes planes with respect to plane $\mathbf{n}_1$. Assuming known homographies $\mathbf{H}_{12}$, $\mathbf{H}_{13}$, then, based on the on-line calibration method, we can evaluate the homography $\mathbf{H}_{14}$. The evaluation results $\mathbf{H}_{14}^{'}$ and the evaluation error are shown in Table 1, at the end of this paper.

In practical applications, the off-line calibration results of the projector are subject to errors. If we only use two arbitrary planes' homographies obtained on-line and the information obtained in the off-line calibration, the achievable accuracy in the resulting homographies will be limited. To illustrate this, we conducted further experiments using five arbitrarily selected stripe planes. Here $\mathbf{T}_k$ and $\mathbf{H}_k$ were obtained in the off-line calibration. Then the entries of $\mathbf{T}_k$ and $\mathbf{H}_k$ were multiplied by the corresponding entries of the $\mathbf{T}$ and $\mathbf{H}$. With the known $\mathbf{T}_k$ and $\mathbf{H}_k$ (k=1,2,3,4), we evaluated the homography $\mathbf{H}_5$, by using the homographies $\mathbf{H}_1, \mathbf{H}_2$ and $\mathbf{H}_1, \mathbf{H}_2, \mathbf{H}_3$ and $\mathbf{H}_1, \mathbf{H}_2, \mathbf{H}_3, \mathbf{H}_4$ to compute homography $\mathbf{H}_5$ and obtained the errors in $\left\|\mathbf{H}_5^{'}\right\|^2 - \left\|\mathbf{H}_5\right\|^2$ as 52.768, -25.361, and – 9.139 respectively. This shows that more than two known stripe planes' information is used to evaluate the other stripe planes' homographies, more accurate results can be obtainable.

6. CONCLUSIONS

In this paper, we presented the calibration method for a 3D vision system using pattern projection. The calibration is divided into an off-line calibration and an on-line calibration. In the off-line calibration stage, we only calibrate the pattern projector. This off-line calibration is only needed if the projector parameters are changed or perturbed before the imaging task begins. The on-line calibration is conducted during the operation and when the camera's intrinsic or extrinsic parameters are changed. In the on-line calibration, we do not need to recalibrate all the light stripe planes' transformations relative to the camera. Instead, only those of any two or more light stripe planes relative to the camera image planes are recalibrated. The other light stripe planes' homographies, with respect to the camera image plane, can be obtained based on the relations derived in the paper between the homographies of the light stripe planes and their transformations with respect to the world coordinate frame. Using such a 3D vision system with the pattern projection for robotic applications, it is possible for a robot to perceive a 3D scene adaptively by controlling the configuration of the vision system on-line to suit varied environments.

Acknowledgment

The work presented in this paper is supported by a grant from the Research Grants Council of Hong Kong [Project No. CityU1136/98E].

References

[1] T. Y. Tsai, An efficient and accurate camera calibration technique for 3D machine vision, IEEE Proceedings of Conf. on Computer Vision and Pattern Recognition, pp. 364-374, 1986.

[2] C. H. Chen and A. C. Kak, Modeling and calibration of a structured light scanner for 3-D robot vision, Proc. of IEEE Conf. on Robotics and automation, pp. 807-815, 1987.

[3] Ian D. Reid, Projective calibration of a laser-stripe range finder, Image and Vision Computing, Vol. 14, pp. 659-666, 1996.

[4] D. Q. Huynh, R. A. Owens and P. E. Hartmann, Calibrating a structured light stripe system: a novel approach, International Journal of Computer Vision, Vol. 33, No. 1, pp. 73-86, 1999.

[5] T. Stahs and F. Wahl, Fast and Versatile range data acquisition in a Robot Work Cell, Proc. of IEEE Conf. on Intelligent Robots and Systems, Raleigh, NC, July 7-10, pp. 1169-1174, 1992.

[6] J. G. Semple and G. T. Kneebone, Algebraic projective geometry, Oxford University Press, New York, 1998.

Table1 Off-line calibration results

Projection center C	(-4.071, -22.262, 212.802)		
$\overline{M}_1$	$\overline{M}_2$	$\overline{M}_3$	$\overline{M}_4$
(0.467, -10.000, 30.000)	(29.029, -0.000, 30.000)	(57.760, -10.000, 30.000)	(86.809, -10.000, 30.000)
n_1	n_2	n_3	n_4
(2.162, 51.783, -205.011)	(2.162, 51.783, -205.011)	(2.162, 51.783, -205.011)	(2.162, 51.783, -205.011)
T_1	T_2	T_3	T_4
-0.123 0.011 -0.992 29.935 0.011 1.000 0.009 29.935 0.992 -0.009 -0.123 29.935 0 0 0 1	-0.081 0.008 -0.997 32.341 0.008 1.000 0.007 32.341 0.997 -0.007 -0.081 32.341 0 0 0 1	-0.038 0.005 -0.999 32.221 0.005 1.000 0.005 32.221 0.999 -0.005 -0.038 32.221 0 0 0 1	0.006 0.002 -1.000 29.491 0.002 1.000 0.002 29.491 1.000 -0.002 0.006 29.491 0 0 0 1
H_{12}	H_{13}	H_{14}	Evaluated homography H'_{14}
2.857 -0.030 270.300 0.030 2.472 15.936 -0.002 0.000 1	7.338 -0.111 917.895 0.117 6.011 56.741 -0.007 0.000 1	33.247 -0.627 4660.856 0.666 26.461 301.195 -0.038 0.002 1	33.247 -0.627 4660.856 0.666 26.461 301.195 -0.038 0.002 1
Evaluation error	$\left\|H'_{14}\right\|^2 - \left\|H_{14}\right\|^2 = -1.922 \times 10^{-7}$		

Silicon Retina Sensing Guided by Omni-directional Vision

Vlatko Bečanović, Giovanni Indiveri[*1], Hans-Ulrich Kobialka, Paul G. Plöger and Alan Stocker[*2]

Fraunhofer Institute for Autonomous Intelligent Systems, Schloss Birlinghoven, 53754 Sankt Augustin, Germany.

Abstract
In the RoboCup mid-sized league, autonomous robots perform at high speed and there is a general need to increase the update rate of the overall optical sensory system. A way of combining a relatively new sensor-technology, that is optical analog VLSI devices, with a standard digital omni-directional vision system is investigated. This is done by the use of a neuromorphic analog VLSI sensor that estimates the global visual image motion. The sensor provides two analog output voltages that represent the components of the global optical flow vector. The readout is guided by an omni-directional mirror that maps the location of the ball and directs the robot to align its position so that a sensor-actuator module that includes the analog VLSI optical flow sensor can be activated. The purpose of the sensor-actuator module is to operate with a higher update rate than the standard vision system and thus increase the reactivity of the robot for very specific situations. This paper will demonstrate an application example where the robot is a goalkeeper with the task of defending the goal during a penalty kick.

Keywords: *neuromorphic, analog VLSI, optical flow, omni-directional vision, RoboCup.*

1. INTRODUCTION

In our lab we exploit analog VLSI (aVLSI) technology in fast-wheeled mobile robotics applications, such as the RoboCup domain, where soccer-playing robots perform at high speed, i.e. in the order of 1 m/s. Our robots use a differential drive for movement, a pneumatic kicker for shooting, two small movable helper arms to prevent the ball from rolling away and a camera-based vision system. The update rate of the overall optical sensory system is increased by the use of neuromorphic analog VLSI sensors that have a continuous time mode of operation. Aspects such

---

[*1] University of Lecce, Department of Innovation Engineering, Via per Monteroni, 73100 Lecce, Italy.
[*2] Institute of Neuroinformatics, University/ETH Zürich, Winterthurerstrasse 190, Y55 G86, 8057 Zürich, Switzerland.

as speed, low weight, low power consumption and small size contribute to a more streamlined design and increased robot performance.

The output from the optical analog VLSI sensors is a usually a low-dimensional analog signal. Sensors with a higher level of complexity typically operate in the kHz range and sensors with more elementary circuits in the MHz range, depending on their internal time constants. The output signal of the sensor gives no information of what hypothetical object is tracked in the scene. Different objects can trigger the output from multiple optical analog VLSI sensors by coinciding with their respective activity regions. In principle, reading out a few analog signals is equivalent to the calculation of certain motion properties, such as relative position or velocity for that particular object. By choosing different lenses, the field of view of a silicon retina can be modified. Multiple data streams can be collected from different sensors to achieve a more robust representation of the visual scene with sensor fusion. Active mechanics, e.g. movable mechanical parts on the robot, that need to be controlled by a very rapid "reflex like" mechanism are implemented as sensor-actuator modules, that is silicon retina sensors embedded with a micro controller module. This shortens the reaction time and reduces the over-all reaction time for the active mechanics.

Experiments performed with a moving ball using optical analog VLSI devices [1] show that those tasks are quite difficult to achieve due to the characteristics of the circuits, which usually have quite a coarse representation (number of pixels) and thus a narrow field of view. The winner-take-all (WTA) function that measures the optical activity of a winning characteristic e.g. contrast or visual motion on the contrary, simplifies the processing task. To investigate the possibility to fuse the information from multiple optical analog VLSI sensors by using an external arbiter, that in our case is a digital vision system equipped with an omni directional camera, could be a possible solution to robustify the sensory information. The omni-directional arbitration system divides the surrounding scene into speciality areas corresponding to the field of view of particular silicon retina sensors. The arbitration system does object tracking. The object tracking corresponds somewhat to performing selective attention and is performed in order to identify novel targets. The object tracking uses a colour threshold and region merging technique in order to determine the centroid of the particular colour marked objects used in the RoboCup environment. Object position prediction is achieved with a statistical approach where certainty of sensory cues is integrated over time. The predicted dynamics of the tracked objects is then, in turn, used to select the amount of autonomy for each particular sensor-actuator module, that is silicon retina devices, and combine the use of those with omni-directional vision.

In this paper we will focus on an example that uses a single optical analog VLSI sensor that calculates global optical flow [2] to predict the motion of a ball during a penalty kick situation. The RoboCup goalie robot of our team will be used to demonstrate the combined use of a digital omni-directional vision system with a silicon retina device.

In section 2 our robot platform and the statistical approach that we used for self-localization is explained. In section 3 the silicon retina device that will be used in our example is described. In section 4 the robot dynamics is described for performing the specific task of defending a penalty kick. In section 5 the work is summarized.

2. OUR ROBOT PLATFORM

The robot platform has actuators in the form of motors to drive and turn the robot and a valve to kick the ball pneumatically. Small robot arms attached to the left and right side of the robot keep the ball in front of the kicker plate. Besides the two optical sensors, camera and retina, it has four infrared distance sensors, a contact sensitive bumper strip with rubber shield, and odometry at the actuated robot wheels. This is augmented by a gyroscope for fast turning movements. All of these peripheral devices are controlled by three 16-bit micro controllers manufactured by Infineon. They are interconnected with a bus interface named CAN, which is a standard in German automobile industry. Motor drive (current) control is performed with one of those micro-controller modules [3]. A second controller supervises all of the analog signals in the soccer robot, e.g., distance reading of infrared sensors and also the A/D conversion of the silicon retina signals. The micro-controller modules communicate to a small notebook PC via CAN-bus. The operating system can be either Windows or LINUX. The cyclic update rate is either 20 [ms] for Windows with dedicated vision hardware, or 33 [ms] for LINUX without dedicated vision hardware. In the future we will move towards a completely LINUX-based platform.

2.1 Vision system

The vision system of our robot consists of a digital camera and a hyperbolical omni-directional mirror. All image analysis is done on-board the notebook PC that also is running the behaviour system. The software package used is a public licensed image-processing package available from the Carnegie Mellon University [4]. This package does colour blob analysis and works well with the colour coded RoboCup environment.

The omni-directional mirror projects the surrounding scene onto a spherical map on the focal plane. This vision system is used to estimate the relative distance and orientation of the robot with respect to the goals and the ball. The angle variables are immediately available from the omni-mirror images without the need any particular processing, while the distance variables only can be estimated after calibrating the system with a quite simple least squares (LS) technique. The mapping between pixel-distances on the image plane and physical distances on the field are reported in the below picture together with the LS estimated model (c.f. Fig. 1).

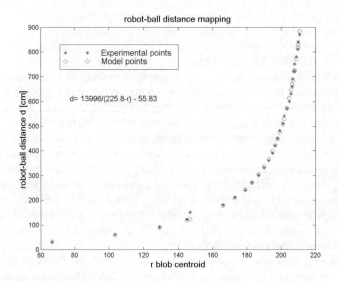

Fig. 1. Distance vs. blob centroid fit for the red ball, raw data indicated as '*' and model points as 'o'. A similar fit is made for the goal distance.

2.2 Ball prediction

Predicting the trajectory of a fast moving ball is crucial for a goalie. In our approach we use the vision system to do a rough approximation of the ball trajectory in order to move the robot in a position where the silicon retina can sense the ball. The sensory input of the silicon retina is then used for fast short-term reaction on movements of the ball.

2.2.1 Estimating the current ball position

The position of the ball is estimated by measuring the center of the biggest blob in the camera picture. These measurements are not very reliable because of varying light conditions on the field causing shadows and reflections on the ball. Especially the distance to the ball is noisy as the center of blob is raised/lowered in the camera picture due to these effects. Furthermore false positives (i.e. red blobs not belonging to the ball) sometime confuse the vision system; therefore the ball is estimated over time. This is done by computing a Gaussian probability density function for each measured ball prediction. The weight of older measurements is decreased by choosing a standard deviation that increases over time. The Gaussian probability density of the most recent ball position has the smallest standard deviation. By adding the probability densities of recently measured positions (and normalizing the result) one gets the density function for the most probable ball position. The probability of the ball being in a certain region of the field is computed integrating the density function over this region.

2.2.2 Predicting the ball using the probability density function

Using the probability density function, the probability of the ball being in a certain position is measured for a number of grid points located near the currently estimated ball position. The grid points, whose probability increase over time, denote the region where the ball can be expected in the next time steps. This statistical treatment is very robust even when facing various kinds of noise.

The performance of the algorithm can be tuned to real-time requirements by choosing an appropriate number of grid points for which these probabilities are computed. It is important to note that this analysis is performed for the standard vision system. For the silicon retina driven sensor-actuator module a smooth temporal filter is applied followed by a simple threshold.

3. IMPLEMENTATION OF THE SENSOR-ACTUATOR MODULE

3.1 The 2-D optical flow sensor

The applied neuromorphic analog VLSI circuit calculates smooth optical flow based on a model of visual motion that includes: the intensity constraint, the smoothness constraint and a bias for slow visual motions [2]. The applied prototype chip has a spatial resolution of 10x10 pixels and potentially provides two output signals at each pixel, representing the components of the local optical flow vector. However, external control voltages bias the sensor such that it provides a global estimate of visual motion. That is, the smoothness constraint is assumed to hold faithfully for the complete image space. Using the 2-D optical flow sensor in such global mode of operation has two advantages: firstly, the output signal is low-dimensional because it consists only of two time-continuous analog voltages. This makes expensive scanning operations through the image space unnecessary and reduces significantly the amount of sensory data. Secondly, the collective computation of global optical flow amongst all sensor pixels increases the robustness of the output signal.

Neuromorphic analog VLSI sensors are inherently prone to fix-pattern noise due to fabrication mismatch. Since the effect of mismatch results usually in a symmetrical error distribution, the collective operation approximately averages out the mismatch errors [6]. Although the output signal is continuous in time, the internal time-constant of the analog circuit limits the read-out rate for a correct estimate of global optical flow to about 1 kHz.

The analog output signals of the chip are linear with respect to the visual motion within a range of ±0.5 volt. This output range can be mapped to different ranges of visual motion according to an external control voltage. As shown in Figure 3a the possible detection ranges cover more than three orders of magnitude with a maximal speed detection of ~5000 pixels/sec. Figure 3b demonstrates the global optical flow estimate of the sensor as a function of the visual contrast of the object for a given visual motion. The output remains constant as long as the contrast of the object remains above 30 % [5].

Fig. 2. The smooth optical flow silicon retina sensor with three layers: lens, chip and supporting printed circuit board.

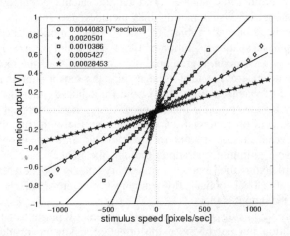

Fig. 3a. The linear output range of the 2-D optical flow sensor can be adjusted to map the range of expected visual motion.

3.2 The optimal mapping of the output range of the optical flow sensor

For an optimal selection of the linear range the maximum focal plane velocity needs to be measured (see also Fig. 3a.). For a RoboCup goalie scenario the maximum speed of the ball is hard to estimate. There are no rules constraining the design of the kicking device of a robot, thus the optimal solution is to choose the range that can detect a minimum velocity that is determined not to allow the standard digital system to react in time, that is, a ball that is too fast for the standard vision system is the minimum velocity in the scope of the silicon retina based sensor-actuator module.

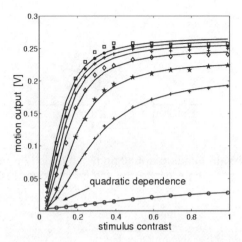

Fig. 3b. The dependence of the optical flow estimate on the visual contrast of the observed moving object. Above 30% contrast, the output is virtually independent of object contrast.

3.3 Integration of the optical flow sensor with the motor controller board

The dual output signal of the silicon retina device is sampled by an A/D-converter on the micro-controller board that controls the actuators of the robot wheels. The silicon retina signal is analysed on the micro-controller in order to determine if action needs to be taken.

3.4 The functionality of the sensor-actuator module

The sensor-actuator module is confined to the motor-controller board of the robot and activated or inactivated through a message over the CAN-bus. This message is sent by the top-level behaviour system. This approach scales nicely and several such modules can be integrated and controlled by the top-level behaviour system, although only one such module is described in this paper.

4. A ROBOCUP GOALIE EXAMPLE

4.1 Robot Model

Given the fast servomotor loop controlling the wheels speed, the guidance control law is designed on a kinematics level only. With reference to the Figure 4, the kinematics equations of the robot can be written as:

$$\dot{r} = u \cos\alpha$$
$$\dot{\alpha} = \omega - \frac{u}{r}\sin\alpha \qquad (4.1.1\text{-}3)$$
$$\dot{\theta} = \frac{u}{r}\sin\alpha$$

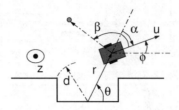

Fig. 4. The robot with heading α,ball angle β,velocity u relative to the goal at absolute reference angle θ and the minimum robot radius d seen from above.

and it can be shown [6] that the guidance law:

$$\omega = \frac{u}{r}\sin\alpha + \gamma\left(\pi/2 - \alpha\right) - hu(d-r)\frac{\cos\alpha}{\pi/2 - \alpha} \qquad (4.1.4\text{-}5)$$

$$u = u_{max}\cos\beta \quad : \quad \gamma, hu_{max} > 0 \text{ (constants)}$$

guarantees that the robot will be asymptotically driven on a circle of radius d aligned with the ball. The convergence is guaranteed as long as r>0 which, in practice, is always the case.

4.2 The sensor-actuator module

When the robot is in place, at the pre-calibrated distance to the ball, the sensor-actuator module is activated. This module will wait for the sensory signal to achieve the limit level determined to be a fast moving ball. Optimally the ball then has a velocity that is faster than the save capability of the goalie with the ordinary vision system.

5. SUMMARY AND CONCLUSIONS

Analog optical VLSI devices could provide an elegant solution for various problems in mobile robotics. The lightweight nature of the sensors, their low power consumption and their substantial on-chip calculation capabilities raises the opportunity to design smaller and faster mobile platforms with advanced scene analysis capabilities. Fast intelligent sensor devises, like silicon retina devices, are especially advantageous for reactive behaviour-based robotics [7], where sensors are influencing actuators in a direct way. Since this is our approach in RoboCup this is another reason to opt for silicon retinas. In general, it can be concluded that silicon retina type analog VLSI devices could be very attractive for solving control problems in mobile robotics. The experiments performed indicate that optical analog VLSI sensors could give a robust enough signal, and that the sensors could have applications to various situations, such as object motion prediction and active ball guidance.

Acknowledgment

This work is funded by the Deutsche Forschungsgemeinschaft (DFG) in the context of the research program SPP-1125 "RoboCup" under grant number CH 74/8-1. This support and cooperation is gratefully acknowledged.

References

1. Bečanović, V., Bredenfeld, A., Ploger, P. G.,"Reactive Robot Control using Silicon Retina Sensors", In proc. of the *IEEE International Conference on Robotics and Automation (ICRA)*, pp. 1223-1228, May 2002.
2. Stocker, A., Douglas, R., "Computation of smooth optical flow in a feedback connected analog network" *NIPS Advances in Neural Optical Systems*, vol. 11, pp. 706-712, 1999.
3. Kubina, S., "Konzeption, Entwicklung und Realisierung – Micro-Controller basierter Schnittstellen für mobile Roboter" *Master thesis at GMD Schloss Birlinghoven*, 2001. (in German).
4. Bruce, J., Balch, T., Veloso, M., "Fast and inexpensive color image segmentation for interactive robots", In Proc. of the *IEEE/RSJ Int. Conf. on Intelligent Robots and Systems (IROS)*, Vol. 3, pp. 2061–2066, October, 2000.
5. Stocker, A., "Constraint Optimization Networks for Visual Motion Perception - Analysis and Synthesis", Ph.D. Thesis No. 14360, Swiss Federal Institute of Technology ETHZ, September 2001.
6. Indiveri, G., Tech. Report, Uni. Lecce, 2001.
7. Brooks, R., "A robust layered control system for a mobile robot", *IEEE Journal of Robotics and Automation*, Vol, RA-2, No. 1, 1986.

Vision Guidance for a Climbing Cleaning Robot

Jian Zhu, Dong Sun, Shiu-Kit Tso,
Department of Manufacturing Engineering and Engineering
Management,
City University of Hong Kong, Hong Kong.

James K. Mills,
Department of Mechanical and Industrial Engineering,
University of Toronto, Canada.

Abstract:
This paper describes a visual sensing application of a climbing robot that provides a cleaning service on the glass wall of high-rise buildings. The vision system, mainly composed of an omnidirectional CCD camera and two laser diodes, is used to perform the real-time measurement of the robot position on the glass surface and location of the dirt to be cleaned. The mathematical model and the measure methodology of the vision system are discussed in this paper. An experiment is performed to calibrate the visual sensor, which is followed by measurement of the position and the location of the dirty area. The experimental results verify the effectiveness of the proposed approach.

Keywords: Climbing robot, visual sensing, glass cleaning.

1. INTRODUCTION

More and more people are interested in developing service robots to relieve human beings from hazardous work [1-3]. It is an important issue for a service robot to obtain the information of its position and orientation, so that it can know where to do the service work, how to avoid obstacles, etc. A climbing service robot has been recently developed at the City University of Hong Kong for cleaning glass curtain walls of the high-rise buildings. A vision system, mainly composed of a CCD camera and two diodes, is used to measure the distance between the robot and the window frame, and the orientation of the robot.

Several types of sensors, such as infrared [4], sonar [5-6], laser [7-8], and ultrasonic sensors [9-10], have been available to the measurement in robots' localization, path planning, obstacle-avoidance, etc. Although certain high-frequency sensors can measure the distance between two objects with good concentration and accuracy, they are not suitable to the climbing robot in measuring the distance between the robot and the window frame. This is because the height of the window frames is usually low and the sensors lean against the glass surface to some extent due to inevitable installation errors, which makes the high-frequency wave difficult to reach the window frame and subsequently affects the measurement of the distance between the robot and the window frame.

In the study, we propose to use a vision system in the climbing robot application. The vision system is mainly composed of a single omnidirectional CCD camera, which has been often used in the robot's localization, visual servoing, and vision guidance [11-15]. In our vision system, two laser diodes are fixed upon the camera as two eyes of the visual sensor. The posture of the vision system can be adjusted to make sure that the laser light reaches the window frame or the dirt to be cleaned. Then, the distance between the robot and the window frame and the location of the dirty area to be cleaned can be measured based on analysis of the coordinates of laser points in the image.

The paper is organized as follows. In section 2, the visual sensing technology is introduced, which is followed in section 3 by the experimental evaluation of measuring the robot's position and the location of the dirty area. Finally, conclusions of the work are given in section 4.

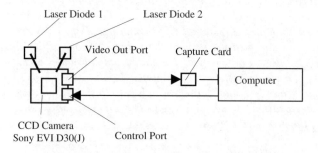

Figure 1. The Vision hardware system

2. VISUAL SENSING TECHNOLOGY

2.1 Vision hardware system

The vision hardware system consists of an omnidirectional CCD camera with the model number Sony EVI-D30 (J), two laser diodes, and a capture card, as shown in Figure 1. The control port of the camera receives the command from the computer, and the video out port of the camera sends the video signal to the capture card. The video signal is displayed on the computer screen, so that the human operator watches the situation around the robot on the glass surface. Two laser diodes fixed on the camera send laser lights to the window frame to generate two laser marks. The posture of the camera and the laser diodes can be adjusted by the computer to make sure that the laser light reaches the window frame. The launching points of the laser diodes are set to be the reference points of the robot. The distances between the reference points and the window frame, and the orientation of the robot relative to the window frame, can be determined by analyzing image pixel coordinates u and v of two laser points in the image plane.

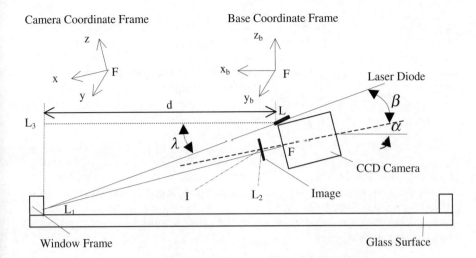

Figure 2. Measurement of the robot position

2.2 Methodology of measuring the position of the robot

Figure 2 (left view) illustrates how the robot position is measured by the vision system. The launching point of the laser diode is represented by L. Point L_1 is the laser mark on the window frame. Point L_2 is the corresponding point of point L_1 in the image plane, where I denotes the center of the image. Treating the focal point of the camera as the origin, denoted by F, a base coordinate frame represented by $F\text{-}x_b y_b z_b$ is established, where x_b coordinate axis is parallel to the glass surface and perpendicular to the window frame, y_b coordinate axis is parallel to both the window frame and the glass surface, and z_b coordinate axis is perpendicular to the glass surface(x_b-y_b plane), as shown in Figure 2. Using F as the same origin, another coordinate frame, named camera coordinate frame, represented by $F\text{-}xyz$, is also established, where the x axis is parallel to line I-F - the main light pivot of the camera – the y axis is the same as y_b axis of the base coordinate frame, and the corresponding z axis is perpendicular to the x-y plane. Denote (u, v) as the coordinates in pixel, and (u_0, v_0) as the pixel coordinates of the central point I. Let d_x and d_y represent distances between two adjacent pixels in horizontal and vertical directions in the images plane, respectively. Define α as the tilt angle of the camera in the base coordinate frame, β as the tilt angle of the laser diode relative to the camera, and γ as the pan angle. Based on analysis in the base and camera coordinate frame, the distance between the robot and the window frame in pixel coordinate u or v can be obtained:

$$d = \frac{a_1 u + a_2}{u + a_3} \cos(\alpha + \beta) \qquad (1)$$

$$d = \frac{b_1 v + b_2}{v + b_3} \cos(\alpha + \beta) \qquad (2)$$

where $a_1 = \dfrac{-x_0}{\cos\beta}$, $a_2 = \dfrac{x_0 u_0 d_x - y_0 f}{\cos\beta \, d_x}$, $a_3 = \dfrac{-u_0 d_x + ftg\gamma}{d_x}$

$b_1 = \dfrac{-x_0}{\cos\beta}$, $b_2 = \dfrac{x_0 v_0 d_y - z_0 f}{\cos\beta \, d_y}$ and $b_3 = \dfrac{-v_0 d_y - ftg\beta}{d_y}$.

in which f denotes the focal distance. The distance d is an important factor to measure the position of the robot relative to the window frame. As shown in Figure 3 (front view), the position of the robot can be known if the distances between the robot and the two window frames, d_1 and d_2, are measured using the proposed visual sensing methodology.

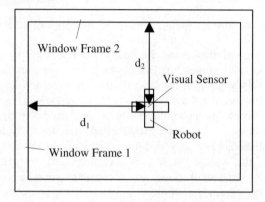

Figure 3.   A robot on the glass surface

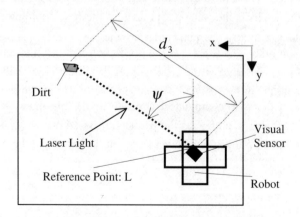

Figure 4.   Measuring the location of the dirt

2.3 Measurement of location of the dirtiness to be cleaned

Figure 4 (front view) illustrates how the location of the dirt to be cleaned is measured by the vision system. The camera and two laser diodes pan to find the dirty area and then acquire its image. The distance between the robot and the dirt can be measured by analyzing u and v coordinates of the laser point in the image, in the same manner to section 2.2. Based on the distance between the robot and the dirt and the pan angle of the camera, the location of the dirty area could be obtained. As shown in Figure 4, d_3 denotes the distance between the robot and the dirt, ψ denotes the pan angle of the camera when acquiring the image, and the reference point L is the launching point of the laser point. Ignoring the pan angle of the laser diode relative to the camera, the coordinates of the dirty area relative to point L are

$$X_d = d_3 \sin\psi \tag{3}$$
$$Y_d = -d_3 \cos\psi \tag{4}$$

3. EXPERIMENTS

Experiments were conducted to measure the distance between the robot and the window frame to determine the robot position, and the location of the dirty zone to be cleaned. The climbing robot with the visual sensor is shown in Figure 5. In the experiment, the posture of the camera, with the pan angle (-100 ~ 100 degree) and the tilt angle (-25 ~ 25 degree), was adjusted by the master computer. The pattern recognition technique was used to identify whether the laser light reaches the window frame to obtain the image of the laser mark in the window frame. The resolution of the image is $u \times v = 2560 \times 1920$ pixels. By analyzing the color matter of the images, the pixel coordinates u and v were obtained.

3.1 Measurement of the robot position

Firstly, the vision system was calibrated by acquiring the images and analyzing the u and v coordinates of the laser points when the robot was at different calibration positions on the glass surface. Table 1 illustrates the experimental results of the tilt angle of the camera and u and v coordinates of the left and right laser points at different distances between the robot and the window frame. With the least square fitting, the distance between the robot and the window frame could be developed from the calibration as a function of u coordinate of the left laser point u_L (pixel) and the tilt angle of the camera α (degree), i.e.,

$$d = \frac{16.67 u_L - 18583997}{u_L - 1286.21} \cos(\alpha + 9.2°) \tag{5}$$

The distance d could also be derived as a function of v coordinate of the left laser point v_L (pixel) and the tilt angle of the camera, i.e.,

$$d = \frac{16.67 v_L - 87786.87}{v_L - 1934.43} \cos(\alpha + 9.2°) \tag{6}$$

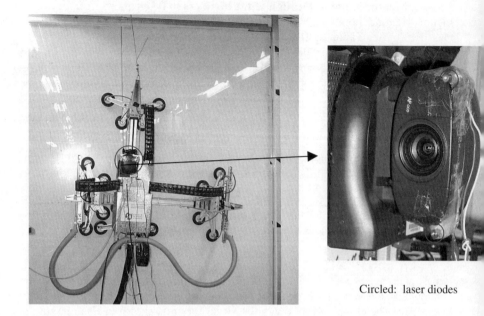

Circled: laser diodes

Figure 5. Climbing robot and vision system

In a similar manner, the distance d could be represented as the function of u or v coordinate of the right laser point, respectively, i.e.,

$$d = \frac{33.09\, u_R + 96310.74}{u_R - 1197.37} \cos(\alpha + 8.9°) \quad (7)$$

$$d = \frac{33.09\, v_R - 136974.29}{v_R - 1948.88} \cos(\alpha + 8.9°) \quad (8)$$

It can be seen from Table 1, when the distance between the robot and the window frame varies from 550 mm to 1500 mm, the u and v coordinates of the left laser point have the variations of 149 and 51 pixels, respectively, and those of the right laser point have the variations of 125 and 68 pixels, respectively. For the left laser point, the average change per pixel is: $\frac{1500 - 550}{149} = 6.38$ mm/pixel for u coordinate, and $\frac{1500 - 550}{51} = 18.63$ mm/pixel for v coordinate.

Table 1. Distance - tilt angle of the camera, u, v coordinates of the laser points

Distance (mm)	Tilt Angle (degree)	Left Coordinate u (pixel)	Left Coordinate v (pixel)	Right Coordinate u (pixel)	Right Coordinate v (pixel)
550	25	1030	1845	1415	1832
600	19	1044	1850	1404	1840
650	16	1051	1857	1396	1846
700	14	1066	1860	1384	1848
750	12	1080	1860	1373	1856
800	11	1092	1868	1363	1862
850	9.5	1100	1870	1356	1866
900	8.5	1110	1878	1346	1866
950	7	1118	1878	1339	1871
1000	6.5	1125	1881	1333	1877
1050	5.5	1132	1883	1327	1879
1100	4.5	1137	1886	1323	1882
1150	3.5	1142	1887	1319	1885
1200	2.5	1148	1889	1313	1888
1250	2	1154	1890	1308	1889
1300	1	1159	1893	1304	1893
1350	0.5	1163	1893	1300	1894
1400	0	1170	1894	1294	1895
1450	-1	1174	1896	1291	1901
1500	-1.2	1177	1896	1288	1900

For the right laser point, the average change is $\frac{1500-550}{125}$ =7.6 mm/pixel for u coordinate, and $\frac{1500-550}{68}$ =14.12 mm/pixel for v coordinate.

Therefore, we conclude that the distance measurement with u coordinate is more accurate than that with v coordinate. In other words, it is better to use equations (5) and (7) to measure the position of the robot on the glass surface, and the distance between the robot and the dirty zone to be cleaned.

In the further experiment, equation (5) was used to measure the position of the robot on the glass surface. When the camera pans 90 degree anti-clockwise, the distance between the robot and the left window frame can be measured. When the camera pans 90 degree clockwise, the distance between the robot and the right window frame can be measured. Experimental results of distances between the robot and the window frame are shown in Table 2. It is seen that the measurement error increases as the distance increases.

When the distance is less than 1000 mm, the measure error is less than 10mm. This implies that when the robot is close to the window frame, the measurement is more accurate.

Table 2. Experimental results of measuring the distance between the robot and the window frame

	To Left Window Frame				
	Tilt Angle (degree)	Left Coordinate u (pixel)	Distance [Reality] (mm)	Distance Measured (mm)	Measure Error (mm)
1	6	1129	1035	1025.2	9.8
2	0	1170	1425	1412.9	22.1
3	0.4	1165	1372	1353.8	18.2
4	5	1134	1073	1063.2	9.8
5	-1.5	1173	1480	1455.6	24.4
6	10	1098	847	840.6	6.4
7	5	1137	1096	1084.3	11.7
8	6.3	1127	1021	1011.1	9.9
9	5	1136	1088	1077.2	10.8
10	5.3	1134	1072	1061.8	10.2

	To Right Window Frame				
	Tilt Angle (degree)	Left Coordinate u (pixel)	Distance [Reality] (mm)	Distance Measured (mm)	Measure Error (mm)
1	3	1145	1168	1154.2	13.8
2	11.2	1084	782	777.6	4.4
3	9.6	1102	866	860.6	5.4
4	3	1146	1176	1162.3	13.7
5	11.2	1084	782	777.6	4.4
6	0	1169	1421	1401.0	20
7	3.5	1143	1148	1136.1	11.9
8	2.6	1150	1212	1197.8	14.2
9	3.5	1141	1134	1120.7	13.3
10	3	1146	1176	1162.3	13.7

	To Upside Window Frame				
	Tilt Angle (degree)	Left Coordinate u (pixel)	Distance [Reality] (mm)	Distance Measured (mm)	Measure Error (mm)
1	2	1156	1270	1254.9	15.1
2	1	1161	1325	1308.6	16.4
3	7	1116	952	943.5	8.5
4	8	1110	915	907.2	7.8
5	13	1068	717	713.0	4
6	7	1115	945	938.1	6.9
7	8	1112	925	917.4	7.6
8	15.2	1059	671	674.1	3.1
9	8	1111	920	912.3	7.7
10	8.5	1108	901	894.7	6.3

3.2 Measurement of the location of the dirty area to be cleaned

The pattern recognition technique was used to find out the dirt to be cleaned and then to obtain the image of the laser mark. Equation (7) was used to measure the distance between the robot and the dirty zone to be cleaned. Based on the measured distance and the pan angle of the camera, the location of the dirt can be known. The experimental results of locating the dirt are shown in Table 3.

Table 3. Experiment results of locating the position of the dirt

	ψ (degree)	Tilt Angle (degree)	Right Coordinate u (pixel)	d_3 (mm)	X_d (mm) Measure Value	X_d (mm) Measure Error	Y_d (mm) Measure Value	Y_d (mm) Measure Error
1	37	11	1364	798.2	480.4	2.6	-637.5	3.7
2	65	17	1400	633.3	574.0	3.2	-267.6	1.6
3	52	22	1410	577.0	454.7	2.5	-355.2	2.1
4	53	11.6	1370	768.6	613.8	3.5	-462.6	2.6
5	54	5	1325	1066.1	862.5	7.1	-626.6	3.4
6	75	2.4	1311	1205.6	1164.5	10.3	-312.0	2.0
7	67	9	1348	890.3	819.5	6.1	-347.9	2.0
8	65	5	1327	1050.1	951.7	7.7	-443.8	2.6
9	78	10.4	1358	830.0	811.9	5.9	-172.6	1.4

4. CONCLUSIONS

This paper describes a visual sensing application in a climbing robot to clean glasses of high-rise buildings. The visual sensing system installed on the robot is composed of an omnidirectional CCD camera, two laser diodes and other associate components. Based on calibration experiments, the relationships amongst the position of the robot, the coordinates of two laser points in the image, and the tilt angle of the camera can be obtained. Using these relationships, the real-time position of the robot and the location of the dirty zone to be cleaned can be measured. Experimental results demonstrate that the developed visual sensing system enables the climbing robot to navigate to do the service work on the glass surface in an effective way.

Acknowledgment

The work described in this paper was partially support by a grant from the Research Grant Council of the Hong Kong Administration Region, China (Project No. CityU 1085/01E), and a grant from City University of Hong Kong (Project No. 7001220).

References

1. C. Balaguer, A. Gimenez, J. M. Pastor, V. M. Padron, M. Abderrahim, Climbing autonomous robot for inspection applications in 3D complex environments, Robotica, Vol. 18, No. 3, 2000, pp. 287-297

2. G. La Rosa, M. Messina, G. Muscato, R. Sinatra, A low-cost lightweight climbing robot for the inspection of vertical surfaces, Mechatronics, Vol. 12, No. 1, 2002, pp. 71-96
3. M. Nilsson, Snake robot-free climbing, IEEE Control Systems Magazine, Vol. 18, No. 1, 1998, pp. 21 -26
4. P. M. Novotny, N. J. Ferrier, Using infrared sensors and the Phong illumination model to measure distances, Proceedings of the 1999 IEEE International Conference on Robotics and Automation, ICRA99, Vol. 2, May 1999, Detroit, MI, USA, pp. 1644-1649
5. A. Grossmann, R. Poli, Robust mobile robot localisation from sparse and noisy proximity readings using Hough transform and probability grids, Robotics and Autonomous Systems, Vol. 37, No. 1, 2001, pp. 1-18
6. V. Koshizen, N. Takamasa, Architecture of a Gaussian mixture Bayes (GMB) robot position estimation system, Journal of Systems Architecture, Vol. 47, No. 2, 2001, pp. 103-117
7. P. Jensfelt, H. I. Christensen, Pose tracking using laser scanning and minimalistic environmental models, IEEE Transactions on Robotics and Automation, Vol. 17, No. 2, 2001, pp. 138-147
8. S. Balakrishnan, N. Popplewell, M. Thomlinson, Intelligent robotic assembly, Computers and Industrial Engineering, Vol. 38, No. 4, 2000, pp. 467-478
9. C. J. Wu, C. C. Tsai, Localization of an autonomous mobile robot based on ultrasonic sensory information, Journal of Intelligent and Robotic Systems: Theory and Applications, Vol. 30, No. 3, 2001, pp. 267-277
10. G. I. Antonaros, L. P. Petrou, Real time map building by means of an ellipse spatial criterion and sensor-based localization for mobile robot, Journal of Intelligent and Robotic Systems: Theory and Applications, Vol. 30, No. 4, 2001, pp. 331-358
11. S. Maeda, Y. Kuno, Y. Shirai, Mobile robot localization based on eigenspace analysis, Systems and Computers in Japan, Vol. 28, No. 12, 1997, pp. 11-21
12. E. Malis, F. Chaumette, S. Boudet, 2-1/2-D visual servoing, IEEE Transactions on Robotics and Automation, Vol. 15, No. 2, 1999, pp. 238-250
13. Y. Ma, J. Kosecka, S. S. Sastry, Vision guided navigation for a nonholonomic mobile robot, IEEE Transactions on Robotics and Automation, Vol. 15, No. 3, 1999, pp. 521-536
14. J. L. Crowley, F. Pourraz, Continuity properties of the appearance manifold for mobile robot position estimation, Image and Vision Computing, Vol. 19, No. 11, 2001, pp. 741-752
15. J. Ferruz, A. Ollero, Integrated real-time vision system for vehicle control in non-structured environments, Engineering Applications of Artificial Intelligence, Vol. 13, No. 3, 2000, pp. 215-236.

Measuring Flank Tool Wear on Cutting Tools with Machine Vision – A Case Solution

Dr Tilo Pfeifer, Dominic Sack, Alexandre Orth,
Laboratory of Machine Tools and Production Engineering (WZL),
Chair of Metrology and Quality Management (MTQ) RWTH, 52056 Aachen, Germany

Dr Marcelo R. Stemmer, Mário L. Roloff,
Intelligent Industrial Systems (S2i), Automation and Systems Dept.
CTC, Federal University of Santa Catarina (UFSC), 88040-900 Florianópolis – SC – Brazil

Abstract
The market has changed significantly over the last years. Nowadays, industries must deal with extremely demanding customers. In order to stay in business, they have to develop quickly customized and specialized products with low prices. In this sense, process monitoring is of crucial importance as it optimises the productivity and reduces the costs by avoiding the production of scrap as well as improving the final product quality. Flank wear is an important parameter in chip forming processes – it allows to estimate the cutting tool's lifetime and to control the product quality. There are many different types of cutting tools, differing one from the other according to the type of machining processes (milling, drilling, etc.), the tool's geometry and its material characteristics. These properties influence directly the optical characteristics of cutting tools. Therefore, the design of a machine vision system for this application is a complex task. This paper describes the development of a image processing system to measure the flank wear and classify the tool wear type (broken tool, flank wear, ...). It applies an image contour classification technique based on Fourier descriptors and neural networks. The information retrieved from this system is essential for an optimised production planning and process control, i.e. optimised tool usage or process control by adaptation of the NC program.

Keywords: Machine vision, processes monitoring, tool wear and neural networks.

1. INTRODUCTION
The current world wide competition is driving companies to improve the production performance and quality level in order to reduce production costs and to avoid the production of scrap. The customers are requiring more individual products with a shorter *"time to market"*. Industrial manufacturers therefore will need flexible production systems with high performance and quality characteristics. In this sense, the antiquate quality assurance by measuring the specification-confor-

mity of a product at the end of the production line is replaced by a preventive quality strategy with in-line-metrology. Milling and turning are very common processes applied in industry today. Therefore, process-monitoring is of crucial importance to optimise the production in quality and costs.

In this context, the "Laboratory of Machine Tools and Production Engineering" (WZL, RWTH Aachen) and the research group "Intelligent Industrial Systems" (S2i, UFSC) defined a cooperation project in the field of tool wear monitoring as one part of the Collaborate Research Center "Autonomous Production Cells" (SFB368). Topic of the project is the development of a image processing system for automated measurement and classification of a cutting tools' flank wear. This paper will focus on the image processing algorithms in this system.

1.1 Cutting Tools

There are many different kinds of cutting tools applied in chip forming processes, a small selection is presented in Fig. 1. The tools differ from each other from the optical point of view in geometry and surface properties. Depending on the type of machining process (milling, turning, drilling, etc.) and the realised manufacturing task, they have different forms and different surfaces (material, coating).

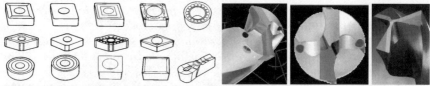

Fig. 1: Samples of cutting tools used in turning and milling processes

1.2 Tool wear

During the machining process, tools are exposed to enormous forces which will lead to wear occurrence. Parameters for the wear magnitude are the cutting conditions (cutting depth, cutting speed) and the duration time. Fig. 2 shows the two most common characteristics of tool wear:
- *Crater wear* (K) on the face of the tool with its parameters crater wear depth and crater wear length.
- *Flank wear* (V_B) on the primary and secondary flank surface of the tool with its parameters flank wear, maximum flank wear and cutting edge offset.

A complete description of tools wear types and wear mechanisms can be found in [2]. Flank wear is the most referred tool wear parameter in process monitoring - today it is usually measured manually by microscopes. The parameter used for wear compensation in the NC program is the cutting edge offset. It can be computed from the flank wear, with information about the cutting edge geometry. Available automated systems, like laser sensors, are just used for the tool calibration as they measure the tool radius, but they do not allow any detailed information about tool wear and the classification of the type of wear.

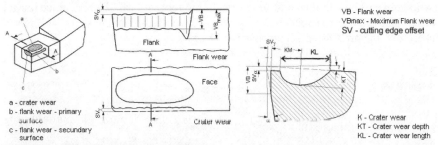

a - crater wear
b - flank wear - primary surface
c - flank wear - secundary surface

Fig. 2: Tool wear and it's parameters

1.3 Goals

The objective of this research project is to develop a vision system for measuring and classification of tool wear by inspection of the tool flank. The system will work intermitting the process, that means that the tool inspection takes place in the tool-magazine of a machine tool or user-supported during the machine setup. The idea is to develop, in the long term, a flexible system for many tool types, but to start with a specific solution for a certain tool type. An important variable for process control, the flank wear (V_B & V_{Bmax}), will be measured.

2 MACHINE VISION SYSTEM

The following sections will present the machine vision system for tool wear measurement and classification. For this project, turning cutting tools of the type SNGN 12 04 08 T01020 from Sandvik Inc. have been used.

2.1 Optics and illumination

The optic system should guaranty a minimum resolution of 10 µm in the measurement realized by the machine vision. The used equipment was a grayscale camera and frame grabber combination (640 x 480 pixels, 8 bits per pixel) and a lens (50mm focus). Based on these parameters, the optic system has been calculated [4]. The overall end resolution of the machine vision is 4.36 µm, providing a precision of ±2,18 µm, which is much more accuracy than required.

For the measurement of the flank wear, the application of an incident illumination technique is necessary because flank wear is a surface defect. Cutting tools have metallic reflecting surfaces with an (after use) undefined micro-topography. Therefore, any illumination will have disturbing effects on the image quality because of reflections and shadows on the surface. The best illumination could be achieved using a directed light with an approximate 35° angle to the surface. An improvement of the illumination system, by means of sensor-data fusion in fields of image acquisition and preprocessing, is proposed for future work (see Section 5).

2.2 Image processing algorithms

Because of disturbing reflections, it is necessary to suppress these effects by preprocessing steps in the software, to guarantee a safe location of the cutting tool and detection of tool wear. The strategy of choice is the comparison of a model image (without wear) and an image of the worn cutting tool in form of a difference image. After that, the wear detection, measurement and classification steps take place. Therefore, the processing is composed of basic steps listed in Fig. 3.

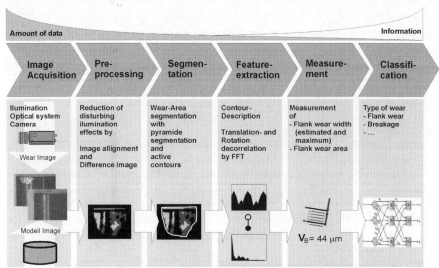

Fig. 3: Algorithm chain for tool wear measurement and classification

2.2.1 Wear image acquisition and model image selection

In a first step, an image of a new unused cutting tool of the same type and geometry must be provided, i.e. from a database. This model image must have been acquired under the same conditions (zoom, illumination) like the image of the worn tool. This condition can be properly guaranteed by an automated system (positioning, illumination and image acquisition). In the sequence, an image of the worn cutting tool is to be acquired for inspection.

2.2.2 Preprocessing: Image Alignment and Difference Image

In order to compare the model and the wear image to a difference image, it is necessary to align both images. The tool under inspection has defined cutting edges, the worn corner is placed to acquire the image, as shown in Fig. 4. The alignment algorithm detects the two main edges in each image – both in the model image and in the worn tool image. After comparing these lines, the algorithm rotates and translates the worn tool image until the position of the tool is the same as the model image. The alignment algorithm works in detail in the following way:
1. Two Regions of Interest (ROI) are defined in both images for the main edge detection: ROI at the upper right corner for the top edge and ROI at down left corner for the left edge. The tool must be located with the main edges in these ROIs for a proper function of the algorithm.
2. The detection of the reference lines is performed as follows: A pyramid segmentation algorithm, in combination with an adaptive threshold, is applied to the main edge ROIs to separate the background from the tool and create a

binary image splitting these regions. Some miss-detected points as a consequence of illumination effects are erased by two morphologic operators in sequence: opening and closing with 3x3 kernel each. The borderline is detected by the morphologic top-hat operator and a common contour detection algorithm. Finally, the line parameters are calculated by minimum distance interpolation. The detected edges are painted in Fig. 4.
3. With the lines found in the ROIs, a reference point - where both lines cross - and the tool orientation is calculated. With these data – both for the model and wear image - the wear image is aligned in relation to the model (Fig. 4).

Fig. 4: Image alignment algorithm for difference image

2.2.3 Wear Detection

Using the difference image of the model and aligned wear image, a more safe detection of the wear area can be performed, as the illumination effects have widely been removed by the image subtraction in the non-wear area. But there are still reflections in the wear area. For wear detection the following algorithm is used:
1. Based on the previously determined reference point, another ROI for the maximum wear dimension is defined for reasons of processing time. The size of this ROI is a parameter of the detection system.
2. The difference image still shows some noise because of the illumination effects and small errors in the alignment algorithm. Therefore, three filters are applied in sequence to smooth the noise effects: minimum (kernel 3x3), medium (kernel 5x5) and gauss (kernel 5x5).
3. A pyramid segmentation is used to distinct the wear region in the image. The "Open"-operator includes small regions which were not properly segmented.
4. The segmented area is still quite jagged because of reflections in the wear area. The segments are combined by the application of an active contour (snake, [4]), which follows 3 energy minimising criteria: a) Equidistant distribution of contour points b) Smoothing of the contour c) Adaptation to edges in images. The initial contour is created containing a set of equally separated points along the edges of the wear area ROI. The snake algorithm compresses this contour to the outline of the wear by iteration.

The wear should be measured in relation to the upper edge of the cutting tool. Therefore, a double white line in the area of the segmented wear is drawn into the segmentation image, representing the upper edge of the tool. The snake includes the upper edge and the wear area: (a) segmentation and snake b) results projected into the source image Fig. 5).

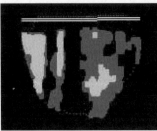 

a) segmentation and snake   b) results projected into the source image
Fig. 5: Wear detection and measurement:

2.2.4 Wear Measurement

The flank wear V_{Bmax} is defined as the vertical distance between the highest and the lowest point, orthogonal to the upper edge. Therefore, this parameter is easily determined. In the sequence, the area of the final contour – respective to the flank wear area AV_B - is calculated by means of the calibration information pixel size. For the average flank wear V_B there are many qualitative definitions, but a accurately quantative definition could not be found in literature. Therefore, the contour area is subdivided here by a horizontal line in two regions, where the upper region contains 80% of the contours' area. The distance between the highest contour point and this line provides the estimation of the flank wear V_B (a) segmentation and snake   b) results projected into the source image Fig. 5).

2.2.5 Wear Type Classification

On the basis of the form of the wear area – respective to the external contour in the form of the snake points - the classification of tool wear type is performed. The Fourier Transformation of the sequence contour pixels, as a periodic complex function, leads to an translation and rotation de-correlated description of the contour. The translation information is found in the 0-coefficient, the rotation information is stored in the phase of the higher level Fourier descriptors. The 10 lowest coefficients without the constant component are used for the classification in a normalised form. The Fourier descriptors are fed to the neural network to associate this descriptors with wear type. The neural network applied is a feed-forward net, which activates the output neuron related to the recognized wear type. The neurons use a tangent-sigmoid activation function.

3   SYSTEM EVALUATION

The image processing methods described above have been implemented in a software tool with a simple user interface (Fig. 6). The algorithms are invoked by buttons, the measurement and classification results are shown in the related edit-/combo-boxes. The interface shows the wear image (upper left), the model image (upper right) and image processing output (lower right).

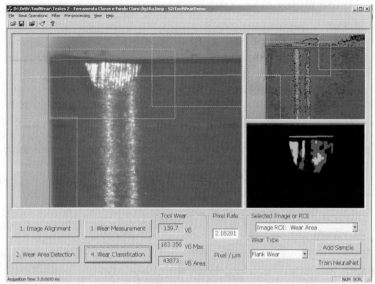

Fig. 6: User interface of the tool wear measurement application

In order to evaluate the system, a set of 9 worn cutting tools (with 2 or 3 images per tool – 24 images at all) was used, including 2 types of wear: flank wear and tool breakage. The neural net was configured to classify just between these two classes. Therefore, the following topology was used: 10 neurons input layer, 4 neuron hidden layer and 2 neurons output layer. The training was performed using the Quickpropagation algorithm (a fast Backpropagation algorithm) and 16 different samples, the learning rate was set to 0.12 and the momentum parameter to 0.65.

The neural network takes in average 43 sec to learn the 16 samples in 2 wear types (Hardware: Pentium III, 550 MHz and 128 Mb RAM). The training algorithm needs approximated 20750 iterations to reach a global error smaller than 2×10^{-3}. The measuring and classification of the tool wear requires 620 ms.

As a reference, the wear was measured in parallel using a microscope with maximum magnification. In the case of tool breakage, the results are almost the same, the difference is 0 - 6 %. In the case of flank wear, the measurement with the microscope is more subjective. The definition of the wear borderline depends on the user, in particular as maximum magnification in some cases is to low for a proper measurement. Therefore, the difference in the results can go up 15 to 30%. But by visual inspection of the segmented images can be concluded, that the vision system realises a proper and accurate segmentation task. The classification shows 100% proper results in the tested case with 2 classes, despite the limited set of samples. Increasing the number of classes will, of course, lower the classification rate.

4 CONCLUSION AND PERSPECTIVES

Based on the obtained results, it can be concluded that the developed machine vision system is more efficient than the manual method using a microscope. More than just to measure the tool wear with a higher precision, the system also guarantees results with repeatability.

The illumination and preprocessing system is the key for improving the vision system. Therefore, this system will be combined with a flexible illumination and image acquisition module using variations in the direction of illumination (Fig. 7) and a software optimisation by sensor data fusion, as proposed in [6]. Another point in future work is the extension of the system different types of cutting tools and the integration of the vision system into an Autonomous Production Cell. Applications and use of the tool wear information are NC-program adaptation to the actual tool wear and estimation and optimisation of the tool life time.

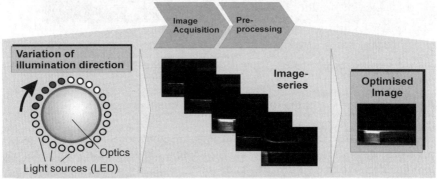

Fig. 7: Variation of illumination and sensor data fusion

5 ACKNOWLEDGEMENTS

This project is funded by the *Deutsche Forschungsgemeinschaft* (DFG) in the context of the Collaborate Research Center "Autonomous Production Cells" (SFB 368, http://sfb368.rwth-aachen.de). We gratefully acknowledge this support as well as Intel Inc. for the computer Vision libraries IPL & OpenCV.

6 REFERENCES

[1] Deschamps, F.; Orth, A.; Stemmer, M.: RAP - Um Sistema De Visão para Robôs Inteligentes Baseado na Interpretação dos Coeficientes de Fourier, Brazilian Automation Congress, Florianópolis, Brazil, 2000
[2] Fremann, J. A.; Skapura, D. M.: Neural Networks: Algorithms, Applications, and Programming Techniques, ed. Addison-Wesley, EUA, March, 1991
[3] König, W., Klocke, F.: Fertigungsverfahren: Drehen, Fräsen, Bohren, ed. Springer Verlag Berlin Heidelberg, Heidelberg, Germany, 1997
[4] Orth, A.: Development of a Machine Vision System to Measure the Flank Wear in Cutting Tools, Master Dissertation, UFSC Florianópolis, Brazil, 2001
[5] Pfeifer, T.; Sack, D.; Orth, A.; Stemmer, M.: Sensor/Actuator Network - The Nervous System Of A New Production Concept: The Autonomous Production Cells, Brazilian Automation Congress (CBA), Florianópolis, Brazil, 9/2000
[6] Pfeifer, T.; Sack, D.; Wiegers, L.: Automatisierte Werkzeugverschleißmessung, werkstatttechnik wt 03/2001

A Floating Point Genetic Algorithm for Affine Invariant Matching of Object Shapes

P.W.M. Tsang,
Department of Electronic Engineering, City University of Hong Kong, Tat Chee Ave., Kowloon, Hong Kong.

W.H. Tsang,
Product Development, Thomson Multimedia (Hong Kong) Ltd.,
13/Fl., Ever Gain Centre, 28 On Muk Street, Siu Lek Yuen, Shatin, Hong Kong

Abstract
This paper presents a novel scheme for matching disjoint edge images of objects that are captured from different viewpoints. Basically, the task of matching a pair of object images is encapsulated as the search for the existence of an Affine Transform (AT) to describe the geometrical changes between the two subjects. As the search landscape is enormous and highly non-linear, a Floating Point Genetic Algorithm (FPGA) is developed to conduct the search. To further reduce the computation time, the search is restricted to regions corresponding to viewing positions that are of relevance in practice. This is accomplished with the use of a Decomposed Affine Transformation (DAT) which is formed by the integration of elementary spatial operators comprised of translation, rotation, scaling and shearing. Experimental results demonstrate that the proposed matching scheme has attained a success rate of over 96% in identifying incomplete edge images with significant reduction in the overall computation time.

Keywords: *Affine invariant shape matching, floating point, genetic algorithm, broken image contours.*

1. Introduction

In model-based object recognition, the identity of an unknown shape can be deduced by comparing it against a library of collection of reference images, each corresponding to a projected view of a known model object. The major difficulty of this approach is that when a real world, three-dimensional object is observed from different viewpoints, its images can be significantly different from each other. As a result, the number of reference images has to be infinitely large to contain every views of every model in the collection. A common direction to overcome this problem is to describe a model object shape with a small set of two-dimensional reference images each representing a distinct aspect on the subject (e.g., front, back, top and bottom view). It can be assumed that within the context covered by each aspect, the projected images resulted from different viewpoints will be related by the Affine Transform. Given an unknown planar shape, it is possible to deduce its identity by matching it against the multi-aspects, two-dimensional projections of each model in the library. The obvious advantage of this approach is that although the model object is three-dimensional, the recognition process can be accomplished with the availability of an effective planar, Affine invariant shape matching scheme.

In most cases, it is suffice to represent an object image with its edge points. Suppose $P = [(x_0, y_0), (x_1, y_1), ..., (x_{N-1}, y_{N-1})]$ and $Q = [(x'_0, y'_0), (x'_1, y'_1), ..., (x'_{N-1}, y'_{N-1})]$ are the edge points of two different projections of the same aspect of an object, their elements can be related by the Affine transform

$$\begin{pmatrix} x'_i \\ y'_i \end{pmatrix}_{0 \leq N-1} = \begin{pmatrix} a & b \\ c & d \end{pmatrix} \begin{pmatrix} x_i \\ y_i \end{pmatrix} + \begin{pmatrix} e \\ f \end{pmatrix} \qquad (1)$$

According to (1), the task of matching can be encapsulated as the search for the existence of an Affine Transform that can convert one edge image to the other. Such an attempt had been made in [1], where tree search was employed to test the similarity between an unknown and a model edge images. Despite the success of this method, the computation time was long, as the space spanned by the Affine parameters is vast. A different direction was taken in [2], by adopting a Genetic Algorithm to provide a fast exploration on the Affine space. The models were restricted to the category of simple star shapes and the method could only be applied for rough classification on an input image. This limitation was overcome with the Genetic Algorithm presented in [3]-[4], capable of identifying closed unknown boundaries with reference to a library of real-world object contours. Later, the method was extended to analyze disjoint edge images [5]. However, due to the enormous search space and the sharp response of the fitness function towards changes in the parameters governing the Affine transform, the successful rate in correct identification of an unknown shape was generally low. Besides, the computation involves in [5] is tedious as the algorithm was comprised of a forward and a backward matching stages.

In view of the above problems, a novel Genetic Algorithm for an Affine invariant matching scheme has been developed and reported in this paper. We have categorized the existing problems into a number of important issues that can be addressed on an individual basis. To start with, it is necessary to reduce the search space by placing suitable constraints on the Affine parameters. This is made possible by the fact that when a particular aspect of an object is recorded by a camera and taken as either a model or an unknown input, there are always restrictions in the coverage of the viewpoint. While the latter can be defined easily, it is difficult to relate it to the corresponding bounds of the Affine parameters.

Apart from that, the transform in (1) does not constitute to a stable representation. An alteration in a single Affine parameter may correspond to spatial changes in the viewpoints along several dimension of freedom. Intuitively, the vector space defining the matching score (e.g. the fitness function in [5]), as a function of the Affine parameters, will be extremely erratic and complicated in general. To rectify this problem we suggest the use of an alternative representation on the Affine Transform so that each variable describes a distinct geometrical change (e.g. rotation, scaling and shearing) on the projected object image.

Second, the effectiveness of encoding the parameters governing the Affine Transform in the form of binary chromosomes is questionable. According to the Schema Theorem, evolution of individuals towards the optimal solution is caused by the exponential growth of schemata that are associated with an average fitness that is above the mean value. A simple experiment will show that in the context of shape matching, correct schema may not strengthen with this mechanism. Suppose O_1 and O_2 that are two identical object contours that are separated by eight pixels along the horizontal direction. A Genetic Algorithm is applied to test whether there exist a value for each translation term 'e' with which O_1 can be shifted to align with the other contour. The fitness value is defined as

$$f(e) = \frac{1}{1 + \frac{1}{N}\sum_{i=0}^{N} dist(p_i, q_i)} \bigg| \{e \mid 0 \leq e \leq 255; \text{horizontal position}\} \qquad (2)$$

Where $p_i \in O_1$, $q_i \in O_2$ is the point nearest to p_i, and $dist(p_i, q_i)$ is the Euclidean distance between the two points. If e is encoded as an eight-bits chromosome, the solution will be the binary string given by '00001000', and the substring '1000' is a schema of the optimal solution. A plot of the fitness function for values of e that contains the schema '1000' for the range [0,60] is shown in Figure 1. Fitness values for $e>60$ are extremely small and not shown in the diagram.

It can be seen that the presence of an indispensable schema has little influence on the fitness value unless it is very closed to the optimal solution. Apparently, the amount of schemata that is active to enable the implicit parallelism mechanism in searching the set optimal Affine parameters is insufficient. This probably explains the low successful rate attained in previous attempts. To address this problem, we proposed the use of a Floating Point Genetic Algorithm (*FPGA*) together with a crossover operation so that correct genotypes can be identified, exchanged and multiplied throughout the evolution process.

Third, early *GA* approaches reported in [3]-[5] had exhibited premature convergence when the initial populations were ill-formed (i.e., small amount of correct schemata). We postulated a new fitness evaluation to encourage "close competition" amongst chromosomes so that genotypes of different nature can have adequate chances to evolve before their survival rights are determined.

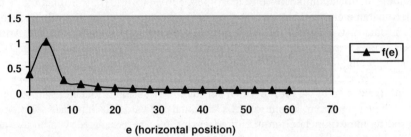

Figure 1: fitness associated with schema '1000'

The paper is organized as follows. In section 2, a brief review on the work reported in [3]-[5] and its shortcomings are described. Description on the Decomposed Affine Transform, together with the architecture, properties and implementation on the proposed Floating Point Genetic Algorithm are given in section 3. Experiment results in the application of the proposed method in matching edge images unknown and model shapes are presented in section 4. Finally, a conclusion is made in section 5 summarizing the essential findings.

2. Genetic Algorithm in Affine invariant contour matching

Since the introduction by Holland [6], there was an increasing trend to incorporate Genetic Algorithm in image analysis and identification. Along this direction, Affine invariant matching techniques were introduced in [3]-[5]. The basic principle of these methods in matching an unknown and a model contours are outlined in Table 1.

Table 1 - Genetic Algorithm Based Contour Matching Technique

Step	Action
1	Set generation count $t = 0$
2	Generate an initial population P consisting of M chromosomes. Each chromosome represents the Affine Transform with the form given in (1)
3	Apply the transform defined by each chromosome to the model contour. Compute the fitness value reflecting the similarity between the unknown and the transformed model contour.
4	If the maximum fitness value in the population exceeds a predefined threshold, go to *Step* 10.
5	Increase generation count by one, i.e., $t = t + 1$
6	If t is greater than the maximum allowable iteration, go to *Step* 10.
7	Select the chromosomes into a mating pool with a probability according to their fitness value.
8	Apply reproduction, crossover and mutation operations to the chromosomes to generate a new population.
9	Go to *Step* 2
10	Object contours are identical. End of process.

In [3] and [4], the fitness value was computed from the normalized area of overlapping between the unknown and the model contours. Although the calculation was straightforward and fast, the contours must be continuous and complete. This assumption, however, was seldom valid in practice. A more sophisticated measure was given in [5] where the fitness value was computed from the product of two scores. The first one was the average spatial distance between the transform model and the unknown contours, computed by overlaying the former onto the distance transform map of the latter. The second term is a measure on the number of points between the pair of contours that are within close proximity.

Despite the moderate success achieved in [3] to [5], they all suffers from the shortcomings described in the introduction of this paper. As the problems were originated from the erratic search landscape and the poor schemata representation, it is unlikely that they can be overcome with minor revisions on the Genetic Algorithm.

3. A novel Affine invariant object matching scheme

The proposed method is a complete overhaul on the works reported in [3] to [5]. First we adopted an alternative form of the Affine Transform so that the six dimensional vector space spanned by the Affine parameters is less erratic and can be significantly reduced by the coverage of the camera viewpoint. Second, a Floating Point Genetic Algorithm is adopted so that useful genotypes are preserved during the evolution process. Third, a new fitness function is defined to encourage close competition between chromosomes so as to avoid premature convergence.

Despite the significant changes that have been made on the matching scheme, the fundamental flow of our method remains similar to that listed in Table 1. In another words, we have adopted the backbone of traditional Genetic Algorithm in our design, but have introduced a more effective and efficient means of selecting good genotypes in the course of evolution.

3.1 Decomposed Affine Transform (*DAT*)

The Affine Transform can be rewritten in the decomposed form as

$$\begin{pmatrix} \hat{x}_{m,i} \\ \hat{y}_{m,i} \end{pmatrix} = \begin{pmatrix} 1 & 0 \\ sh & 1 \end{pmatrix} \begin{pmatrix} s_x & 0 \\ 0 & s_y \end{pmatrix} \begin{pmatrix} \cos\theta & \sin\theta \\ -\sin\theta & \cos\theta \end{pmatrix} \cdot \begin{pmatrix} x_{m,i} \\ y_{m,i} \end{pmatrix} + \begin{pmatrix} t_x \\ t_y \end{pmatrix}_{i=1,2,\ldots,n} \quad (3)$$

where sh represents the shearing factor, s_x and s_y are the scaling factors, t_x and t_y are the translation and θ is the angle of rotation. The subscripts x and y parameters denote the horizontal and vertical directions, respectively. Although equation (3) involves the same number of parameters as in equation (1), it allows an individual parameter to be bounded according to physical constraints. For example, we can safely assume that the scale factors (i.e. the ratio between the scene and model image sizes) should be greater than 0.2 or else the scene image will be too small to be identified. According to Interval Analysis [7], the volumetric ratio between the vector space of the original and Decomposed Affine Transform can be over 1000 times if the bounds are properly assigned. This implies a significant reduction in the computation time required to locate an optimal state in the confined search space.

3.2 Floating Point Genetic Algorithm and its operators

In a Floating Point Genetic Algorithm, the chromosomes are formed by genes represented in real values. To represent the Decomposed Affine Transform, the chromosome is comprised of six real numbers, each corresponding to one of the parameters, as shown in Figure 2.

Figure 2: Structure of a chromosome in FPGA

Apart from a different chromosome representation as compare with binary *GA*, evolution in *FPGA* is governed by a new set of genetic operators.

a) Arithmetical crossover

The operation is a linear combination of a corresponding pair of genes in two parent chromosomes. Suppose $g_{p,j}^t$ and $g_{q,j}^t$ is a pair of genes in chromosome p and q to be crossover at location j, their offspring are given by

$$g_{p,j}^{t+1} = k_m \cdot g_{q,j}^t + (1-k_m) \cdot g_{p,j}^t \text{ and} \qquad (4)$$

$$g_{q,j}^{t+1} = k_m \cdot g_{p,j}^t + (1-k_m) \cdot g_{q,j}^t \qquad (5)$$

where $k_m \in [0.5, 1)$. It can be seen that the two offspring genes will fall into a range that is somewhere between the values of their parents. As a result, genetic information is inherited with different weightings from the parents but will not be disrupted as in the case of binary *GA*.

b) Mutation

In *FPGA*, mutation is performed by adding or subtracting a random value from a gene. Suppose u_j and l_j are the upper and lower bounds of $g_{p,j}^t$. After mutation

$$g_{p,j}^{t+1} = w_j \left(g_{p,j}^t \pm k_c \Delta \right) \qquad (6)$$

where k_c is a random sign and $\Delta = (u_j - l_j)$. $w_j(\bullet)$ is a wrap-around function given by

$$w_j(a) = (a + u_j - 2l_j) \bmod (u_j - l_j) + l_j \qquad (7)$$

The genetic operations in *FPGA* have integrated a certain degree of the *Hill-Climbing* behavior, as new genes are derived from interpolating the states of their parents. However, from a global perspective, the evolution is still conducted by the principle of natural adaptation and survival of fitness.

3.3 A new fitness function

Suppose $A = \{a_1, a_2, ..., a_{n_A}\}$ and $B = \{b_1, b_2, ..., b_{n_B}\}$ denote the edge points on the unknown and the transformed model images. Their difference can be determined by measuring the average distance between their edge points as given by

$$h(A,B) = \frac{1}{n_A} \sum_{i=1}^{n_A} d_B(a_i) \qquad (8)$$

Where $d_B(a) = \min_{b \in B} \|a - b\|$ denotes the minimum distance between a point $a \in A$ to the nearest point in B. Based on this distance measurement, a fitness function can be defined for each chromosome (representing an Affine Transform) as

$$AS(A,B) = \frac{1}{1 + h(A,B)} \qquad (9)$$

The above function had taken a major role in the calculation of the fitness function in [5]. As revealed by a plot of $AS(A,B)$ against $h(A,B)$ in Figure 3, the fitness function is a nonlinear and erroneous reflection on the distance. It can be seen that the fitness value drops rapidly from its maximum value of unity to 0.25 when the average distance is only 3 pixels. Consequently, chromosomes will be discarded before they have a chance to evolve, thus inhibiting competition and leading to premature convergence. To overcome this problem we have proposed a new fitness function:

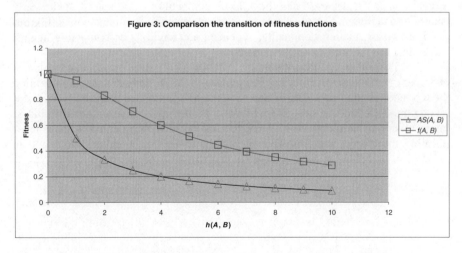

Figure 3: Comparison the transition of fitness functions

$$f(A,B) = \left(1 + \frac{h(A,B)^2}{k_f}\right)^{-\frac{1}{2}} \tag{10}$$

A plot of $f(A,B)$ against $h(A,B)$ for $k_f = 9$ is shown in Figure 3. It can be seen that a more linear relation is established between the two functions. As a result, chromosomes that are more remote from the optimal solution (as reflected by larger distance value) will have sufficient room to mature and justify their survival.

4. Results

The proposed scheme has been evaluated by repetitively matching 600 unknown edge images to a library of 10 model contours. All the images are digitized with a resolution of 256×256. A population size of 100 is adopted and the matching process is classified as successful if the maximum fitness function exceeds 0.95. The crossover rate, mutation rate, k_f and maximum generation are taken to be 0.55, 0.2, 9 and 100, respectively. A few examples are presented to demonstrate the feasibility of the method.

To start with, selected stages in the matching of a scene and a model image of a cutter (marked in black and gray lines) are illustrated in Figures 4a to 4h. It can be observed that the best individual in the initial population (Figure 4a) has a moderate fitness value of 0.418 representing transformation of the model to a small area around the tip of the scene contour. If no constraint is placed on the lower bound of model size as in the methods reported in [5], the model should evolve around this local optimum and ultimately shrink into a single point. This erroneous situation has been prevented in our method and the correct search path towards the global solution is soon located in the 4th generation (Figure 4b). From that point onward, the evolution process continues and the correct match is found in the 25th generation with a high fitness value of 0.98. The second set of example shown in Figures 5a to 5h demonstrates the "close" competition experienced at the early stages of matching a pair of spanner contours. In the first four generations (Figures 5a-5d), two orientation dominating schemata are found to be competing against each other to gain survival in the population.

Two other set of results in the matching of a pair of scissors and a pair of hammer contours are given in Figures 6a-6h and 7a-7h. In both cases, competition between dominating schemata of opposing nature is exhibited at the early stages of evolution. This kind of struggling between genotypes is quite common found in most of our experiments. With very few exceptions, the method is capable of locating the correct solution. The overall successful rate is around 96% and in general around 23-33 generations are taken to arrive at the optimal solution.

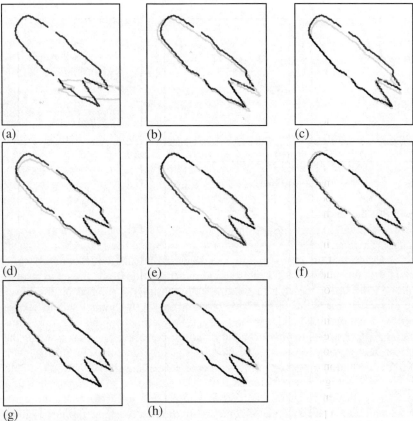

Figure 4: The model and scene contours of a cutter. (a) 1st generation, fitness=0.418140. (b) 4th generation, fitness=0.597644. (c) 11th generation, fitness=0.617070. (d) 13th generation, fitness=0.662378. (e) 14th generation, fitness=0.752007. (f) 17th generation, fitness=0.847959. (g) 20th generation, fitness=0.916653. (h) 25th generation, fitness=0.980600.

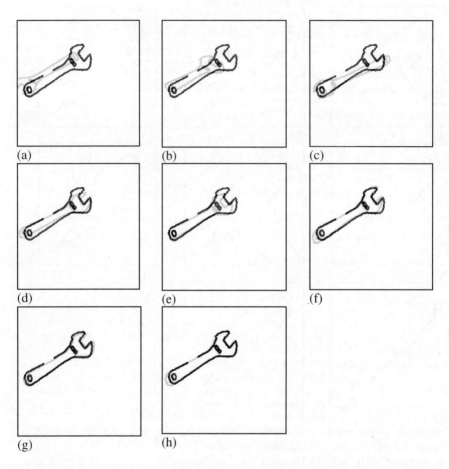

Figure 5: The model and scene contours of a spanner. (a) 1st generation, fitness=0.551018. (b) 2nd generation, fitness=0.566140. (c) 6th generation, fitness=0.668157. (d) 12th generation, fitness=0.751039. (e) 19th generation, fitness=0.826097. (f) 21st generation, fitness=0.854947. (g) 23rd generation, fitness=0.914502. (h) 34th generation, fitness=0.963580.

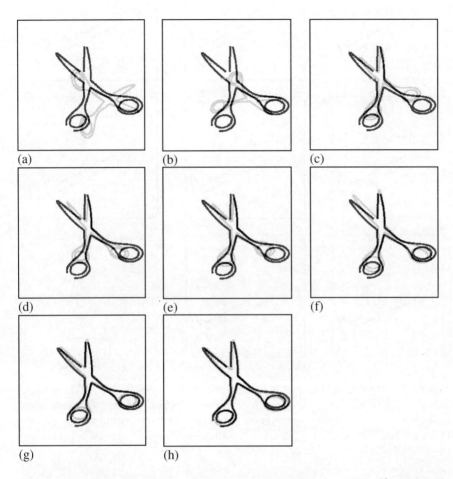

Figure 6: The model and scene contours of a scissors. (a) 1st generation, fitness=0.407167. (b) 3rd generation, fitness=0.593223. (c) 9th generation, fitness=0.712355. (d) 11th generation, fitness=0.730547. (e) 19th generation, fitness=0.813159. (f) 23rd generation, fitness=0.839247. (g) 29th generation, fitness=0.932681. (h) 31st generation, fitness=0.951276.

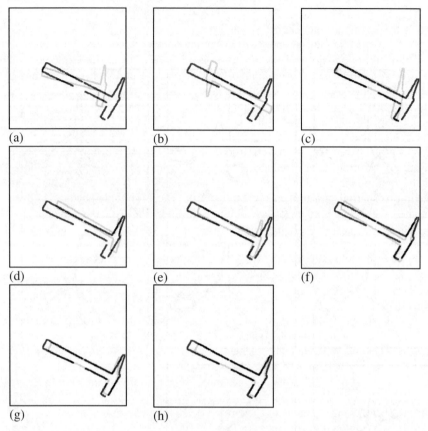

Figure 7: The model and scene contours of a hammer. (a) 1st generation, fitness=0.361653. (b) 3rd generation, fitness=0.467454. (c) 4th generation, fitness=0.542728. (d) 9th generation, fitness=0.619009. (e) 14th generation, fitness=0.737457. (f) 18th generation, fitness=0.760428. (g) 24th generation, fitness=0.921594. (h) 35th generation, fitness=0.950060.

5. Conclusions

In this paper, a novel scheme for Affine invariant matching of partial or broken edge images of an object is presented. Based on the approach taken in [3]-[5], the similarity between a pair of object contours can be determined by whether they are related by a legitimate geometrical transform. This allows the matching task to be conducted as an optimization problem that can be technically addressed with a search process. In the past, the success of this approach was limited by the long search time and the high uncertainty in locating the global solution.

In view of these unfavorable factors, we have proposed to adopt an alternative representation of the Affine Transform so that the search space could be significantly reduced by assigning bounds on the Affine parameters according to the constraints on the camera viewpoints.

As for the search process, we had employed the "Floating Point Genetic Algorithm" for two major reasons. First, the search landscape can be explored with very fine resolution based on a simple, single gene chromosome representation. Second, the latter allows the adoption of a crossover operation that localizes the child chromosomes to the vicinity of their parents. In contrary to binary *GA*, where schema may be destroyed after the crossover operation, genetic information is redistributed and preserved in the child population. To alleviate some of the inherent problems in *GA*, we have introduced a new fitness function to encourage "close" competition and to reduce premature convergence in early generations. This has led to high and consistent success rates that are difficult to achieve in the matching of partial and broken object contours. The proposed scheme has been evaluated with large amount of test samples and the results are encouraging. For the majority of cases, the similarity between pairs of contours is successfully determined in less than 35 generations. The encouraging findings demonstrate the feasibility of the approach and its strong potential to be adopted as a foundation for further research in occluded object recognition.

References

[1] W.J.Rucklidge, "Efficiently locating objects using the Hausdroff distance", *Int. J. of Comp. Vis.*, vol.24, no.3, pp. 251-270, 1997.
[2] A. Toet and W.P. Hajema, "Genetic contour matching", *Pattern Recognition Letters*, vol. 16, pp. 849-856, 1995.
[3] P.W.M.Tsang, "A Genetic Algorithm for Affine Invariant Object Shape Recognition", *Proc. Instn. Mech. Engrs.*, vol. 211, *part I*, pp. 385-392, 1997.
[4] P.W.M. Tsang, "A genetic algorithm for aligning object shapes", *Image and Vision Computing*, vol. 15, pp. 819-831, 1997.
[5] P.W.M. Tsang, "A genetic algorithm for affine invariant recognition of object shapes from broken boundary", *Pattern Recognition Letters*, vol. 18, pp. 631-639, 1997.
[6] J.H. Holland, *Adaptation in Natural and Artificial Systems*, Univ. of Michigan Press, Ann Arbor, Mich., 1975.
[7] Wing Hon Tsang, "A novel floating-point genetic algorithm for matching rigid near planar objects under suppression of specular edge", *Thesis (Ph.D.)*, City University of Hong Kong, 2001.

Programmable Focussing Mechanism for the Vision System of a Gold Wire Bonding Machine

Ajit S Gaunekar, Gary P Widdowson, Wang Guangneng,
Luo Xiao Ming and Chen Xiong Hui
R&D Department, ASM Technology Singapore Pte. Ltd.
2 Yishun Avenue 7, Singapore 768924
Email: ajit@asmpt.com

Abstract
This paper deals with the development and performance evaluation of a high precision opto-mechanical device, which is used to move a lens relative to another fixed lens, thus altering the focus of the optical system. The device employs flexure bearings and a voice coil motor (VCM) as actuator. It is used in the computerised vision system of a wire bonding machine, which bonds gold wire on stacked dies.

Keywords: Lens focus, flexure bearing, voice coil motor (VCM), vision system, wire bonder.

1. INTRODUCTION

Of all semiconductor interconnection technologies, wire bonding is the most flexible and the most widely used [1]. Fig. 1 shows a schematic diagram of the bond-head of a ball-bond type wire-bonding machine. Gold wire (17-75 μm in diameter) connecting the chip to the substrate, is bonded onto bond-pads on the respective surfaces, using a ceramic capillary tip mounted into an ultrasonic transducer. Bonding is achieved through a "rocking" motion of the bond body about a suitably located pivot point on the bond-head housing, through a small angle (< 10 degrees). At each touchdown of the capillary tip, ultra-sonic energy is fed to the interface between the wire and the bond-pad (heated to 200-300 degrees Celcius) to form the bond. The typical bonding rate is 12-15 wires per second implying about 24-30 cycles per second for the bond-body. The bond-head is mounted on a X-Y table which moves in a horizontal plane at a peak acceleration of 100-120 m/s^2 in each direction, possibly simultaneously. Present–day wire-bonders are expected to bond on pads with a pitch of around 40-50μm with an accuracy of +/- 4 μm and a repeatability of +/- 2 μm at their best.

Before bonding begins, a computerised vision system using pattern recognition algorithms, locates precisely the position and orientation of the substrate and the die bonded onto it. This includes locating all the leads on the substrate and bond pads on the die and then transforming and correcting the taught locations for each

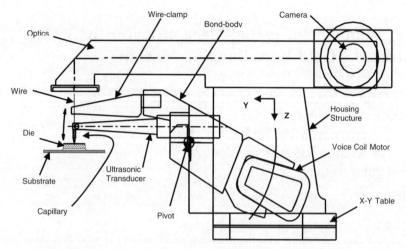

Fig. 1. Schematic of a Voice Coil Motor driven bond-head of a gold wire bonder

bond. The vision system is also used for post-bond inspection in order to monitor bond placement accuracy and wire sweep.

2. NEED FOR A FOCUSSING SYSTEM

Fig. 2 shows a typical fixed focus optical assembly of the vision system. It comprises:
a) a set of lenses for magnification and focus,
b) arrays of LEDs for illumination,
c) relay components such as mirrors, beam splitters, filters and
d) a camera

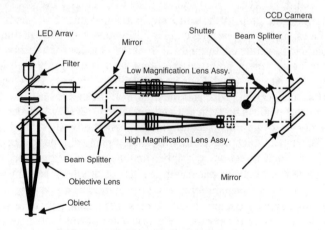

Fig. 2. Wire bonder dual path optics

Two optical paths, one for low magnification and the other for high magnification with a reduced field of view, are provided. A shutter mechanism actuated by a torroidal actuator toggles between the two paths, as and when required. The low magnification path is used for imaging the substrate leads and the high magnification path for the die pads.

Such fixed focus optics, although sufficient for bonding on conventional devices, is inadequate to serve the newly emergent stacked die architecture wherein two or more dies are mounted on top of each other onto a single substrate. In such a configuration, the need to bond on die-pads at different heights necessitates real time focussing while imaging the die-pads. This may be achieved by either moving the objective lens or one of the relay lenses in the high magnification path. The latter option has been adopted in the present case and a Programmable Focussing Mechanism (PFM) has been designed and developed to serve the purpose.

3. REQUIREMENT SPECIFICATIONS

The motional and optical requirement specifications of the PFM are listed below.

Travel range	3mm
Typical stroke response	0.5 mm in 30 ms
Repeatability in axial position	10 µ
Maximum process duty cycle	10 %
Maximum process "ON" time	5 s
Absolute inter-lens de-centration/tilt	50 µ/ 0.2°
Repeatability in image position	1µ

4. OVERALL DESIGN APPROACH

Fig. 3 shows a schematic of the PFM design. It consists of two flexure bearing stacks, separated adequately in order to provide sufficient moment stiffness. Housed in the gap between the two stacks of flexures, is an actuator in the form of a cylindrical Voice Coil Motor (VCM). A Linear Variable Differential Transformer (LVDT) provides position feedback, enabling the VCM to be operated in closed loop servo mode for very precise control over the axial position of the lens mounted on the moving member.

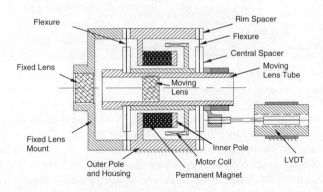

Fig. 3 Programmable Focussing Mechanism (PFM) schematic layout

Since all the moving parts are axi-symmetric about the lens axis, except for the very lightweight LVDT core, the centre of gravity of the moving parts lies very close to the lens axis. Thus the actuating force due to the VCM, the restoring force of the flexures and the inertial force during acceleration are very nearly co-linear. This helps to keep the moment load on the flexures to a manageably low value, ensuring thereby that the lens executes rectilinear motion with negligible tilt.

5. FLEXURE BEARING

The PFM employs circularly symmetric flexure bearings for precise axial motion. Flexures with circular symmetry do not suffer from parasitic lateral motion, but instead give rise to a small angular rotation about the translation axis. These are in the form of flat metallic discs, fractions of a millimetre thick. In general, each disc may have a specified number of slots (usually but not always, three), of either spiral or straight or arc shape or a combination thereof [2,3,4]. They are machined using wire Electro-Discharge Machining or etched using conventional photolithography or any other suitable method, yielding a number of flexing "arms" which bear the load of the moving member.

In the present case, each flexure bearing stack consists of one such disc (Fig. 4a) sandwiched between rim spacers (Fig. 4b) and central spacers (Fig. 4c) used for clamping the non-flexing sections of the flexure discs. The central hole in each flexure bearing fits snugly over the lightweight tube, which houses the moving lens. Very high ratios of radial stiffness to axial stiffness, typically about three orders of magnitude, can be realized using such flexure bearings.

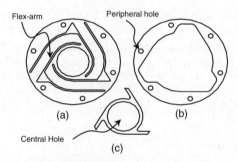

Fig. 4. Flexure bearing components

6. VOICE COIL MOTOR

As seen in Fig. 3, an axially magnetized ring-shaped permanent magnet is glued into the main housing. The permanent magnet is made of a high energy density material such as Neodymium Ferrous Boron. A ring-shaped pole piece of magnetically permeable ferrous alloy, is glued on the magnet. The main housing, which is also made from magnetically permeable iron alloy, acts as the outer pole. Thus the annular air gap between the inner pole piece and the main housing contains a radial magnetic field. The motor coil is appropriately positioned in the magnetic air gap. When the coil is energized with an electrical current, an axial force is induced on it. The direction of the current determines the direction of the actuating force. This force of the VCM is used against the restoring force of the flexure bearings, to move and position the lens at the desired location.

7. CONTROLLER ARCHITECTURE

The controller/driver board shown in Fig. 5, is used to implement the command from the host PC of the wire bonder, in driving the PFM. Its functions are: i) moving the lens from one specified position to another, ii) holding the lens at a specified position and iii) driving the lens in an oscillating motion during operation in the Synchronised Imaging (SI) mode (discussed below).

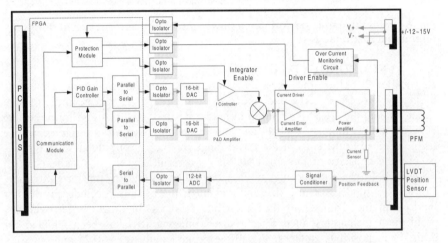

Fig. 5. Block diagram of the controller architecture

The board communicates with the host PC via a PCI interface. In addition to normal control functions, the control board has some other features, like fault protection (short circuit, over-power and over-temperature protection), self-detection (PCI interface, power supply and lens home position) and self-tuning. The self-tuning function enables the control board to drive the lens to various positions with optimal performance in terms of minimum settling time, no over-shoot, and negligible position error.

8. MOTION PERFORMANCE

Tests have been conducted to evaluate the motion performance of the PFM. The dynamic response and repeatability in axial position have been measured. The PFM meets the specifications in this regard throughout the travel range. Fig. 6 shows the typical motion profile of the PFM and the current drawn for a bi-directional move from home position to +/-1.3 mm respectively. The settling time of 14 ms is well within the requirement specification.

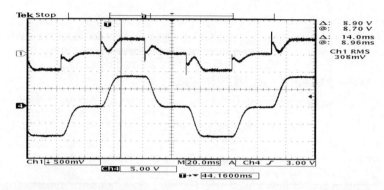

Fig. 6 Typical dynamic performance (lower trace) and current drawn during bi-directional response test

9. SYNCHRONOUS IMAGING

As mentioned before, the worst case duty cycle and the maximum "ON" time during the process are 10% and 5 seconds respectively. However, in the manual 'teach' mode, the VCM is required to hold a desired axial position for anywhere between a few minutes to half an hour. Balancing the restoring force of the flexures, close to the ends of the travel range for such a long duration of time, is beyond the capacity of the small VCM. As a result, the motor overheats when made to hold close to the maximum specified travel of 1.5 mm on either side. Constraints imposed on the size and weight of the module preclude any further increase in the diameter of the motor and flexures. In order to get over this problem, a technique called Synchronised Imaging (SI) has been adopted for desired lens positions greater than half the stroke length, i.e. +/-0.75 mm on either side of the home position.

When required to focus beyond +/-0.75 mm, the lens is first moved to the desired position and held there for a short time interval (~8-10 ms) adequate enough for image capture by the grabber board. The controller then de-energizes the motor whereby the PFM essentially acts as an under-damped, single degree of freedom spring mass system in free vibration. The lens thus springs back to the home position and overshoots beyond, until the restoring force of the flexures (now acting in the opposite direction), brings it to rest and swings it back towards the home position. As the lens is on its way back, the controller energizes the VCM again to bring it back under control to be able to re-position it accurately and delay it there again long enough for image capture once more. This cycle (Fig. 7) is repeated at a rate of about 24 cycles per second, faster than the persistence of human vision. Thus, in the manual "teach" mode, the image appears to be steady to the eyes of the operator. A higher refresh rate for improved image quality is possible with a higher frequency SI mode operation since system response and delay for image capture together account for only about 25 ms, at worst. However, this will increase the duty of the flexures and may compromise overall reliability.

The typical motion profile and current drawn during SI mode operation at 1.3 mm is depicted in Fig. 7. The r.m.s. current drawn is about 0.61 A, substantially lower than the 1.2 A needed to hold the lens steady in that position. This reduction

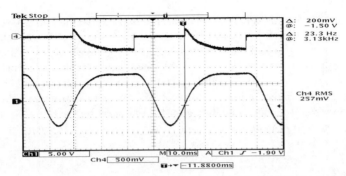

Fig. 7 Motion profile (lower trace) and current drawn during SI mode operation for a stroke of 1,3 mm

of current to 50% of its static hold value translates to a 75% reduction in copper loss. Besides, forced convection heat dissipation rate in the SI mode is substantially higher than natural convection in the static mode.

10. PATTERN RECOGNITION TEST

The PFM has been tested for long term image repeatability alongside an existing fixed focus optical module for comparison. Representative results of tests conducted for different axial positions of the PFM are depicted in Fig. 8. Also shown are results for a conventional fixed focus optical assembly tested alongside the PFM.

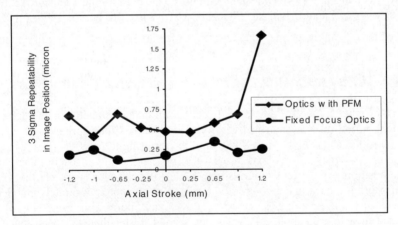

Fig. 8 Repeatability of image position of PFM and fixed focus optics

The PFM meets the optical specification to about 1 mm travel stroke on either side of the home position. However, as the stroke increases, the spread of data spills outside the specified band. This is on account of a coupled effect of two factors: i) variability in lens position during SI mode operation due to random effects and ii) a rapid drop in the radial stiffness of the flexure bearings with increasing stroke.

To ameliorate this situation, it has been proposed that the home position of the fixed lens be adjustable to an extent that relaxes the total travel range of the moving lens to about half the present value. Thus we have two possible approaches to the problem at hand: i) For "hands-off" applications with less stringent optical repeatability specifications, the PFM can operate in the SI mode without the need for any manual intervention whatsoever. ii) For more stringent wire bonding applications viz. fine pitch bonding, adjustment option for the home position of the moving lens will obviate the need for SI operation, while maintaining very high imaging precision.

Conclusion
A high precision programmable focussing mechanism has been developed and tested for motional and optical accuracy in operation. The use flexure bearings and a cylindrical voice coil motor has resulted in a compact design and high repeatability in lens positioning. In order to avoid excessive heating of the motor, a special technique called Synchronized Imaging is used for focussing needs beyond half the maximum stroke range. At the time of writing, further tests were being conducted to determine the long term reliability and robustness of the unit.

References
1. G.G. Harman: Wire Bonding in Microelectronics: Materials, Processes, reliability and Yield, Mc Graw-Hill, 1997.

2. A. S. Gaunekar, T. Goeddenhenrich and C. Heiden,: Finite Element Analysis and Testing of Flexure Bearing Elements, *Cryogenics* **36**, No. 5, 1996, pp. 359-364.

3. T.E. Wong, R.B. Pan, H.D. Marten, C. Sve, L. Galvan, and T. S. Wall: " Spiral Flexure Bearing ", *Cryocoolers* **8**, *Plenum Press, New York,* 1995, pp. 304-310.

4. T.E. Wong, R.B. Pan and A.L. Johnson: "Novel Linear Flexure Bearing", 7th *Intl. Cryocooler Conf.,* 1992, pp. 675-698.

Using a Panoramic Camera for 3D Head Tracking in an AR Environment

Björn Giesler, Tobias Salb, Rüdiger Dillmann,
IAIM, University of Karlsruhe (TH), Germany
E-mail:{giesler,salb,dillmann}@ira.uka.de

Tim Weyrich,
ETH Zürich

Abstract
For Augmented Reality (RA), using a pair of transparent 3D glasses, a precise and fast method for head tracking is required to determine the user's position and direction of gaze in all six degrees of freedom. The methods currently available require expensive external sensors and have small working areas and/or other limitations. We propose a method that uses a panoramic camera that is mounted directly on the user's head, combined with cheap, easily mountable passive artificial landmarks. The panoramic camera uses a paraboloid mirror, which allows for interesting algorithmic simplifications. The system has been tested both in simulation and in reality and shows promising results.

Keywords: *Augmented reality, head tracking, panoramic camera.*

1 INTRODUCTION

Augmented Reality (AR) is the layering of Virtual Reality elements (such as 3D models or markers) over a viewer's image of the real world. This can be achieved in a number of ways, the most popular of which is to have the user wear a pair of semi-transparent 3D goggles. In Karlsruhe, we are constructing an AR system for human-robot interaction, to make it easy for the user to immediately see the interpretation of his or her actions by the computer. To be able to overlay elements of simulation with the view through the 3D glasses, it is necessary for the system to very precisely keep track of the viewer's head position and direction of gaze [7].

The tracking must meet real-time requirements to prevent lagging of the virtual elements. Since it is very difficult to model human head motions, the benefit of using motion-prediction techniques is very limited. For real-world usability, the tracking should also be able to cover as wide an area as possible, should be very easy to set up, require little or no modification to the environment and, last but not least, should not be prohibitively expensive.

Some existing approaches to this problem make use of external camera systems, such as the commercially available *POLARIS* [5]. Such systems, while often very precise, have a limited working space and are therefore more suitable for usage in stationary applications, such as surgical aids [6]. Others require extensive modification to the environment, such as the University of North Carolina's *HiBall* tracking system [8].

Our novel approach consists of a panoramic camera that is affixed to the AR glasses and tracks artificial environmental features. This approach, using only a single camera, requires relatively little computation and therefore comes close to real-time requirements; it is also inexpensive. We are currently tracking artificial targets that can be mounted very quickly and easily, so very little environmental modification is necessary.

2   PROPERTIES OF THE SYSTEM COMPONENTS

We have decided to use a system that uses artificial landmarks instead of natural environmental features, because we consider a 6-degree of freedom (DOF) position reconstruction using natural features to be still too difficult and most of all too costly in terms of processing time to meet with the required real-time constraints. Therefore, we can design both the features that we are recognizing and the sensors that we recognize them with to be perfectly adapted to each other.

2.1   Properties of the Panoramic Camera

A panoramic camera is a system using a CCD camera taking pictures of a convex mirror that reflects a distorted view of the environment. This mirror can be a half-sphere, cone, paraboloid or any other regular convex body. Using panoramic cameras for position reconstruction is not in itself a novel approach; such cameras have been successfully used, for example, for position estimation of mobile robots (with a conical mirror: [1], [2]; with a spherical mirror: [3]). However, so far work has been limited to two-dimensional reconstruction. For three dimensions, a novel approach is needed, and the camera should fulfill certain conditions.

Position reconstruction from a single image and known landmarks is essentially triangulation, since information about distance to the landmarks is not known. For triangulation, it is necessary to take multiple bearings from a single point in space. That means that the *rays of sight* that hit the landmark centers have to emanate from, or meet in, a single point.

For two-dimensional position reconstruction, it is sufficient to locate environmental features that lie in the plane of reconstruction, or in a plane parallel to that. A conical mirror is perfectly suited for this application; as can be seen from **figure 1**.

To make three-dimensional position reconstruction possible from a single image, it is necessary to find environmental features distributed in all three dimensions around the camera; it must be possible to take bearings toward multiple non-coplanar environmental features.

A parabolical mirror has this important property (see **figure 2**): All rays of sight that hit the mirror parallel to the mirror's symmetry axis are reflected in such a way that they all seem to originate in the focal point of the paraboloid. Therefore, we can relate all bearings to the focal point.

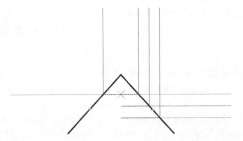

Figure 1: Deflection of rays of sight by a conical mirror.

A parabolical mirror does have the disadvantage that it distorts images of objects in a non-trivial way; while a circle reflected in a conical mirror becomes an easily-matchable ellipse, a paraboloidal mirror distorts it into an egg shape. Furthermore, the center of gravity of the pixel set that forms an object's image does not coincide with the object's center of gravity in cartesian space. It is therefore not trivial to match an object in the camera image or even its center of gravity with numerical methods.

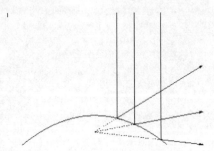

Figure 2: Deflection of rays of sight by a parabolical mirror

However, **figure 2** shows also that in a picture taken of a parabolical mirror, there is a simple correlation between azimuth and elevation of a point in space and the image coordinates of its picture. Since we are reconstructing the position by triangulation, we can perform all our calculations in *ray space*, that is, the vector space spanned by all rays of sight emanating from the mirror's focal point. Therefore, the reconstruction of object shapes should take place in ray space as well. We achieve this by conic matching; the method is described in section 3.1.

The camera currently in use has a mirror covering an angular area of 360° azimuth and 60° elevation. The mirror's image is taken by a 640x480 pixel CCD and encoded as a PAL signal. The resolution is very low considering the large section of the environment that is depicted; **figure 3** shows an image taken with the camera and its cartesian reconstruction. However, simulation and experiments both show that the low resolution mainly poses problems for actual landmark recognition; the camera resolution has only marginal effect on the accuracy of the position reconstruction if sufficiently many landmarks are successfully recognized.

2.2 Properties of the Artificial Landmarks

To meet with the real-time requirements outlined in section 1, landmarks should be designed in such a way that we can find them very easily and quickly. We are currently using circular landmarks in the primary colors red (lighter zone) and blue (dark zone), as shown in **figure 4**. The landmarks are not uniquely coded in any way, because owing to the limited resolution of the panoramic camera they should be extremely simple in structure to be easily recognizable even if their image is only a few pixels large.

The landmark colors make it easy to find the exterior and interior of the landmarks by examining only the red and blue channels of the RGB image stream (thus working on two separate binary images) and finding blobs. This results in two sets of blobs, one for each channel. We then calculate the size of each blob's bounding box and compare the bounding boxes in the red channel with those in the

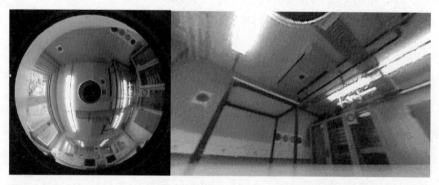

Figure 3: A 360° image and its cartesian reconstruction. The image resolution is 640x480 pixel; this can be seen in the coarse reconstruction results, especially towards the top of the picture.

blue channel. If a 'red' bounding box lies within a 'blue' one or vice versa, and the sizes of the bounding boxes fulfil, within certain limits, the known size ratio of the inner and outer circles, a landmark has been found. **Figure 5** shows the basic process of landmark identification.

3   LANDMARK RECOGNITION AND POSE RECONSTRUCTION

The outlined process makes it possible to quickly find landmarks in the camera image. However, their *precise* location and shape in ray space is still not known. The camera's limited resolution necessitates the development of some scheme to achieve sub-pixel accuracy. As stated in section 2.1, the camera's parabolic mirror makes simple center-of-gravity calculation difficult; but in the course of this work, a method has been developed that is both more precise and makes excellent use of the camera's properties.

Figure 4: Circular landmarks in red (light area) with a blue (dark zone) border (left circle) and blue with a red border (right circle). The landmarks are easily recognizable and have a defined color border circle that is important in conic matching.

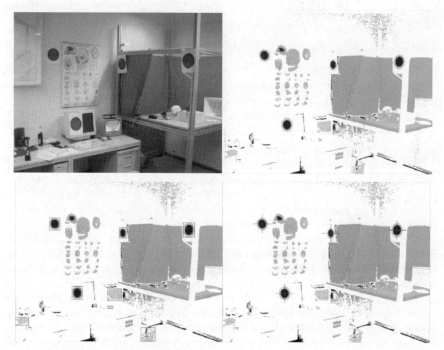

Figure 5: The process of finding landmarks in the camera picture. Top left: Original image. Top right: Finding red and blue areas. Bottom left: Calculating bounding boxes. Bottom right: Detecting landmarks where correct bounding box ratios are found.

3.1 Conic Matching

Since working with the 2D representation of the landmarks is not practical, due to the distorting properties of the parabolic mirror, we have decided to use the landmarks' 3D shape. Therefore, the border pixels are detected where a landmark's outer ring adjoins to the inner circle. It is known that these pixels must (within some margin of error) lie on a cone with an elliptical cross-section whose origin is the optical center of the camera. Therefore, rays from the optical center are constructed that go through the border pixels, and a least-squares error minimization algorithm is used to construct a conic through the optical center that approximates the rays most closely. The conic's center ray is then used as a bearing for triangulation.

3.2 Different Landmark Centers

As can be seen in **figure 6**, one final ambiguity must be resolved: The conic matching process delivers a very good estimate of the landmark's perimeter, but not of the landmark's center. If the landmark is not seen straight-on, there are two possible surface normals and therefore two possible centers: Let H be the set of all planes whose section with a given conic C is circular. Then H consists of two plane lots $_i\{H_i = H \parallel H_i, H \quad H_{\cap} \quad \notin i \in \{1,2\}$ created by two planes H_1 and H_2 going through the coordinate origin. There are especially two "candidate planes", $H_{c1}, H_{c2} \in H$, that contain the landmark's center. The parallel projections of H_{c1} and H_{c2} onto a plane perpendicular to C are identical and elliptical; since all we see is this projection, we cannot tell which of the two planes contains the landmark. Since we are only interested in the center for bearing, this normally would not matter; however, we do not really see a *parallel* but a *perspectivic* projection of $H_{c1} \cap C$ or $H_{c2} \cap C$, and these are not identical but slightly distorted; therefore, the center is slightly shifted as well (see **figure 7**).

To solve this ambiguity, we use a heuristic that employs the fact that most walls in human-habited areas are perpendicular to each other; therefore, if multiple landmarks can be seen, their candidate planes can be determined by selecting a combination of candidate planes over all landmarks that are either coplanar or perpendicular (within a margin of error).

Up to this point, all calculation takes place in ray space, that is, the three-dimensional rotational coordinate system rooted at the optical center of the camera. Due to lack of stereo information, it is not possible to determine the distance to the landmark at this point; therefore, all rays are represented as infinite lines. This method delivers a high sub-pixel accuracy while operating directly on the camera image, which is given in ray space. No conversion to the cartesian coordinate system is necessary, which makes this approach both fast and accurate.

Figure 6: Border pixel detection and conic matching. The first and second images show the distorted camera picture; the jagged edges are due to the interleaved video signal. The third image shows the landmark in "real-world", i.e. cartesian coordinates. The two crosses mark the two possible normal vector roots. Dark outer zone is blue, lighter inner zone is red.

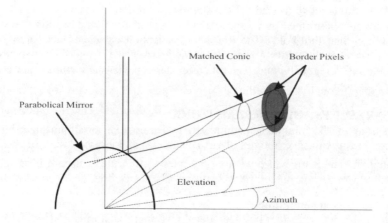

Figure 7: Schematic of conic matching. Elevation and azimuth of the rays toward the border pixels are extracted from the mirror image and are used to create a conic using a least-squares method. The resulting conic describes the landmark in ray space except for its distance, which is not needed for triangulation.

Unfortunately, the method only works for landmarks that are close enough for a sufficiently large number of border pixels to be detected. Therefore, we use a simple center-of-gravity approach as a fallback if the landmark's image occupies less than 16x16 pixels. If the image is so small, the distorting effect of the parabolic mirror can be neglected.

3.3 Pose reconstruction

After the landmarks have been detected and identified, the viewer's pose can be reconstructed: As stated above, the angles of the landmark normals are now known, and therefore the angle that the camera takes relative to each landmark is known as well. When the positions of the landmarks in cartesian space are known, the reconstruction becomes trivial. Theoretically, three recognized landmarks would suffice to determine the camera's pose, but if more can be found in the image, the resulting over-determined system can be used to enhance accuracy.

The pose is reconstructed via translatoric and rotatoric adaption. The adaption steps are also performed in ray space and are repeated until the resulting error lies below a certain threshold[1]. The translatoric adaption calculates the approximate section point of the viewing rays of the landmark centers, and then matches the viewer's position to that point. The rotatoric adaption uses the method described by B.K.Horn in [4].

4 RESULTS AND FURTHER WORK

The system has been implemented initially in a simulation environment, to be able to experiment with several types of virtual cameras and use the results in choosing the optimal real camera. Therefore, we present two sets of experimental results; one for the simulation system and one for the actual camera.

4.1 Results in simulation

In [9], a simulation system has been developed that transforms a 3D scene into a simulated camera image. This system has made it possible to test the effects of changes in camera resolution and interlaced/non-interlaced sensors on the accuracy of position reconstruction. **Figure 8** shows an image obtained from the camera simulation.

The simulation environment allows the recording and playback of trajectories in the simulated 3D scene. In playback, the position reconstruction works on the simulated camera image. **Figure 9** shows a recorded trajectory from two different angles. The graphs on the right display the rotational and translational errors corresponding to the reconstructed trajectory.

The graphs show that the rotational error on this (fairly typical) run is always well below one degree, and the translational error below 5mm. These results were achieved by simulating a NTSC camera of 640x480 pixels interleaved resolution. Moving toward a simulated camera of 1024x768 pixels and "progressive scan" technology improves the medium errors to 4.25mm (10.9% improvement) and

---

[1] It can be shown that the employed method does not converge in some rare cases. Therefore, the reconstruction process is aborted after a maximum of 48 repetitions.

0.14° (4.4% improvement). This shows that an increase in resolution alone does not significantly increase the accuracy of the method.

4.2 Results with a real camera

The actual panoramic camera (right) is an NTSC camera yielding pictures at 640x480 pixel interleaved resolution, equalling the camera simulation. The results with this real camera mirror those achieved with the simulated one; however, image noise (which was disregarded in simulation) poses a large problem; the color-separated red and blue images are too noisy to effectively recognize the more distant landmarks. **Figure 10** shows the results of a test run with this real camera; the dashed lines mark a time segment when the camera was moved out of the range of several landmarks. Naturally, in

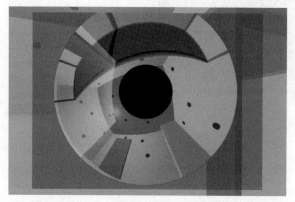

Figure 8: A simulated camera image. The image is superposed with a conventional, cartesian view of the simulation environment, i.e. the view that a user carrying the camera upon his/her head would see. The image segment representing this view in the simulated panoramic camera is marked as a bright segment in the camera image.

this time the error increased well below the tolerable measure, which reflects in the medium accuracy over this test run.

4.3 Further work

The camera module used at the moment is far too noisy to be really useful for position reconstruction in larger areas. Therefore, our main focus right now is twofold: First, we are examining image enhancement techniques to investigate whether images from the existing camera can be improved sufficiently; secondly, other more adequate camera modules are being evaluated.

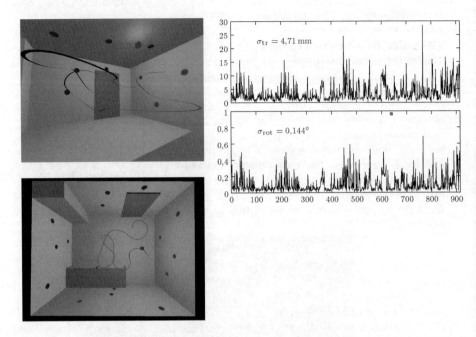

Figure 9: A recorded trajectory in the simulation environment (left) and the translational and rotational errors of the position reconstruction working on the simulated camera image.

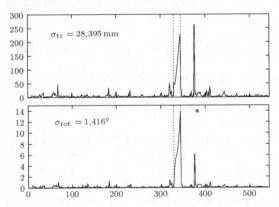

Figure 10: Rotational and translational errors derived from a trajectory followed with the actual camera. The dashed lines mark a time segment where only three landmarks were seen.

Once a sufficiently good camera has been found, the system will be enhanced to support automatical calibration, so that the landmark positions do not have to be initially known. Afterwards, we will use the system as the main head tracking component in the AR system used at the University of Karlsruhe for robot programming by demonstration.

5 REFERENCES

[1] M.O. Franz, B. Schölkopf, H.A. Mallot and H.H. Bülthoff, A.Zell. *Navigation mit Schnappschüssen*. In: Proceedings of the 20th DAGM Symposium, Springer Verlag, Berlin, 1998.

[2] M.O. Franz, B. Schölkopf and H.H. Bülthoff. *Homing by Parameterized Scene Matching*. In: Proc. of the 4th European Conference on Artificial Life, MIT Press, Cambridge, 1997.

[3] J. Gaspar and J.S. Victor. *Visual Path Following with a Catadioptric Panoramic Camera*. In: Proceedings of the 7th International Symposium on Intelligent Robotic Systems (SIRS'99), University of Coimbra, 1999.

[4] B.K. Horn. *Closed-form solution of absolute orientation using unit quaternions*. In: Journal of the OSA, Issue 4, April 1987.

[5] Northern Digital Inc. *Product website for the Polaris Optical Tracking System*. At http://www.ndigital.com/polaris.html

[6] T. Salb, J. Brief, O. Burgert, S. Haßfeld and R. Dillmann. *Intraoperative presentation of surgical planning and simulation results using a stereoscopic see-through head-mounted display*. In: Proceedings of Stereoscopic Displays and Virtual Reality Systems VII, part of SPIE / Photonics West 2000, San Jose, CA, January 2000.

[7] T. Salb, O. Burgert, T. Gockel, B. Giesler and R. Dillmann. *Comparison of tracking techniques for Intraoperative Presentation of medical data using a see-through head-mounted display*. In: Proceedings of Medicine Meets Virtual Reality 9, Newport Beach, CA, January 2001.

[8] G. Welch, G. Bishop, L. Vicci, S. Brumback, K. Keller, D'. Colucci. *The HiBall Tracker: High-Performance Wide-Area Tracking for Virtual and Augmented Environments.*In: Proceedings of the ACM VRST 99, University College London, December 20-22, 1999.

[9] T. Weyrich. *Entwicklung eines Kopfverfolgungssystems auf der Basis einer Panoramakamera und künstlicher Landmarken*. Master's Thesis, University of Karlsruhe (TH), 2001.

Industrial Applications

The use of mechatronics in industry has been one of its core applications. From the beginning, the discipline has taken two general directions – one for production mechatronics, the other for design mechatronics. This section looks at six different applications that are mainly used in industry, some of which have many applications elsewhere.

The first paper considers a very important problem in the garment industry – how do you classify defects in the manufacturing of fabric? It presents a new method for fabric defect classification by using a wavelet frames-based feature extractor and a minimum-classification -error based neural network. The authors claim that a 93.4% classification accuracy has been achieved.

Sensors play a dominant role in achieving autonomous and intelligent behaviour, by allowing a system to learn about the state of its physical environment and thereby interact with it intelligently. In the second paper, the integration of a vision system and a force/torque sensing system in a flexible manufacturing workcell environment is presented. The workcell primarily includes two six-degree-of-freedom robotic systems for carrying out cooperative tasks under computer control.

The third paper considers an important problem in many manufacturing situations – the tendency for a workpiece to vibrate when it is being machined. Active vibration control appears to be a suitable solution for the noise and vibration problem in boring operations. By using small piezo-ceramic stack actuators, large forces can be applied to the boring bar while at the same time the space where the actuator is located can be kept small. This paper presents the results of a project using such methodology.

The automobile industry is one of the heavy users of both production and design mechatronics. The fourth paper in the section presents a fully automated raw foundry brake disk visual inspection system. Three different computer vision techniques are used via complete piece rotation. The whole system could be applied to any component with circular symmetry, and is a blending of mechanics, automation, computer vision and robotics.

The fifth paper presents a completely different application: robot assemble of the two halves of a mobile phone casing. Joining of the front housing and back chassis of a typical mobile phone is made possible where notch-locked assembly is involved. An assembly strategy based on three force-based compliant motions is described.

The final paper looks at a technology that will become more important in the future – micro-engineered mechanical structures, MEMS. One fundamental challenge lies in the fact that at micro-scale, micro-mechanical structures are fragile and easy to break. This paper presents the development of a polyvinylidence fluoride (PVDF) multi-direction micro-force sensing system that can potentially be used for force-reflective manipulation of micro-mechanical devices or micro-organisms over remote distances.

Fabric Defect Classification using Wavelet Frames and Minimum Classification Error-Based Neural Network

Grantham Pang, Xuezhi Yang and Nelson Yung
Dept. of Elec. & Electronic Engineering,
The University of Hong Kong.

Abstract
This paper presents a new method for fabric defect classification by using a wavelet frames feature extractor and a minimum classification error-based neural network. Channel variances at the outputs of the wavelet frame decomposition are extracted to characterize each non-overlapping window of the fabric image, which is further assigned to a defect category with a neural network classifier. In our work, a Minimum Classification Error (MCE) criterion is used in the training of the neural network for the improvement of classification performance. The developed defect classification method has been evaluated on the classification of 329 defect samples from nine types of defects and 82 non-defect samples, where an 93.4% classification accuracy was achieved.

Keywords: *Fabric inspection, defect classification, wavelet frames, neural network, minimum classification error.*

1. INTRODUCTION

Fabric Automatic Visual Inspection (FAVI) is becoming an attractive alternative to human vision inspection in modern textile industry. Based on the advances in image processing and pattern recognition, FAVI can potentially provide an objective and reliable evaluation on the fabric production quality. The classification of fabric defects is an important part of FAVI. Since different types of fabric defect have different effects on the fabric product, the classification of fabric defects is necessary for the grading of fabric products. Based on defect classification, the statistics for each type of defects can be obtained, which further indicates the potential problems in certain components of the weaving machine. Moreover, on-line classification of fabric defect provides necessary information for the real-time quality control of the weaving process.

Currently, most of the FAVI systems only achieve the detection of fabric defects. However, the classification of fabric defects still remains a research topic. The large variations within each type of defect and the similarity among different types of defect are the major obstacles in fabric defect classification. Previous works on defect classification can be divided into two categories. In the first category, fabric defects are classified in terms of their shape characteristics [1,2].

The success of this approach relies on the accurate detection of the defect region. The second category is based on texture analysis. Since different type of defect locally causes distinct type of texture, the classification of defects can be formulated as a texture classification problem. To achieve that, autocorrelation function [3], gray level difference method [4] and local integration [5] have been used to extract statistical texture features for defect classification.

In this paper, a new method for defect classification is presented, which is illustrated in Fig. 1. The method has the following features:
- Wavelet frame decomposition [6] is used for feature extraction in the defect classification. Compared to the single-scale statistical texture features, channel variances at the output of the wavelet frame decomposition characterize the fabric texture at multiscale, and are able to provide more efficient discriminations among different types of defective textures.
- A neural network, which is trained by Minimum Classification Error (MCE) training method [8], is employed for defect classification. Compared to the traditional backpropagation algorithm, the MCE training method yields a network which is more consistent with the objective of minimum classification error.

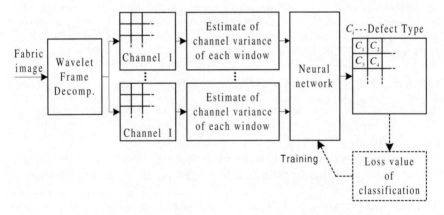

Fig. 1. The developed fabric defect classification method

The developed defect classification method has been evaluated on the classification of 329 defect samples containing nine types of defects, and 82 non-defect samples, where an 93.4% classification accuracy was achieved.

2. FABRIC DEFECT CLASSIFICATION USING WAVELET FRAMES AND MCE TRAINING

The developed defect classification method consists of a feature extraction module and a classification module. In the feature extraction module, feature vectors consisting of channel variances at the outputs of the wavelet frame decomposition are extracted to characterize each non-overlapping window of the fabric image. In the classification module, a neural network classifier is used. Minimization of the classification error is achieved by using the MCE training method, which has been illustrated in Fig. 1 using dashed lines.

2.1 Feature extraction based on wavelet frame decomposition

Fig. 2 illustrates the filter bank implementation of a 2-D wavelet frame decomposition, where $H(z)$ and $G(z)$ denote the z-transform of the low-pass filter $h[n]$ and high-pass filter $g[n]$ respectively. $I(x,y)$ denotes an image and (x,y) is the spatial indices. $\{W_r^1(x,y), W_r^2(x,y), W_r^3(x,y)\}$ are the wavelet coefficients at scale r, in the diagonal, horizontal and vertical orientation respectively. $R_r(x,y)$ represents the residue signal at scale r.

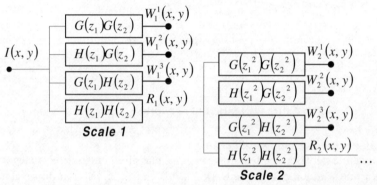

Fig. 2. Filter bank implementation of 2-D wavelet frame decomposition

Corresponding to a window in the fabric image, the channel variances are estimated as the mean energy of the wavelet coefficients in the window [6]

$$w_r^d = \underset{(x,y)\in window}{Mean} [W_r^d(x,y)]^2, \text{ for } d = 1,2,3 \qquad (2.1.1)$$

The channel variances at each channel of the wavelet frame decomposition form the feature vector to characterize the image window

$$\mathbf{x} = [w_1^1, w_1^2, w_1^3, \cdots, w_D^1, w_D^2, w_D^3], \qquad (2.1.2)$$

where D is the depth of the wavelet frame decomposition.

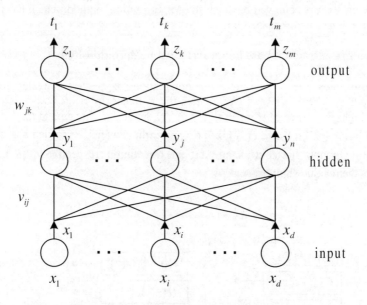

Fig. 3. A three-layer neural network

2.2 Classification using neural network

By using a neural network, the feature vector x is classified into a defect category. Fig. 3 illustrates a three-layer network. $\mathbf{x} = \{x_1, x_2, \ldots, x_d\}$ is a d-dimensional feature vector presented in the input layer. y_j is the activation output of the jth node in the hidden layer, which is obtained using the equation

$$y_j = f(nety_j) \text{ with } nety_j = \sum_{i=1}^{d} x_i v_{ij}, \qquad (2.2.1)$$

where v_{ij} denotes the weights connecting the ith node in the input layer to the jth node in the hidden layer. $f(\cdot)$ is a nonlinear activation function, and the sigmoid function can be used. Similarly, z_k, the activation output of the kth node in the output layer, is obtained using the equation

$$z_k = f(netz_k) \text{ with } netz_k = \sum_{j=1}^{n} y_n w_{jk}, \qquad (2.2.2)$$

where w_{jk} denotes the weights connecting the jth node in the hidden layer to the kth node in the output layer.

For the training of the neural network, the backpropagation algorithm [7] is employed. The backpropagation algorithm sets a target output $\mathbf{t} = \{t_1, t_2, \ldots, t_m\}$ for the network, where m is the number of classes. An error J_{mse} on a training pattern x is then defined as the sum-squared difference between the target output t and the actual output $\mathbf{z} = \{z_1, z_2, \ldots, z_m\}$ as follows

$$J_{mse} = \frac{1}{2} \sum_{k=1}^{m} (t_k - z_k)^2. \qquad (2.2.3)$$

Based on the gradient descent method, the backpropagation algorithm adjusts the weights of the network to minimize the error J_{mse}

$$v_{ij}(r+1) = v_{ij}(r) - \mu \frac{\partial J_{mse}}{\partial v_{ij}}; \ w_{jk}(r+1) = w_{jk}(r) - \mu \frac{\partial J_{mse}}{\partial w_{jk}}, \qquad (2.2.4)$$

where $\mu > 0$ is the learning rate, and r indexes the presentation of training patterns.

2.3 MCE training-based neural network for defect classification

A drawback of the traditional backpropagation algorithm is that the decision rule of the classifier is not directly incorporated into the error criterion J_{mse}. As a result, the weights that minimize the error J_{mse} may not be consistent with the objective of minimum classification error. One way to improve the backpropagation algorithm is the use of MCE training method proposed by Juang and Katagiri [8]. Instead of using the criterion J_{mse}, a MCE criterion J_{mce} is given below.

Given a training pattern $\mathbf{x} \in C_q$ presented to the network, where C_q denotes the qth class, a misclassification measure d is defined for $\mathbf{x}$ as follows:

$$d = -netz_q + \ln \left[\frac{1}{m-1} \sum_{p \neq q}^{m} e^{\eta \cdot netz_p} \right]^{1/\eta}, \qquad (2.3.1)$$

where η is a positive constant which controls the contribution from the competing classes. Note that, when η approaches ∞, Eqn. (2.3.1) becomes

$$d = -netz_q + \max_{p, p \neq q} \{netz_p\}, \qquad (2.3.2)$$

According to the decision rule of the neural network, the misclassification measure d enumerates how likely the training pattern x is misclassified.

An error J_{mce} on the training pattern x is then defined by a smoothed zero-one function of the misclassification measure d. The sigmoid function is an example.

$$J_{mce} = l(d) = \frac{1}{1+e^{-\alpha d}}, \text{ where } \alpha > 0. \quad (2.3.3)$$

Following the gradient descent method shown in Eqn. (2.2.4), the weights of the network are adjusted to minimize the error J_{mce}. The gradients of J_{mce} with respect to the weights v_{ij} and w_{jk} are derived as

$$\frac{\partial J_{mce}}{\partial w_{jk}} = \frac{\partial J_{mce}}{\partial d} \frac{\partial d}{\partial netz_k} \frac{\partial netz_k}{\partial w_{jk}}; \quad (2.3.4)$$

$$\frac{\partial J_{mce}}{\partial v_{ij}} = \frac{\partial J_{mce}}{\partial d} \frac{\partial d}{\partial nety_j} \frac{\partial nety_j}{\partial v_{ij}}, \quad (2.3.5)$$

where

$$\frac{\partial J_{mce}}{\partial d} = l'(d), \quad (2.3.6)$$

$$\frac{\partial d}{\partial netz_k} = \begin{cases} -1 & k = q \\ \dfrac{e^{\eta \cdot netz_k}}{\sum_{p,p \neq q}^{m} e^{\eta \cdot netz_p}} & k \neq q \end{cases}; \quad \frac{\partial netz_k}{\partial w_{jk}} = y_j \quad (2.3.7)$$

$$\frac{\partial d}{\partial nety_j} = \left[-w_{jq} + \frac{\sum_{p,p \neq q}^{m} w_{jp} e^{\eta \cdot netz_p}}{\sum_{p,p \neq q}^{m} e^{\eta \cdot netz_p}} \right] f'(nety_j); \quad \frac{\partial nety_j}{\partial v_{ij}} = x_i. \quad (2.3.8)$$

To summarize, the MCE criterion has been used in the training of a neural network. The next section presents an evaluation of this method, and compares it with the traditional backpropagation training method.

3. EVALUATIONS

3.1 Data collection

The developed defect classification method was evaluated on the classification of nine types of typical fabric defects on plain, twill fabrics, as shown in Fig. 4. Fabric without defect should be classified into the nondefect class. In total, eighty-three fabric images containing nine types of defects were used for the evaluation. Feature vectors were extracted to characterize the non-overlapping image windows of size 32x32 pixels. Fourty-two fabric images were used for training, where 336 defect samples and 84 nondefect samples were collected. The remaining fourty-one fabric images were used for the test, where 329 defect samples and 82 nondefect samples were collected.

3.2 Evaluation results and comparison

In the feature extraction based on the wavelet frame decomposition, the Haar wavelet is used, and the feature vectors contain features from scales 1, 2 and 3. The classification rate using the developed method are summarized in Table 1. The method achieves 93.4% accuracy in the classification of the test samples. For a comparative study, the neural network trained by the traditional backpropagation algorithm was also implemented, and its classification rate is 91.4%. Hence, the neural network trained by the MCE method is more consistent with the objective of the minimum classification error.

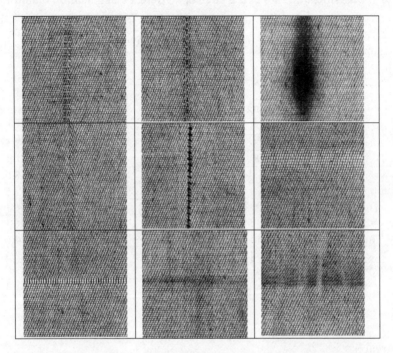

Fig. 4. Fabric images containing defects: Upper row (from left to right): BrokenEnd, SlackEnd, Dirty Yarn; Middle row: Wrong Draw, Netting Multiples, ThinBar; Lower row: Mispick, Thick Bar, Thick Bar type B

Comparing with results by other researchers, Brzakovic and Vujovic [1] gave a classification accuracy of 85% for web material inspection. The same classification accuracy was given by Bradshaw [2] in his classification of defects into four categories: vertical, horizontal, local and slubs. Tolba and Abu-Rezeq [3] reported on a 100% accuracy, but the result was based on classification into only three categories (vertical, horizontal and area defects). Also, only 22 test samples were used. Karayiannis et al. [4] gave an 85% classification accuracy over eight classes of defects (light vertical, dark vertical, light horizontal, dark horizontal, light area, dark area, wrinkle and non defect) but the number of test samples is not mentioned.

4. CONCLUSIONS

This paper presents a new method for fabric defect classification using a wavelet frame decomposition and a MCE training-based neural network. In an evaluation of the method using a test set consisting of 329 defect samples and 82 non-defect samples, an accuracy of 93.4% is achieved. Since non-defect fabric is classified as a distinct class, the method can also be used for the detection of fabric defects.

Table 1. Classification rate of the developed defect classification method

Defect Type	Classification rate (%)			
	MCE training method		Traditional backpropagation training method	
	Training set	Test set	Training set	Test set
Broken End	100	90.0	100	95.0
Slack End	100	97.5	100	97.5
Dirty Yarn	100	100	100	100
Wrong Draw	96.8	75.0	96.8	65.6
Netting Multiples	100	100	100	87.5
Thin Bar	100	100	100	100
Mispick	100	100	100	100
Thick Bar	100	92.8	100	94.6
Thick Bar B	100	100	100	82.5
Nondefect	100	87.8	100	87.8
Overall	99.7	93.4	99.7	91.4

References

[1] D. Brzakovic and N. Vujovic, "Designing a defect classification system: a case study", *Pattern Recognition*, Vol. 29, No. 8, pp. 1401-1419, 1996.

[2] M. Bradshaw, "The application of machine vision to the automated inspection of knitted fabrics", *Mechatronics*, vol. 5, pp. 233-243, 1995.

[3] A. S. Tolba and A. N. Abu-Rezeq, "A self-organizing feature map for automated visual inspection of textile products", *Computers in Industry*, vol. 32, pp. 319-333, 1997.

[4] Y. A. Karayiannis R. Stojanovic, P. Mitropoulos, C. Koulamas, T. Stouraitis, S. Koubias and G. Papadopoulos, "Defect detection and classification on web textile fabric using multiresolution decomposition and neural networks", *Proc. IEEE Int. Conf. on Electronics, Circuits and Systems*, vol. 2, pp. 765-768, 1999.

[5] D. Rohrmus, "Invariant web defect detection and classification system", *Proc. IEEE International Conference on CVPR*, vol. 2, pp. 794-795, 2000.

[6] M. Unser, "Texture classification and segmentation using wavelet frames", *IEEE Trans. on Image Processing*, vol. 4, no. 11, pp. 1549-1560, Nov. 1995.

[7] D. Rumelhart, E. Hinton and J. Williams, "Learning internal representation by error propagation", in D. Rumelhart, J. L. McClelland, and the PDP Research Group Eds., *Parallel Distributed Processing*, pp. 218-364, MIT Press, 1986.

[8] B.H. Juang and S. Katagiri, "Discriminant learning for minimum error classification", *IEEE Trans. on Signal Processing*, vol. 40, no. 12, pp. 3043-3054, Dec. 1992.

Multi Sensor Fusion in a Flexible Workcell Environment

Devendra P. Garg
Professor & ASME Fellow

Manish Kumar
Graduate Student

Duke University, Box 90300
Durham, NC 27708-0300
USA

Abstract
The capability to acquire, process, and integrate information from diverse sources, is a hallmark of intelligent behavior. It is this characteristic that is essentially responsible for the ability of humans and animals to survive and improve. Sensors play a dominant role in achieving this autonomous and intelligent behavior, by allowing a system to learn about the state of its physical environment, and thereby interact with its environment. In this paper, the integration of a vision system and a force/torque sensing system in a flexible manufacturing workcell environment is discussed. The workcell primarily includes two six-degree of freedom (DOF) robotic systems for carrying out co-operative tasks under computer control. Fuzzy logic, Simulink, and Matlab's State Flow toolbox are used for achieving real-time, autonomous and intelligent behavior of the two robots.

Keywords: *Multiple robots, sensor fusion, camera calibration, force/torque sensing.*

1. INTRODUCTION

Modern-day robots operate in a manufacturing world that is inherently uncertain, which arises in the perception and modeling of environments, in the motion of manipulators and objects, and in the planning and execution of tasks. A sensor is a device used to gain, update or refine the information about the environment. To extend their capabilities to uncertain environments, sensor systems must be developed as to be able to interpret the observations of the environment dynamically (in terms of a task to be performed), accounting for this uncertainty and obtaining a refined model of the robot world. Much of the earlier research concentrated on single sensor measurements that inherently incorporated uncertainty and were occasionally incorrect and spurious. This problem could be overcome by the use of multiple sensory systems that have the advantages of redundancy, complimentarity, diversity and timely information.

The area of multi-sensor fusion has been of increasing interest among researchers and its applications in robotic systems have been published in recent literature. Xiao *et al.* [1] have considered a problem of controlling a robot manipulator for a class of constrained motions. Similarly, Matia and Jiminez [2], have implemented mathematical and artificial intelligence techniques in vehicle survival in a dynamic environment, while the robot carries out a specific task. The use of an adaptive or learning approach, such as artificial neural networks, fuzzy logic and genetic algorithms, can prove to be extremely beneficial in real world applications where the explicit model cannot be obtained at all. Mahajan *et al.* [3] have developed an intelligent multi-sensor integration and fusion model that uses fuzzy logic. Similarly, Sun *et al.* [4] have used a neural network-based adaptive controller with an observer, for trajectory tracking of robotic manipulators with unknown dynamic nonlinearities.

2. SENSOR FUSION

The integration of sensory data obtained from several sources into a single element of information is referred to as "sensor fusion". Sensor fusion is especially beneficial when sufficient information about the environment cannot be obtained from a single sensory mechanism for real-time, reliable decision making. A use of sensory signals from different sources provides corroborative information, which allows the maximization of information needed for typical monitoring and control tasks, increases accuracy, and resolves ambiguities in the knowledge about the environment. Moreover, multiple sensors introduce redundancy into the system whereby the decision making process can be made fault-tolerant. Also, since each sensor is in tune with different properties of the environment, a variety of properties can be sensed using multiple sensors, thereby enriching the available information about the environment.

Multi-sensor data fusion addresses the synergistic combination of information made available by various knowledge sources in order to provide a better understanding of a given scene. To achieve these targets it is expedient to have a common method of representing or describing the information contained in the measurements from different sensors. Consistent and coherent methods for combining diverse information are required in a manner that avoids distortion or biases caused by malfunctioning sensors or rogue measurements. Methodologies have been presented based on structural paradigms such as architectures, logical structures or as estimation based algorithms. The variety of methods generally proposed for multi-sensor fusion are: Weighted Average, Bayesian Theorem, Dempster Shafer Evidence Theory, Adaptive Decision and Kalman Filter, which fall into the Statistical Approach category. The methods like Expert System, Rule Based System and Adaptive Learning are not based on Statistical Approach and are called Information Theoretic.

3. FUZZY LOGIC

Fuzzy logic uses linguistic if-then statements involving fuzzy sets, fuzzy logic and fuzzy inference. A fuzzy set consists of a universe of discourse and a membership function that maps every element in the universe of discourse to the membership value between 0 and 1. Fuzzy inference of fuzzy reasoning determines the rule outcome from the given rule input information. Finally, the fuzzy sets generated by fuzzy inference converted to a real number using a defuzzifier and the process is called defuzzification.

Fuzzy logic is robust, can be tweaked easily to improve or alter the system performance, eliminates the need of expensive, accurate sensors, can handle reasonable number of inputs and outputs, and can control nonlinear systems. The knowledge of fuzzy logic control has been applied to dynamic control of robot manipulators to deal with nonlinearities and strong coupling of robot dynamics. The most popular PID controllers are based on the dynamic model of the system. Since robotic manipulators are highly nonlinear, fuzzy logic seems to be a very useful control technique, which can be applied here. Lim and Hiyama [5] have presented a fuzzy logic control strategy for robotic manipulators that incorporates a proportional plus integral (PI) controller with a simple fuzzy logic. Similarly, Yi and Chung [6] have investigated the robustness and stability of a fuzzy logic controller applied to a robotic manipulator with uncertainties, such as friction, unmodeled dynamics, external disturbances, etc.

4. CAMERA CALIBRATION

Camera calibration computes the transformation from the camera space to any given world space. The usual approach to camera calibration has generally been to map camera co-ordinates to arbitrary world co-ordinates. The problem of mapping camera co-ordinates directly into robot co-ordinates is typically referred to as the Camera Robot (CR) problem.

In recent years much effort has been made towards solving the CR problem or the hand-eye calibration problem. This problem is also known as the 'AX = XB' problem and it was first formulated in this manner by Shiu and Ahmad [7]. In this form, the calibration problem becomes finding X, the unknown Homogeneous Transformation Matrix (HTM) that relates the end effector co-ordinate frame to the camera standard coordinate frame, while A and B are other known HTMs. Other researchers have used: least mean square approach [8], quarternion representation of the HTMs and singular value decomposition [9], nonlinear optimization algorithm [10], and finally, both the closed form and the minimization method. Nagchaudhuri *et al.* [11] have employed neural network learning capability to learn the CR transformation, which is easily implementable in a laboratory experiment. The results have been compared with the pseudo inverse method of obtaining the transformation. The authors have used the same data to train a neural network via the back error propagation (BEP) approach to obtain the CR transform. Three kinds of neural networks have been used for camera calibration namely: back propagation, Elman network and radial basis functions. However, the best approach was found to be a neural network trained via the back propagation method.

5. MULTIPLE ROBOT CONTROL

A large number of situations in industry require the use of multiple robots working in co-ordination with each other. For example, tasks such as assembly of components and flexible material handling operations could be more efficiently carried out using two manipulator arms instead of one. A large object, which might be very difficult to be handled by a single manipulator, could easily be carried and maneuvered by two arms or two manipulators together. Multiple manipulators make manufacturing systems more flexible and these systems become capable of handling more complex operations.

Many control strategies have been proposed for the control of multiple robots working in continuous co-ordination that differ in focus, approach and the variables they control. Obviously, position is the most popular variable to be investigated. However, pure position control, including velocity and acceleration can prove to be inefficient. When multiple robots grasp an object, under this type of control, any misalignment or positional errors could yield undesirable forces exerted on the manipulator. These interactive forces can cause difficulties in controlling the position of the robot's end effector accurately. Moreover, the forces acting on the object can cause some damage to the object itself. For these reasons, position control by itself is not suitable for manipulating objects via multiple robots. Similarly, a pure force control is likely to lead to errors in position since there would be no position feedback, rendering this method unusable for many robotic applications. Hence, in multiple robotic manipulator applications, the control scheme must include elements of both position and force control.

Earlier researchers have used various control strategies to achieve position/force control of the robotic manipulator. Some researchers have been using master/slave method, hybrid position/force control and symmetric hybrid position/force control. Recently, a number of control strategies, such as fuzzy logic, neural networks, general redundancy optimization method, and two-level hierarchical fuzzy logic for hyper-redundant co-operating robots have been published in technical literature. Much of the recent research efforts have been devoted to optimize load sharing and to estimate internal forces acting on the object. Nagchaudhuri and Garg [12] have analyzed the problem of load sharing and internal forces from a fundamental outlook that exploits the geometry and weight distribution of the common object. Garg *et al.* [13] have proposed a strategy for force balance and energy optimization for co-operating manipulators. They employed a linear programming technique to calculate external forces [14], such that the power used in the direction of motion was minimized. Kosuge and Oosumi [15] have proposed a decentralized control algorithm of multiple robots handling a single object in co-ordination. The motion command of the object is given to one of the robots, referred to as a leader, and the other robots (referred to as followers) estimate the motion of the leader. Each robot is controlled by its own controller without having explicit communication among robots. Different from the conventional leader/follower or master/slave type of robot control algorithms, the proposed control algorithm specifies the internal force applied to the object, while handling the object is due to co-ordination based on robot dynamics.

6. WORKCELL SETUP

The central core of the experimental setup illustrated in Figure 1 consists of two ABB IRB 140 six-degree of freedom industrial robots each having a payload capacity of 5 kilograms. The robots are controlled by individual S4Cplus controllers with proprietary ABB control algorithm based on PID strategy. The controller contains the electronics required to control the manipulator, external axes and peripheral equipment. The controller computer is loaded with BaseWare Operating System, which controls every aspect of the robot, such as motion control, development and execution of application programs, communication, etc. The robots can be programmed using the ABB RAPID programming language using the teach pendant. The controller can be configured to deal with digital inputs and outputs to and from external devices such as camera, gripper, etc., for synchronization.

The two robots have been stationed on pedestals and are fitted with ATI gamma Force/Torque (F/T) sensors that communicate the six dimensional force/torque data to the host computer through National Instruments' Data Acquisition Board. These F/T sensors are interfaced to ATI QC – 5 pneumatic tool changers which allow the robot to choose from four different Schunk grippers. The flexible work cell is also equipped with two Cognex Insight 1000 vision sensors mounted above the work table that provide the vision data of the workspace. The vision systems use Ethernet to communicate and transmit data to a host PC. The workcell is controlled by a Dell Dimension Intel 1.4 GHz Pentium 4 digital computer with 384 Mega Bytes of RAM. Both the S4Cplus robot controllers and both the cameras are connected to this host PC through 100 MB of Ethernet network.

7. FUZZY LOGIC CONTROL

Fuzzy logic has been employed in this research to co-ordinate the movement of the second robot with respect to the movement of the first robot, when both the robots manipulate a common object. When the first robot moves, because of the object it carries, it generates internal forces and transmits those forces to the other robot via the common object they carry. The extent to which the other robot should move in order to compensate for the internal forces generated depends on the magnitude and direction of force, torque and rate of change of force. Hence, there are three inputs to the fuzzy inference engine. Different fuzzy inference systems (FIS) have been designed to determine the magnitude and direction of movement of the second robot for different directions. The present research involves an investigation of the co-ordination problem in the X-Y plane. Hence, there are two different fuzzy inference systems (FIS), one each for determining the extent of motion in 'X' and 'Y' direction. Each FIS has three inputs: force, rate of change of force in the direction in which the FIS determines the motion, and torque along the 'Z' direction. The output of the FIS is a numerical value with proper sign that determines the magnitude the robot has to move in that direction in millimeters.

Fig. 1. Photograph of the Flexible Workcell located in the Robotics and Manufacturing Automation (RAMA) Laboratory. Vision system is not shown.

The FIS has been designed using the MATLAB's Fuzzy Logic Toolbox. The FIS designed in the present research uses Mamdani fuzzy rules. The FIS uses the 'min' method for ANDing the variables and the 'centroid method' for defuzzification to obtain a crisp output variable. Gaussian functions have been used throughout for representing the membership functions.

The main program combines all of the different modules of the program on one platform, utilizes the information gathered from various sources, such as controllers, vision sensors, force/torque sensors, and uses the fuzzy logic to achieve the decision regarding the movement of the robots. The main program is a Simulink diagram with different blocks representing communication blocks, vision calibration blocks, fuzzy logic blocks, Stateflow [16] block and other differentiation and mathematical blocks.

8.   RESULTS

Two different cases were examined and the results derived from them were compared. In each of these cases, the first robot was given a command to move in a pre-specified manner while the second robot followed based on the force torque data obtained from the F/T sensor mounted on its wrist and the output of the fuzzy logic controller. Case I consists of the first robot moving along the negative Y direction and Case II consists of first robot moving along the positive Y direction. In both of these cases, the first robot was commanded to move 1 mm in each sample time.

The variation of forces in all the three (X, Y and Z) directions with time were plotted. The variation of the torques in the three directions was also plotted with time. Figure 2 shows the variation of forces for Case I. In all of the three plots shown in Figure 2, the forces remain approximately constant during the initial phase. During this time, the robot moves from the starting position to the position where it grasps the object. After that, a sudden change in force marks the event of the robots grasping the object and lifting it up. When the robots had lifted the object to a height specified in the program, the first robot is given the command to move 1 mm in the negative Y direction at each sample time.

The fuzzy logic controller takes control of the decision of the movement command of the second robot based on force, rate of change of force and torque data. There are two different fuzzy logic controllers employed to calculate the distance required to move in the X and Y directions. Once the fuzzy logic controller takes control (this happens shortly after time step 100), the forces in the X and Y directions oscillate at the rate given by the sample time. Hence, when the first robot moves, the force at the other end of the robot increases and when the second robot moves a certain distance in the appropriate direction, as obtained by fuzzy logic output, the forces decrease. The force in the X direction can be seen to be varying close to 0, as the first robot moves in the Y direction. The co-ordinate frames of both robots are not exactly aligned, hence the movement in the Y direction in the first robot causes the force in the X as well as Y direction. However, as the alignment difference is not much, the force in the X direction is small. The force in the X direction is also created because when the first robot moves, the second robot is at rest. The plot of the force in the Y direction starts at a higher value after the fuzzy logic controller begins to feed the movement command to second robot. But, it decreases (as well as oscillates about a decreasing force). This shows that initial gripping position of the robot gripper caused a higher internal force that became smaller with the help of the fuzzy controller. The force in the Z direction remains almost the same all throughout the event except when the robot grips the object, which is as expected because the robot does not move in the Z direction to cause any internal force in that direction.

9.   CONCLUSION

In this research paper a modern non-conventional techniques (viz. fuzzy logic) has been proposed to achieve the desired goal of controlling two six-degree of freedom industrial robots. Two Cognex Insight Vision systems have been calibrated with the help of a number of methods e.g., pseudo-inverse method and neural networks (back propagation, radial basis functions and Elman Network). The calibration results from these methods have been compared and successfully used for vision applied to the two 6-DOF industrial ABB robots available in the Robotics and Manufacturing Automation (RAMA) Laboratory. The simulations and the experiments performed show that the fuzzy logic control system was able to decrease the internal forces considerably.

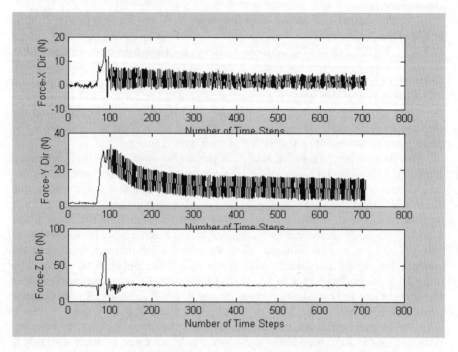

Fig. 2. Forces in the X, Y and Z directions versus number of time steps.

Acknowledgement
The financial support provided by the National Science Foundation under research award numbers 99-08177 and 99-11195 is gratefully acknowledged. Thanks are also due to Mr. Piyush Jain for his assistance with preparing the final draft of this paper.

References
1. Xiao, D., Ghosh, B.K., Xi, N. and Tarn, T.J., "Sensor-Based Hybrid Position/Force Control of a Robot Manipulator in an Uncalibrated Environment", *IEEE Transactions on Control Systems Technology*, Vol. 8, No. 4, 2000, pp. 635-645.
2. Matia, F. and Jimenez, A., " Multisensor Fusion: An Autonomous Mobile Robot", *Journal of Intelligent and Robotic Systems*, Vol 22, 1998, pp 129-141.
3. Mahajan, A., Wang, K. and Ray, P.K., "Multisensor Integration and Fusion Model that Uses a Fuzzy Inference System", *IEEE/ASME Trans. on Mechatronics*, Vol. 6, No. 2, June 2001, pp. 188-196.

4. Sun, F., Sun, Z. and Woo, P.Y., "Neural Network based Adaptive Controller Design of Robotic Manipulator with an Observer", *IEEE Transactions on Neural Networks*, Vol. 12, No. 1, January 2001, pp. 54–67.
5. Lim, C.M. and Hiyama, T., "Application of Fuzzy Logic to a Manipulator", *IEEE Transactions on Robotics and Automation*, Volume 7, No. 5, October 1991, pp. 688-691.
6. Yi, S.Y. and Chung, M., J., "A Robust Fuzzy Logic Controller for Robot Manipulators with Uncertainties", *IEEE Transactions on Systems, Man and Cybernetics*, Volume 27, No. 4, August 1997, pp. 706 – 713.
7. Shiu, Y. and Ahmad, S., "Finding the Mounting Position of a Sensor by Solving a Homogeneous Transform Equation of Form AX = XB", *Proceedings of IEEE International Conference on Robotics and Automation*, 1987, pp. 1666-1671.
8. Tsai, R.Y. and Lenz, R.K., "A New Technique for Fully Autonomous and Efficient 3D Robotics Hand Eye Calibration", *Proceedings of The Fourth International Symposium in Robotics Research*, 1987, pp. 287-297.
9. Chou, J.C.K. and Kamel, M., "Finding the Position and Orientation of a Sensor on a Robot Manipulator using Quarternion", *International Journal of Robotics Research*, Vol. 10, No. 3, June 1991, pp. 240-254.
10. Horaud R. and Dornaika F., "Hand – Eye Calibration", *International Journal of Robotics Research*, Vol. 14, No. 3, June 1995, pp. 195-210.
11. Nagachaudhuri, A., Thint, M. and Garg, D.P, "Camera Robot Transform for Vision Guided Tracking in a Manufacturing Work Cell", *Journal of Intelligent and Robotic Systems,* Vol. 5, 1992, pp. 283-298.
12. Nagchaudhuri, A. and Garg, D., "Load Sharing and Internal Forces in Multiple Cooperating Manipulators: A New Perspective", *Proceedings of the 23rd Annual Pittsburgh Modeling and Simulation Conference*, Vol. 23, Pt. 4, May 1992, pp. 2025-2032.
13. Garg, D. and Ruengcharungpong, C., "Force Balance and Energy Optimization in Cooperating Manipulation", *Proceedings of the 23rd Annual Pittsburgh Modeling and Simulation Conference*, Vol. 23, Pt. 4, May 1992, pp. 2017-2024.
14. Kwon, W. and Lee, B., "A New Optimal Force Distribution Scheme of Multiple Cooperating Robots Using Dual Method", *Journal of Intelligent and Robotic Systems*, Volume 21, 1998, pp. 301-326.
15. Kosuge, K. and Oosumi, T., "Decentralized Control of Multiple Robots Handling an Object", *Proceedings of the International Conference on Intelligent Robots and Systems*, Volume 1, 1996, pp 318 –323.
16. *"User's Guide: StateflowToolbox "*, Mathworks Inc., 2001.

Active Control of Internal Turning Operations Using a Boring Bar

L. Pettersson, L. Håkansson, I. Claesson and S. Olsson,
Department of Telecommunications and Signal Processing
Blekinge Institute of Technology
372 25 Ronneby
Sweden

Abstract
Vibrations in internal turning or boring operations are usually a cumbersome part of the manufacturing process. The manufacturing industries are having problems with these kinds of cutting operations. When cutting in pre-drilled holes the cross sectional area of the boring bar is limited, at the same time, as it is long. Since a general boring bar is long and slender, it is sensitive to external excitation and thereby inclined to vibrate. The vibration problem affects the surface finish, in particular. The demand for smaller and smaller tolerances of the surface finish has prompted the manufacturing industry to seek for a solution to the boring bar vibration problem. The tool life is also likely to be influenced by the vibrations involved in a cutting operation. Another problem in boring operations is the high noise level in the cutting process. The noise level in the environment of the operators is today more and more regulated, especially in the western world. Active vibration control will reduce the amount of vibrations during cutting operations. Since the noise is induced by the vibration of the boring bar, the noise level will also be reduced due to the cancellation of the noise source. Preliminary results show reduction of vibrations in the boring bar by up to 30dB.

Keywords: *Active vibration control, boring bar, piezo-actuator.*

1.   INTRODUCTION

Boring operations have a history of being a cumbersome metal cutting process in the workshop. Several boring operations are simply not possible to perform and in many boring operations it is impossible to meet the desired tolerances. The majority of the problems in boring operations are vibration related. Common boring bars are long and slender and as a consequence they are sensitive to excitation forces, such as the forces in metal cutting. When the bar vibrates the surface finish is deteriorated. The tool life is also likely to be influenced by the amount of vibrations induced in the cutting operation. Severe noise in the operator area is also frequently a result of the boring bar vibrations. In boring operations the motion of the boring bar has components in both the cutting speed and the cutting depth directions [1]. The vibrations are usually dominated by the first two resonance frequencies of the boring bar, one resonance in each direction. The vibration level in the cutting speed direction is generally larger than the vibration level in the cutting depth direction [1]. Fig. 1 shows a boring operation. Thus active vibration control in the cutting speed direction is a good first step towards a solution in reducing noise and vibration.

The active control of boring bar vibration involves a secondary source, driven in such a way that it will interfere destructively with the original vibration induced by the cutting operation. A complication in cutting operations is that the original excitation cannot be observed directly, thus the control system must be based on a feedback approach. An actuator is used as a secondary source and an accelerometer is needed to provide the control algorithm with sufficient information regarding the boring bar vibration. The bar is modified to fit a piezo-ceramic actuator and an accelerometer. The actuator and accelerometer are embedded and sealed to protect them from cutting fluid and the chips from the metal cutting operation.

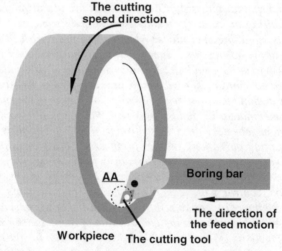

Fig. 1. A schematic picture of a boring operation.

During the machining of a workpiece, variations in the spectral properties of boring bar vibrations are likely to occur. Variations in the spectral properties are caused by changes in the excitation of the bar and/or changes in its structural response. A solution to the controller problem is to use an adaptive FIR filter controller that is able to handle the nonstationary environment introduced by the cutting process. Such a control algorithm is the feedback filtered X LMS algorithm. The reason for choosing this algorithm was due to a so-called forward path or secondary path. The forward path is the difference between the output of the controller and the input of the accelerometer. An estimate of the forward path is needed to reduce the effects of A/D converters, amplifier, transfer path in the boring bar and D/A converters. The forward path is estimated in an initial phase before the actual control of the bar vibration by using an ordinary LMS algorithm.

One of the objectives in this research project was to find a control solution that fits in a standard lathe. Since both the actuator and the accelerometer are embedded into a standard boring bar, the only modification that is required on a standard lathe to enable the active control of tool vibration is the installation of signal cables to the actuator and accelerometer.

This paper discusses single channel feedback control of the bar's vibrations during boring operations.

2. MATERIALS AND METHODS

2.1 Experimental Setup

The cutting trials have been carried out in a Mazak SUPER QUICK TURN - 250M CNC turning center, see Fig. 2a, with 18.5 kW spindle power, maximal machining diameter 300 mm, and with 1007 mm between the centers. The boring bar is based on a standard WIDAX S40T PDUNR15 bar, see Fig. 2b, it is modified to fit a piezo-ceramic stack actuator and an accelerometer to enable active vibration control. The amplifier used was custom designed for piezo-ceramic actuators.

2.2 Work Material, Cutting Tool and Cutting Data

The material of the workpiece used in the cutting trials was chromium molybdenum nickel steel. After a preliminary set of cutting trials, a combination of cutting data and tool geometry was selected, see Table 1. The diameter of the workpiece was deliberately chosen large (> 150 mm), in order to render the workpiece vibrations negligible.

Geometry	Cutting speed, v (m/min)	Depth of cut, a (mm)	Feed rate, s (mm/rev)
DNMG 150508-SL 7015	117	0.4	0.22

Table 1. Cutting data and tool geometry

2.3 Active Boring Bar

A considerable part in an active control application is to select a proper location for the actuator. The actuator must introduce vibrations into the vibrating modes of the bar. The actuator is embedded in the length direction below the centerline of the boring bar. When the actuator applies load on the bar in its length direction, due to the expansion of actuator, the boring bar will bend. A schematic picture of the active boring bar is shown in Fig. 3. The active boring bar also incorporates an accelerometer, embedded as close as possible to the cutting tool.

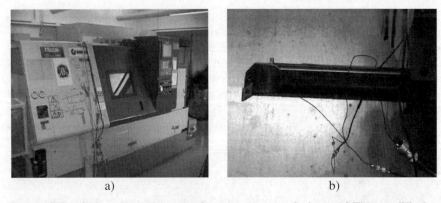

Fig. 2. a) The lathe where the experiments were carried out. b) The modified active boring bar used in the experiments.

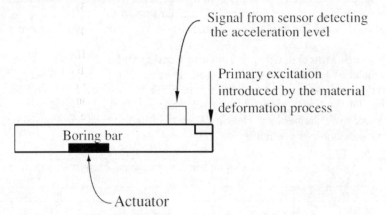

Fig. 3. A schematic figure of the active boring bar with embedded actuator and accelerometer.

2.4 Control System

In boring operations it is impossible to measure the primary excitation, i.e. the excitation induced by the cutting process. A system based on a feedforward solution is consequently not possible, hence the control system must be based on a feedback approach. Another important point in selecting a proper algorithm is that the error estimate is not the difference between the output signal from the adaptive filter and the desired signal. The only signal available is the signal from an accelerometer sensing the acceleration of the boring bar. The error estimate in our case is the vibrations induced by the actuator that are summed with the vibrations originated from the cutting process. The output signal from the adaptive filter will be filtered by a D/A converter, amplifier, actuator and the transfer path in the boring bar before it is picked up by the accelerometer. The transfer path the output signal from the adaptive filter has to pass before it is sensed by the accelerometer is called forward path or secondary path and is present in active control applications.

A suitable control algorithm must be able to handle these kinds of conditions and such an algorithm is the filtered X LMS algorithm.

The use of an error signal as input to the control algorithm causes the algorithm to act as a feedback controller. A block diagram of the feedback filtered X LMS algorithm is shown in Fig. 4. The unit delay z^{-1} handles the fact that we are dealing with adaptive digital filter in a real time environment, C denotes the actual forward path and C^* is the estimate of the forward path. $y(n)$ is the output signal from the adaptive FIR filter, $d(n)$ is the original excitation, $y_c(n)$ is the vibrations induced by the actuator and $e(n)$ is the signal from the accelerometer. The search for a minimum in the mean square sense is performed by the filtered X LMS algorithm, which is given by [2]

$$y(n) = \mathbf{w}^T(n)\mathbf{x}(n) \qquad (1)$$

$$e(n) = d(n) - y_c(n) \qquad (2)$$

$$x_{C^*}(n) = \mathbf{C}^{*T}\mathbf{x}(n) \qquad (3)$$

$$\mathbf{w}(n+1) = \mathbf{w}(n) + \mu \mathbf{x}_{C^*}(n)e(n) \qquad (4)$$

where $x_{C^*}(n)$ is the filtered reference signal vector. The difference between the estimate of the forward path and the true forward will affect both the stability and the convergence rate of the adaptive controller [3]. Using the error signal as input to the filtered X LMS algorithm will complicate the relation between the mean square error and the filter coefficients, i.e. the mean square error will not be a quadratic function of the filter coefficients.

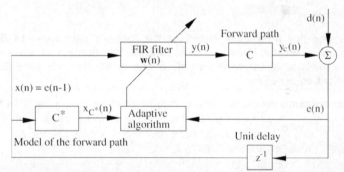

Fig. 4. A block diagram of the filtered X LMS algorithm used in the adaptive feedback control system.

The forward path is estimated in an initial phase using an ordinary LMS algorithm. During the estimation, the actuator is feed with pseudo random noise in order to minimize the hysteresis effects of the actuator. The hysteresis is highly variable based on different load and acceleration values. In the active control of boring bar vibrations the estimate of the forward path based on a FIR filter with 40 coefficients, the adaptive filter had 20 coefficients and the sample rate in the DSP was 8 kHz.

3. RESULTS

To evaluate the performance of the active control of boring bar vibrations, power spectral densities have been estimated of the vibrations with and without active vibration control. The spectra are based on the acceleration signals from the accelerometers during a continuous boring operation in both the cutting speed and the cutting depth directions.

The forward path is estimated in an initial phase. The 40 estimated FIR filter coefficients are shown in Fig. 5a. An estimate of the amplitude and phase function of the forward path, based on 60 seconds time data during the identification process, is shown in Fig. 5b.

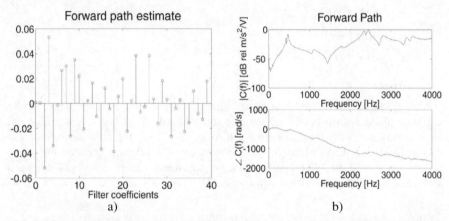

Fig. 5. a) The FIR filter coefficients of the forward path used to filter the reference signal in the filter X LMS algorithm. b) Estimate of the amplitude and phase function of the forward path based on 60 seconds time data from the identification process.

The performance of the active control solution in boring operations is illustrated in Figs. 6 and 7. Fig. 6 shows the vibrations in the cutting speed direction with and without active control and Fig. 7 shows the vibrations in the cutting depth direction with and without active control.

The active control application is able to attenuate the vibrations in the boring by approximately 25 dB at the first two resonance frequencies. Several harmonics are also attenuated significantly. At the first harmonic in both the cutting speed and the cutting depth directions the attenuation was approximately 30 dB.

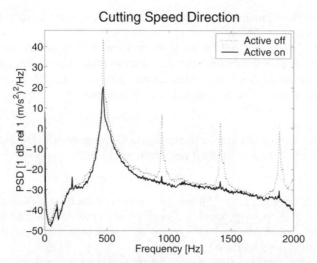

Fig. 6. Power spectral density of boring bar vibrations with and without active vibration control in the cutting speed direction.

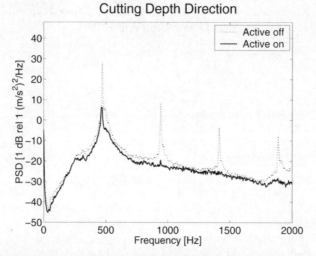

Fig. 7. Power spectral density of boring bar vibrations with and without active vibration control in the cutting depth direction.

4. SUMMARY

Active vibration control appears to be a suitable solution for the noise and vibration problem in boring operations. By using small piezo-ceramic stack actuators large forces can be applied to the boring bar at the same time as the space where the actuator is located can be kept small. The filtered X LMS algorithm showed stable behavior during the cutting experiments. A leaky version of the algorithm has been implemented but not tested yet. A leakage factor will make the algorithm even more robust.

This solution is based on a single piezo-ceramic stack actuator but future work will include a solution with two actuators to be able to control the vibrations in the cutting speed and the cutting depth directions separately. Investigating different algorithms is also an important task in this project.

Acknowledgement
We thank our industrial partner, Staffansboda Compagnie AB, Fänestadsvägen 8, 330 12 Forsheda, Sweden, for their participation.

References
1. Linus Pettersson, Lars Håkansson, Anders Brandt and Ingvar Claesson, Identification of dynamic properties of boring bar vibrations in a continuous boring operation, Journal of Mechanical Systems & Signal Processing, Academic Press, 2001, Submitted for publication.
2. I. Claesson and L. Håkansson, Adaptive active control of machine-tool vibration in a lathe, IJAV-International Journal of Acoustics and Vibration, 3(4), 1998.
3. S.M. Kuo and D.R. Morgan, Active Noise Control Systems, Telecommunications and Signal Processing, Wiley, 1996.

Automatic Foundry Brake Disk Inspection through Different Computer Vision Techniques, using a New 3D Calibration Approach

Pedro Martín Lerones[*], José Llamas Fernández[*], Jaime Gómez García-Bermejo[$], Eduardo Zalama Casanova[$]
[*] C.A.R.T.I.F; Parque Tecnológico de Boecillo. Parcela 205; 47151-Boecillo (Valladolid), Spain; *{pedler, joslla}@cartif.es*
[$] ETSII; Dep. of Automatic Control, University of Valladolid Paseo del Cauce, s/n; 47011-Valladolid, Spain; *{jaigom, eduzal}@eis.uva.es*

Abstract
We present a fully automated raw foundry brake disk visual inspection system in which three different computer vision techniques are used via complete piece rotation. For the characterisation of the piece, a 3D calibration is performed through a new method, which does not require any previous mechanical alignment and leads to more accurate results than conventional ones. The whole system could be fitted to any kind of piece with a circular symmetry, and is an accurately synchronised blending of mechanics, automation, computer vision and robotics. Some results for industrial implementation are presented.

Keywords: *Automobile industry, automatic surface acquisition, camera calibration, computer vision, image analysis, measured values.*

1. INTRODUCTION

Raw foundry brake disk manufacturing processes produce pieces which have different faults, so the piece surface is finished off with a grinder. A human operator checks the brake disk on the production line, and puts it in the proper container. This process may be carried out by a fully automated dimensions characterisation and surface and ventilation slots fault detection [1].

A calibrated 3D structured light-based technique is being used for circularity defect examination, over-grinding and piece characterisation, as a first process step. For this purpose, a number of traditional calibration methods are available. Because the precision and versatility required, Tsai's method is one of the most commonly used [2]. There are also specific 2D-2D calibration methods for laser plane-based projection pattern [3]. Any traditional method is characterised by a mathematical model followed by parameter valuation from control points measured over a known geometry sample model [4]. However, usually mathematical models are not complete, because only the relationship between world co-ordinates (fixed and tied to a laser plane) and image co-ordinates (fixed on image screen and measured in pixels) is explicitly expressed. The relationship between object co-ordinates (tied to inspected piece and used to express scene points) and world co-ordinates, has to be done by manual mechanical adjustments [3,5]. This leads to calibration inaccuracies, so big as to lower production and cause poor mechanical fitting quality. To avoid these disadvantages, McIvor has proposed a new calibration method [6,7] that allows the user to obtain a direct relation between image and object co-ordinates. None the less, it can only be used in the case of uniform linear motion objects. It's extension to the rotation motion case, with huge practical interests, has been our main goal [8]. The results obtained show that the proposed calibrating method is more accurate than traditional ones.

Conventional 2D vision techniques for ventilation slots inspection and a 3D uncalibrated structured light-based technique for pores, hard masses and featheredges inspection are used after dimensional characterisation.

In this context, the three techniques described above consist of two different experiments (a robot is used in the second one), during which every disk is examined, and whose checking order depends on the disk type. The system could be fitted to a new model or any other 3D rotational inspection. The whole system is an accurately synchronised blending of mechanics, automation, computer vision and robotics.

The remainder of the text is divided as follows: in section 2 we present the description of a fully automated visual inspection, with the new calibration approach. Section 3 shows the industrial process, system set-up details and some results. Finally, section 4 outlines conclusions and future work.

2. DESCRIPTION OF AUTOMATIC INSPECTION

Brake disks faults to be inspected are: circularity and defective grinding (+/-1mm tolerance), hard masses and featheredges (greater than 2mm thick), ventilation slots obstructions (because of small metal pieces lodged in them) and hole jump obstruction (up to 30%), veining because of wrong moulds union (up to 15%) and pores greater than 1.5mm wide (inner gasbags are likely to be the origin).

Traditionally the brake disk inspection is carried out by humans. This requires manual disks handling, which leads operators to considerable physical efforts (10 Kg in weight for each piece, on average, and a new one every 4 s) and visual tiring as well, with the subsequent results that defects will go undetected. This has enormous implications for the automotive industry in maintaining the high quality standards required.

2.1. Computer vision techniques and placement

The developed system combines the following computer vision techniques: a calibrated 3D structured light-based vision technique for circularity defect examination, over-grinding and piece characterisation; conventional 2D vision techniques for ventilation slots inspection; and finally a 3D uncalibrated structured light-based technique for pores, hard masses and featheredges examination.

The first technique uses a very high-speed digital CCD camera with 955 fps. Calibration is essential for 3D measurement data accuracy and reliability [2,8]. In this paper, a new non-mechanical alignment calibration method is presented. It presents the works from McIvor in a linear case [6,7] to the rotating case.

The second techniques involves conventional lighting (halogen light, in the V spectrum area) and relies on frames sequence acquisition with an analogic high-speed CCD camera (120 fps). These images are developed to a threshold and the evolution through time of the obtained area (referring to the ventilation slots) is then analysed.

The third technique uses a laser structured-light (plane pattern) whose projection on the piece is measured by a very high speed CCD digital camera (955 fps). This subsystem need not be calibrated, so the processing speed is high. Also while taking an entire image sequence, laser traces variations are examined via a discrete gradient calculation.

Physically, the entire computer vision system is divided into two components. The first one consists of two different subsystems and the rotation motion for inspection is via a turning table, at constant speed:
- A 3D calibrated computer vision subsystem (first described technique) with a high-speed digital camera with computer and framegrabber inside it, and a 670nm line pattern laser focused on the piece (Fig.1-A). As a first step, this subsystem allows the disk under production verification of the brim and crown brake disk diameter and height measurement and the determination of circularity defects. As a second step, without considering calibration, this subsystem (using the third above-mentioned technique) allows featheredges, hard masses (Fig.1-B) and pores detection on the external disk surface to be analysed.
- In case of backward-ventilated brake disks, a non-calibrated 2D computer vision subsystem (second technique used) with an analogic camera with the proper framegrabber and host computer, and an halogenous light whose spot is just perpendicular and over disk hole jump. It is used for ventilation slots inspection (Fig.1-C), ribs examination, veining and hole jump obstruction determination.

In the same way, the second computer vision system consists of another two different subsystems, whose inspection purposes 6-DOF robot provides the rotation motion at wrist constant speed:
- A 3D non-calibrated computer vision subsystem with computer and framegrabber inside it, and a 670nm line pattern laser focused on the piece. This subsystem (using the third technique) allows featheredges, hard masses and pores detection on the internal disk surface.
- In the case of ventilated brake disks production, a non-calibrated 2D computer vision subsystem (second technique used) with an analogic camera (and same framegrabber and host computer) as the 2D subsystem in the first component. An halogen light is also used for ventilation slots inspection, ribs examination, and whose spot is just over the ventilation slots and parallel to the rotation motion.

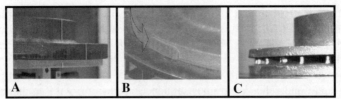

Fig. 1. Laser lighting on brake disks, hard mass and ventilation slots

2.2. New 3D calibration approach

A precise and fully automated new calibration method is proposed for 3D-computer vision systems, through a laser line pattern projection over the piece to which a rotation motion is applied. In this way, the mathematical model that gives the direct relation between image co-ordinates and object co-ordinates will be established.

For this purpose, we first have to establish a object to world co-ordinates relationship in the case of a piece rotation motion. We need an auxiliary 'rotation axis reference system' (RARS), whose Y-axis is the rotation and the other two are parallel to the corresponding piece co-ordinates system [8]. The RARS origin is the intersection between its Y-axis and the turning table used to apply rotational motion to the piece. We can also easily obtain a degree of perpendicularity between these two elements. As the object co-ordinates are obtained, the RARS is tied to the piece. Then the object to world co-ordinates relation is:

$$X_w = T_0^1 \cdot T_1^2 \cdot X_p \qquad (2.2.1)$$

Where X_w, X_p express world and object co-ordinates respectively, T_0^1 world to RARS interchange matrix, and T_1^2, RARS to object interchange matrix. These last ones are an homogeneous expression.

Take into account the world to image co-ordinates relation, expressed as follows:

$$X_i = A \; R_t \; X_w \qquad (2.2.2)$$

Where X_i are row-column co-ordinates of the image reference system (homogeneously expressed). Matrix A includes image co-ordinates origin and scale factors. Matrix R_t reflects rotation and translation factors for world to image reference when moving.

From the above combination of equations, we have the direct desired relation between image and object co-ordinates. By lineal grouping of unknown factors with respect to known ones, it leads to two independent equation subsystems, one for the camera (2 equations, 35 elements) and other for the laser (1 equation, 11 elements).

The following step is for the application of a parameter estimation through the minimum square algorithm. To do this, the object and image co-ordinates have to be known. Accordingly, 45 control points were enough to solve the posed system (second previous subsystems combination), but for a better estimation that number must be higher. In order to do this, a specific black two-steps pyramidal structured sample piece has been developed, whose dimension is similar to the pieces to be measured and over which 232 control points are selected (Fig. 2). These points are silk-screen printed white circles. An estimation is made using 200 points, and an estimation of the quadratic square error is obtained with the remaining 32 points. The results show that the proposed calibrating method is more accurate than the traditional ones (section 3.2).

Fig. 2. Calibration sample piece

Control points image co-ordinates and estimation (through a proposed mathematical model) and error calculation for a robust validation are included in a specifically developed software application. This program is divided into 7 steps; the first to the fifth step includes several morphology operations [9, 10] and guide to image co-ordinates achievement and their correspondence to object co-ordinates. The sixth step consists of a minimum square algorithm estimation, and the last step is the error calculation. To test this software, 955 images are acquired from the entire sample piece. Afterwards, via a 8-pixel FIR technique, the laser trace is obtained. Then the image can be generated.

2.3. Brake disk dimensional characterisation

Firstly, a simple laser trace image is obtained and compared to the characteristic one (measured on each disk reference). In this way we determine if the brake disk is the proper one under production, to avoid breaking down the fingers that clutch the piece during rotation.

Next, a complete image sequence is taken over 990 frames. Following the first technique described in section 2.1, the obtained laser traces are re-sampled. In this way, the radius and height medium values for the brake disk brim and crowns are calculated. Sub-pixel precision allows the geometrical profile characteristics to be distinguished. The obtained data compared to the real disk parameters point to dimensional errors.

2.4. Ventilation slots inspection

The second technique described on section 2.1 is used for this inspection. An entire image sequence (140 frames) corresponding to a complete piece rotation is taken and these frames are binarized and labelled later. Every ventilation slot shows an area whose maximum is evaluated within a fixed image number. Afterwards, the average over all maximums is set and it is used to compare the remaining ventilation areas, leading to the determination of the percentage of obstructions.

Defects in ribs, disk hole jumps (in backward-ventilated case) and veining are detected as a percentage of ventilation slots obstructions.

2.5. Surface inspection

The third technique described on section 2.1 is used to inspect the whole brake disk surface. Within both system positions (external and internal checking), an entire image sequence is taken (990 frames). Light changes on traces emphasize any brake disk edges. Then, when all laser traces are obtained and the average, as a geometrical average, is worked out. Featheredges are detected as longer traces than the average.

Hard masses, pores and small hits are detected through the 8-pixel FIR technique and horizontal sweeping, as well as via 2-pixel FIR followed by vertical sweeping.

3. INDUSTRIAL PROCESS AND RESULTS

Although only fifteen different pieces, representing three different types (solid, ventilated and backward-ventilated) of disks, were tried, we believe that this system has important industrial applications to every new model of disk, ensuring that the client can continuously adapt to any new market requirements.

3.1. Process and system set-up

The general set-up of our system consists of a commercial blending of mechanical, robotics, automation and computer vision components (Fig.3). Each vision position (section 2.1) combines a laser source that projects a light plane (30° angle, 670 nm wavelength, Class III-A), a 5° spot halogen light and two CCD cameras that operate according to the optical triangulation principle: a *DALSA CA-D6* digital high-speed area scan camera with a *Viper-Digital* framegrabber (from *CORECO*), and a *JAI-M30* analogic area scan camera with *IC-RGB* framegrabber (from *IMAGING* and common to both analogic cameras). The computers are: Pentium III 500MHz PCs with WIN 2000 OS. In the first position, f=16mm for the digital camera and f=50mm for the analogic one. In the second position, f=12mm and 25mm for both respective cameras. These two positions are on the same steel structure.

Brake disks are fed into our system through a belt, and their presence is detected through a photoelectric detector. A pneumatic cylinder (1m/s speed, 6 bar pressure) then guide the pieces to first inspection point on the rotation table (Fig. 3-A) and it is turned via an engine-inverter conjunction and gearhead (1/30 factor) to 60rpm. During rotation, dimensional characterisation, external surface checking and ventilation inspections in backward-ventilated disks are performed (Fig. 3-B). Depending on the disk model (geometrical profile), nine sets of specially-designed three-finger clutch the disk during rotation. If this inspection is successful, a 6-DOF robot (*ABB IRB-4400F* 60Kg) takes the piece for internal surface inspection, rotating it at 55rpm (Fig.3-C). During a 6-axis movement, internal surface checking and ventilation inspection is performed. Also, depending on the model of the disk in production, ten specially-designed three-finger sets are used to clutch the disk during rotation.

A *TSX Premium* automaton (from *TELEMECANIQUE*) manages the whole system through an *IBS* network. From both computer vision and inspection results, the brake disk is discarded into one of three different containers: faulty (via the robot, if first inspection was not passed), recovery (human expertise criteria needed) or correct (through a 30° steel slide to avoid disks hitting previously inspected ones, see Fig. 3-D).

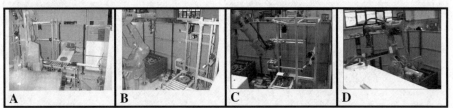

Fig. 3. Different system in process views set-up. Reproduced with LINGOTES ESPECIALES, S.A. permission.

3.2. Brake disk inspection results

Industrial results are focused on the following three aspects: calibration for dimensional characterisation, inspection of entire disk surface and ventilation slot inspection.

For calibration results, we will compare those obtained by the method described in section 2.2 to the traditional techniques (see [5] as reference). A specially-designed sample piece is mounted on the turning table and follows the same process as the brake disk in the first computer vision position. Immediately after, Table 1 is produced, in which the quadratic square error is shown, for both image co-ordinates errors (RC Errors) and object co-ordinates errors (XYZ Errors).

Table 1. Image to object co-ordinates errors on estimated points

	RC Errors	XYZ Errors
Traditional method quadratic medium square error	0.860 pixel	0.803 mm
Proposed method quadratic medium square error	0.371 pixel	0.672 mm

Points on the sample piece lead to the following table being produced (Table 2):

Table 2. Image to object co-ordinates errors on check points

	RC Errors	XYZ Errors
Traditional method quadratic medium square error	1.789 pixel	1.041 mm
Proposed method quadratic medium square error	0.396 pixel	0.694 mm

The results obtained show the proposed calibrating method is more accurate than the traditional one (found error is about 80% of the error of traditional methods) in the same conditions [8].

The same as for calibrating procedure, a specific software application has been developed for brake disk surface and ventilation slots inspection (if needed). The screen output for dimensional characterisation, hard masses, featheredges and pores on the external surface (the same for internal ones) is shown on Fig. 4-A. A colour coding patter indicates where faults occur. For the inspection of ventilation slots (Fig. 4-B), an obstruction is dark coloured, a critical zone is lighter in colour and the rest indicates areas correct ventilation.

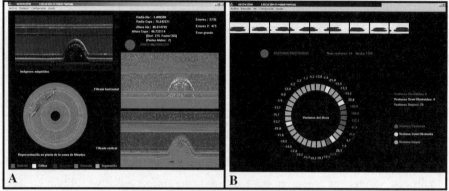

Fig. 4. Inspection software's output

4. CONCLUSIONS AND FUTURE WORK

A fully-automated brake disk inspection system for the automotive industry has been presented. Structured light vision is used, which is somehow uncommon in the industry. Another innovative aspect is the application of very high speed CCD cameras. A 3D calibration is performed through a new method, which does not require any previous mechanical alignment.

Because the system is implemented in a foundry, every computer vision element is covered in specially-designed stainless steel casing, surrounded by a 1bar pressure air curtain, to avoid contamination by metallic dust. Optimisation in robot trajectories and higher production speed is also a new achievement. The system could be fitted to any new model or any other kind of piece with a circular symmetry and geometry. Adapting this system to other industries will also be a real challenge.

Acknowledgment

We want to acknowledge the financial support of LINGOTES ESPECIALES, S.A., Valladolid (Spain), as well as the EU FEDER program.

References

1. F.A. Rodrigues Martins, J. Gómez García-Bermejo, E. Zalama Casanova, J.R. Perán González: A system for automatic surface scanning, Sept 2001, The Eighth IEEE Conference on Mechatronics and Machine Vision in Practice, Hong Kong.

2. R.Y. Tsai: A versatile camera calibration technique for high-accuracy 3D machine vision metrology using off-the-self TV cameras and lenses, 1987, IEEE Journal of Robotics and Automation, Vol. 3, No. 4, pp 323-344.
3. P. Saint-Marc, J.L. Jezouin, G. Medioni: A versatile PC-based range finding system, 1991, IEEE Transactions on Robotics and Automation, Vol. 7, No. 2, pp 250-256.
4. R. Jain, R. Kasturi, B.G. Schunck: Machine Vision, 1995, MIT Press & McGraw-Hill.
5. J. Gómez García-Bermejo, F.J. Díaz Pernas, J. López Coronado: Obtención conjunta de las informaciones tridimensional y cromática. Primera aproximación a la caracterización óptica de superficies, 1997, Informática y Automática, Vol. 30, No. 3, pp 19-33.
6. A.M. Mc Ivor: Calibration of a Laser Stripe Profiler, Third International Conference on 3D Digital Imaging and Modelling, 3DIM99, October 1999, IEEE Press, pp 92-98.
7. R.J. Valkenburg, A.M. Mc Ivor: Accurate 3D Measurement Using Structured Light System, February 1998, Image and Vision Computing, Vol. 16, No. 2, pp 99-110.
8. I. Riñones Mena, P. Martín Lerones, J. Gómez García-Bermejo, J. Llamas Fernández, E. Zalama Casanova: Un nuevo procedimiento de calibración para medición tridimensional automática de piezas en rotación, July 2000, Revista Electrónica de Visión por Computador (REVC)-Electronic Paper.
9. S. Ullman: High-level Vision. Object Recognition and Visual Cognition, 1997, MIT Press.
10. G. Medioni, M.S. Lee, Ch.K. Tang: A Computational Framework for Segmentation and Grouping, 2000, Elsevier.

Force-guided Compliant Motion in Robotic Assembly: Notch-locked Assembly Task

Kong Suh Chin[a], Mani Maran Ratnam[a], Rajeswari Mandava[b]
[a]School of Mechanical Engineering,
Engineering Campus,
University Science Malaysia,
13400 Nibong Tebal,
Seberang Perai Selatan, Penang.

[b]School of Computer Science,
Universiti Sains Malaysia,
11800 Minden,
Penang.

[a]E-mail: eric16919@stud.usm.my

Abstract
This paper presents how a force-guided robot can be implemented to perform compliant motions for a notch-locked assembly. A study on assembly operation of front housing and back chassis of a typical mobile phone is carried out where notch-locked assembly is involved. An assembly strategy based on three force-based compliant motions is proposed in order to perform the automated assembly of the front housing and back chassis of a mobile phone. The assembly strategy based on force-based compliant motion is performed and the experimental results are discussed. The implementation and the setup of a force-guided robot and the end of arm tools are also presented. The system is optimized for high-speed performance while considering the constraint and limitation involved.

Keywords :*Force-guided robot, advanced manufacturing, automated assembly and notch-locked assembly.*

1. INTRODUCTION
Robots offer many benefits in manufacturing simply because they are machines. They are not as susceptible to fatigue, boredom, discomfort, or similar factors that negatively impact a human worker's job performance in harsh, noisy, hot or hazardous environments.

With the development of more sophisticated automation concepts, such as the flexible manufacturing system (FMS), industrial operations are usually best automated through the application of multi-purpose machine, such as industrial robot, rather than special-purpose machine. In recent years, the use of robots in industry has increased rapidly. There are approximately 116,000 industrial robots currently operating in United State factories. Over the huge number of robot populations, only approximately 10% is used in assembly applications [1]. This is due to the complexity of assembly tasks involved. These assembly tasks not only consist of simple mating processes but also consist of higher level of assembly work such as snap fit and screw fastening [2-3]. Many concepts and methods have been developed to accomplish the assembly tasks, which leads to the application of force-guided robot in automated assembly [4-6]. Implementation of a force-guided robot to perform an automated assembly still needs more investigation using actual products before it can be really implemented in a production line.

This paper describes how a force-guided robot can be used to perform in an automated assembly. The assembly of back chassis and front housing of a typical mobile phone using notch-locked assembly is demonstrated.

2. NOTCH-LOCKED ASSEMBLY

Notch-locked assembly is used in the mobile phone industry to assemble the back chassis and the front housing of a mobile phone. This assembly task is currently done by human operators in a local manufacturing factory [7].

In order to automate the assembly process, a study to identify the compliant motion for a notch-locked assembly task, as a critical assembly task, is required before it can be implemented effectively in actual applications.

2.1 Manual assembly

This assembly task generally involves the insertion of a back chassis known as the battery compartment into the front housing of a mobile phone. Fig. 1(a)-(f) shows the details of the assembly operation and sequence for notch-locked assembly of the back chassis and front housing. Fig. 1(a)-(b) shows the approaching stage where the back chassis is approached to the front housing. The back chassis is allowed to slide on the front housing until the front end of the back chassis is snapped into the notch of the front housing, as shown in Fig. 1(c)-(d). The back end of the back chassis is pressed and snapped into the front housing to complete the assembly tasks of the notch-locked assembly, as shown in Fig. 1(e)-(f).

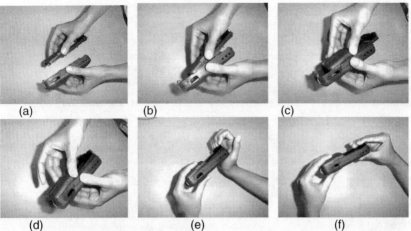

Fig. 1. Notch-locked assembly by human operator: (a) approaching stage. (b) Contact at front end. (c) Sliding on front housing. (d) Snapping into notch. (e) Pressing back end. (f) Assembly accomplish.

2.2 Assembly strategy

Notch-locked assembly can be performed automatically using force-guided compliant motions. Three force-guided compliant motions are identified to perform notch-locked assembly: stopping, aligning and sliding skills [7].

This approach combines three force-guided compliant motions in a specific sequence in fine motion planning to perform the notch-locked assembly. The front housing is placed on the assembly stage, while the back chassis is gripped by the gripper at the end-of-arm. The back chassis approaches the front housing under position control, while the force readings are continuously monitored, as shown in Fig. 2(a). An increase of sensing force opposition to the motion is expected when the back chassis comes into contact with the front housing.

The detection of a predetermined force level is used as a stopping skill to halt the manipulator's motion during the approaching stage. Using the sliding skill, the back chassis is made to slide on to the front housing, as shown in Fig. 2(b), while maintaining a certain contact force orthogonal to the front housing. A sudden drop of the force level is expected when the end corner of the back chassis is successfully snapped into the notch of the front housing. This condition is used to stop the previous sliding motion, as shown in Fig. 2(c). The gripper is opened and the manipulator is moved to the back end of the front housing, as shown in Fig. 2(d). The back-end of the back chassis is pressed into the front housing where a sudden drop in force level is detected during the pressing motion. The sudden drop in force level again shows that the back-end of the back chassis is snapped into the notch of the front housing, as shown in Fig. 2(e).

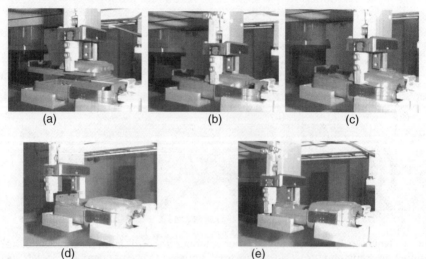

Fig. 2. Automated assembly for notch-locked assembly: (a) Approaching. (b) Sliding. (c) Snap front-end. (d) Approaching back end. (e) Snap back end.

3. EXPERIMENTAL SETUP

In order to perform a force-guided compliant motion for a notch-locked assembly, a force-guided robot is implemented. A conventional industrial robot, the Seiko Epson SSR-X Accusembler robot with SRC-42M controller, is integrated with a six axis JR3 force sensor in between the robot wrist and the end-effector tool, as shown in Fig. 3. Since the force sensor has a small tolerance (0.02 mm in the x-axis), a compliant device was developed and mounted to the sensor to reduce the stiffness and the risk of damaging the sensor and grasping product during assembly.

A PC was used as console to the robot controller and the force sensor using the RS232 and RJ-11 cables shown in Fig. 4. Serial programming was used to facilitate the interfacing between devices and the acquisition of data computed in robot controller. A proportional-based external force control, with a hybrid framework, was designed and fitted into the closed architecture industrial robot used in our applications. This control scheme is simple, easily implemented, satisfies the industrial constraints and experimental goals in a notch-locked assembly. All the data of force and position during compliant motion were processed online and analyzed offline.

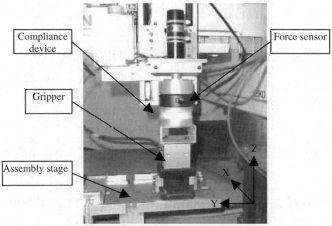

Fig. 3. End-of-arm tool setup.

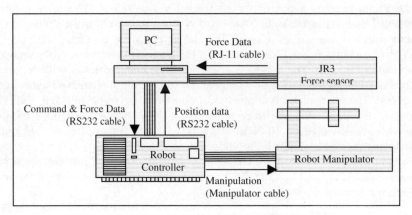

FIG. 4. Design architecture of force-guided robot.

4 EXPERIMENTAL RESULT

The approach is tested in the experimental setup for automated assembly and the results are shown in Fig. 5.

The robot manipulator performs home positioning for self-calibration and moves to the target position above the front housing during the first 20 seconds. The robot manipulator is commanded to move in the negative z-axis with step size of 0.2 mm while force is fed back to the force controller. Fig. 5(b) shows that the position in the z-axis increases in the negative direction while it remains constant along the other axis. Contact between back chassis and front housing is detected at 28.9 second.

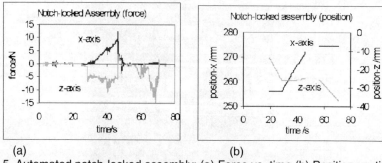

Fig. 5. Automated notch-locked assembly: (a) Force vs. time (b) Position vs. time.

The motion of the manipulator along the z-axis is stopped after the contact force in the z-axis exceeds a predetermined force level of 0.5 N. The force control is enabled to maintain the sliding force at 5 N along the z-axis, while sliding in the x-axis with the step size of 0.5 mm. The sliding motion is continued until a sudden drop of a predetermined force of 8 N in the x-axis is detected at 46.76 second, as shown in Fig. 5(a). The gripper is opened and the manipulator is made to move to the back-end of the back chassis where the force control is enabled again at 56.1 second. The manipulator is commanded to move in the negative z-axis with a step size of 0.2 mm, while the force is continuously monitored. A sudden drop of a predetermined force of 10 N in the z-axis is detected at 70.6 second and the snapping motion is stopped. The sudden drop in force of 10 N in the z-axis shows that the back-end of the back chassis is completely snapped into the notch at the back-end of the front housing. This represents the complete process of the notch-locked assembly.

5. OPTIMIZATION OF THE NOTCH-LOCKED ASSEMBLY

In a real production line, the cycling time of assembly is important since it determines the volume of the production. A high-speed assembly is needed to ensure high productivity. The assembly system must be optimized in order to produce optimum performance for real applications. However, there are few factors that limit the optimization process.

The most critical factor to be considered is the maximum payload of the compliant device. This factor may cause damage to the compliant device and needs attention in the consideration of an optimization process. For safety purpose, the maximum payload for a compliant device is recommended at a safety payload level of 20% less than the maximum payload. The maximum payload and the recommended safety payload of the compliant in each axis are shown in Table 1.

Table 1. Maximum payload for compliant device

Axis	Maximum payload /N	Safety payload /N
X	15	12
Y	15	12
Z	50	40

The optimization process must be achieved to obtain the fastest assembly time within the allowable force, as given in Table 1. From the force-guided compliant motions performed previously, three factors that can be optimized are identified. They are: sliding force, step size and arm speed. Sliding force refers to the contact force that is orthogonal to the contact surface in the sliding motion. Step size refers to the length used in the hybrid framework to move the robot in every single control loop of the force controller program. The arm speed refers to the speed of the manipulator to move from a set point to another set point in positioning controller within the step size specified in the control scheme.

Two responses are used to study the performance of the different parameter setup. They are snap force and snap time. Snap force refers to the minimum force used to perform the snap fit task and snap time refer to the time to accomplish the snap fit task in the notch-locked assembly. In a notch-locked assembly, the lowest snap force with the lowest snap time is preferred. The experimental results for various factors specification and responses are presented in Table 2.

Table 2. Optimization of various factors vs. response.

Factor			Response	
Sliding force /N	Step size /mm	Arm speed /mm/s	Snap force /N	Snap time /s
7	0.5	100	15.93	22.19
5	0.5	100	10.41	18.9
2	0.5	100	7.68	13.52
1	0.5	100	6.3	12.47
0.5	0.5	100	6	9.94
0.5	1	100	8.63	7.08
0.5	1.5	100	9.87	5.82
0.5	2	100	10.8	5.27
0.5	2.5	100	12.04	4.73
0.5	2	1	11.92	15.88
0.5	2	5	12.65	4.83
0.5*	2*	10*	10.93	3.47
0.5	2	15	11.41	3.85
0.5	2	20	10.16	5.93
0.5	2	50	9.72	8.51
0.5	2	1000	10.61	9.73

* Optimum parameter setup.

Friction force is proportional to the sliding force which is orthogonal to the contact surface. The friction force occurs in the opposite direction of the snapping motion. A higher snap force may be needed to overcome the friction force in order to accomplish the snapping task. This explains how the various sliding forces may result in different snap forces.

The optimum sliding force for the notch-locked assembly is 0.5 N because it results in the lowest snap force for safety purposes and the lowest snap time for a high-speed assembly. The sliding force is limited to 0.5 N, and not lower, to ensure the two parts are always in contact with each other and that it is always greater than the noise of the force data. Step size is one of the factors to determine the total travel time of a manipulator motion. Incrementing the step size decreases the snap time but increases the snap force in a notch-locked assembly. This means that the bigger the step size used, the faster the assembly time achieved. However, the maximum step size is limited by the constraints of the allowable maximum payload of the compliant device. A step size of 2.5 mm causes an overshoot in the allowable force in the x-axis, which may damage the compliant device. Therefore, the maximum step size of 2 mm is used in the hybrid framework to move the manipulator in the sliding motion within the allowable force in the x-axis. The arm speed of the SSR-X Cartesian robot can be set from 1 mm/s to 1000 mm/s.

A high arm speed of the manipulator results in a short travel time, which means a fast assembly time. The optimum arm speed is obtained at 10 mm/s, where it produces the lowest snap time of 3.47 second within the allowable force. Arm speeds of more than 10 mm/s cause system errors, which delay the snap time, because these higher arm speeds cannot be obtained with a small step size of 2 mm for every single loop in the hybrid framework system. Therefore, an optimum setup is achieved at a sliding force of 0.5 N, step size of 2 mm and arm speed at 10 mm/s.

6. CONCLUSION

A force-guided robot has offered a reliable solution to perform a notch-locked assembly in a real production line. With fine motion planning, these force-based compliant motions are programmed in a specific sequence and predetermined conditions to perform a notch-locked assembly, previously done by human operators in a manufacturing plant. The experimental result shows that a notch-locked assembly can be done automatically using the proposed assembly strategy. The system is optimized to perform the notch-locked assembly with the fastest assembly time, within the allowable conditions of the assembly tasks. This research will be extended to investigate various strategies for compensating and overcoming the defeats and uncertainty in real assembly tasks.

Acknowledgement
This work is sponsored by the IRPA grant, University Science Malaysia.

References
1. Robotic Industries Association, "Robotic Industry Statistic 1st Quarter 2002", http://www.roboticsonline.com/public/articles/articles.cfm?cat=201
2. Robert W. Messler Jr., Suat Genc, Gary A. Babriele, "Integral attachment using snap-fit features: a key to assembly automation. Part 1 – Introduction to integral attachment using snap-fit features", Assembly Automation, Volume 17, No. 2, 1997, pp 143-155.
3. Takeshi Tsujimura, Tetsuro Yabuta, "Adaptive force control of screwdriving with a positioning-controlled manipulator", Robotics and Autonomous System 7(1991) pp 57-65.
4. Ke-Lin Du, Xinhan Huang, Min Wang, Jianyuan Hu., "Assembly robotics research: A survey", Int. Journal of Robotics and Automation, vol. 14, no. 4, 1999.
5. H. Qiao, B.S. Dalay, R.M. Parkin, "A novel and practical strategy for the precise chamferless robotic peg-in-hole insertion", Robotica, 13(1), 1995, pp 29-35.
6. K.L. Du, "Impedance learning for peg-in-hole assembly, in research on assembly robotics and control techniques", Ph.D. Dissertation, Huazhong Univ. of Sci. & Tech., China, April 1998.
7. Kong Suh Chin, Mani Maran Ratnam, Rajeswari Mandava, "Force Guided Robotic in Automated Assembly: A Case Study in Mobile Phone Assembly", International Conference of Mechatronic, Kuala Lumpur, Malaysia, Feb. 2001, pp 186-200.

A PVDF-Based Micro-Newton Force Sensing System for Automated Micro-Manipulation

Carmen K. M. Fung[1], Wen J. Li[1], and Ning Xi[2]
[1]Dept. of Automation and Computer Aided Eng., The Chinese University of Hong Kong, Hong Kong SAR
[2] Dept. of Electrical and Computer Engineering, Michigan State University, USA

Abstract
Despite the enormous research efforts in creating new applications with MEMS, the research efforts at the backend, such as packaging and assembly, are relatively limited. One reason for this is the level of difficulty involved. One fundamental challenge lies in the fact that at the micro-scale, micro-mechanical structures are fragile and easy to break - they typically will break at the micro-Newton (mN or $10^{-6}N$) force range, which is a range that cannot be felt by human operators. In this paper, we will present our ongoing development of a polyvinylidence fluoride (PVDF) multi-direction micro-force sensing system that can be potentially used for force-reflective manipulation of micro-mechanical devices or micro-organisms over remote distances. Thus far, we have successfully demonstrated 1-D and 2-D sensing systems that are able to sense force information when a micro-manipulation probe-tip is used to lift a micro-mass supported by 2mmx30mmx200mm polysilicon beams. Hence, we have shown that force detection in the 50mN range is possible with PVDF sensors integrated with commercial micro-manipulation probe-tips. We believe this project will eventually make a great impact to the globalization of MEMS foundries because it will allow global users to micro-assemble and micro-manipulate surface micro-machined devices from their laboratories, and therefore, reduce the time from design to production significantly.

Keywords: *Micro-manipulation, micro-force feedback, micro-sensor, micro-actuator, micro-manufacturing.*

1. INTRODUCTION

MEMS devices have been steadily finding their usefulness in our daily lives in the past decade. However, a major obstacle for the advancement of MEMS technology in the commercial sector is the availability of a technique for automated batch packaging and assembly. The development of such a technology will directly impact the throughput and long-term reliability of many MEMS devices. One reason for this is the level of difficulty involved. An intrinsic difficulty lies in the fact that at the micro-scale, micro-mechanical structures are fragile and easy to break – they typically will break at the micro-Newton (μN or 10^{-6}N) range force, which is a range that cannot be felt by a human operator hoping to assemble micro-structures.

And while material properties data can be used to predict fracture strength, there is no existing micro-manipulation system that can provide in-situ μN force data during assemblage of commercial MEMS devices. The consequence of this is that devices are often damaged during assembly, decreasing overall yield and driving up cost. For instance, the well-known surface micro-machining commercial foundry technology MUMPs™ (Multi-User MEMS Processes), run by Cronos Integrated Microsystems, has a 3-polysilicon and 2-sacrificial-layer process that can now be used to produce many micro-mechanical devices with scientific and commercial applications, including micro-mirrors, micro-optical bench, micro-RF switches, and micro-sensors [1]. However, after MUMPs fabrication, many surface micromachined devices need to be micro-assembled or micro-manipulated to realize a final device or be experimentally tested. Case in point, in a micro-piezoresistive cantilever sensor, as shown in Fig. 1, manipulation is required to lift it from the horizontal plane, for mechanical tests and calibration (the sensor is a micro-mass platform suspended by 2 cantilevers 2μmx30μmx200μm in dimension). The micro-cantilevers have fracture strength in the order of μN, so an operator hoping to lift the platform for calibration will often break the cantilevers unintentionally due to excessively applied force through the commercial micro-manipulator probe-tips.

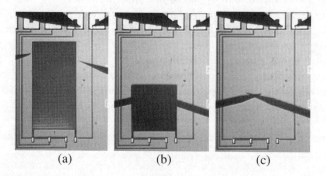

(a)   (b)   (c)

Fig. 1. MUMPs micro-structures are typically tested by using commercial probes without any force sensors for lifting and moving. (a) to (c) is a sequence of pictures showing the lifted micro-device may be damaged suddenly due to excessive force applied by a human operator. In (d) the mass-platform disappeared from microscope view due to breakage of the beams, which "sprung" the structure to a different physical location.

Micro-manipulation and control are rigorously being investigated worldwide currently. Our on-going project aims to integrate PVDF sensors on commercial probe-tips used for contact micro-manipulation and assembling, which will allow a large sensing and actuation range. The sensors' force data can be calibrated to assist a human operator in exercising manipulation forces below the fracture limit of micro-mechanical structures under manipulation. The data can also be used to establish a micro-manipulation model via an on-line learning scheme. As a result, contact/impact forces can be regulated to maintain safety margins and improve yield and reliability during micro-assembly - factors that will eventually make automated batch micro-assembly feasible.

We have recently demonstrated both 1D and 2D sensing systems that are capable of detecting an impact force when a manipulation probe hits a silicon substrate or when it lifts up a micro-structure. Our current results are presented in the following sections.

2. POLYVINYLIDENCE FLUORIDE (PVDF) SENSOR

Piezoelectric materials create electrical charges when mechanically stressed. A commercial PVDF strips from Measurement Specialties, Inc. (MSI, Shenzen, China) was used in our sensing system. In this project, we have investigated the possibility of using PVDF as force sensors because the charge generated by PVDF is almost linearly proportional to the force on its surfaces - the current generated by the PVDF can be related to an applied force. Moreover, PVDF is an ideal piezoelectric rate-of-force sensor because of its low-Q response, ease of use, and compliance - properties that are lacking in most non-polymeric piezoelectrics.

The voltage output $V(s)$ of a PVDF sensor due to an applied force $F(s)$ in Laplace domain can be written as [2]:

$$\frac{V(s)}{F(s)} = \frac{d_{33}}{A\varepsilon_{33}^T/h} \frac{\tau s}{1+\tau s} \qquad (2.1)$$

where A is the area of the crystal plate, h the thickness of the plate, ε_{33}^T is the mechanical strain in the 3 direction due to tensile stress T in the 3 direction (which represents the thickness direction), τ is the time constant of the PVDF sensor and is calculated as $\rho h C_p/A$, where $\rho h/A$ is R_p, the resistance of the PVDF sensor and ρ is the resistivity of PVDF, and C_p is the capacitance of the PVDF. The above transfer function is a high-pass filter type, so an undesired characteristic of the PVDF sensor is that its lower limit of frequency response is $> 1/\tau$, indicating that the measurement of constant force is not possible (no DC response). However, with proper electrical circuit design, a few mHz input can still be detected [2]. Our current effort in optimizing a sensor design for maximum sensitivity at low frequency force input is based on Eq. (2.1) above. As rate-of-force sensors, PVDF have already proven to be effective in controlling force damping in *macro* robotic manipulators [3]; we are currently investigating its applications in the micro-world.

3. 1-D PVDF-BASED SENSING SYSTEM

In [4] we have demonstrated an 1-D system (see Fig. 2 and Fig. 3) capable of sensing μN force, i.e., force signals from manipulating a micro-structure could be detected.

In that work, the micro-manipulator was manually controlled for positioning the probe-tips during micro-manipulation. We have since then integrated the manipulator with a computer controllable controller and have investigated in more detail the performance of the 1-D system. The results are presented in this section.

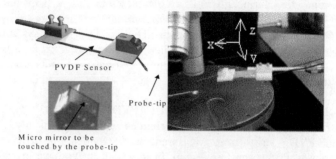

Fig. 2. 1-D PVDF force sensor probe. This orientation of the PVDF plate allows sensing of force in the z-direction.

Fig. 3. The computer controllable micro-manipulation station.

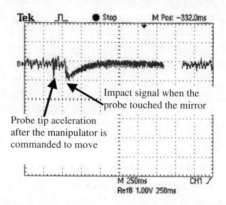

Fig. 4. Signal from touching a micro mirror at speed v = 2000μm/sec and displacement d = 252μm.

The force of the probe-tip impacting the micro-mirror can be detected by the PVDF sensing system. In this experiment, a controller was used to move the probe-tip to approach the mirror (in the positive x-direction). Once the probe-tip touched the micro-structure, an impact signal can be detected (Fig. 4).

Some basic experiments were performed to calibrate the sensor signal while lifting a micro-structure by using the experimental set up shown in Fig. 5. After the probe-tip is positioned under a micro-mass, suspended by 2 cantilevers (as shown in Fig. 6), the probe-tip was commanded to move upward to lift the micro-structure to a certain displacement (in the positive z-direction), then the probe was made to stop at that position.

Fig. 5. (a) Modified 1-D PVDF Sensor Probe. (b) micro-mass-plate lifted by probe-tip.

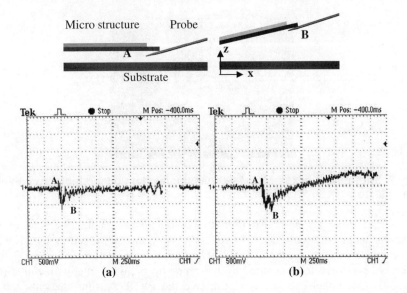

Fig. 6. Signal from lifting a micro-mass plate at v = 1000µm/sec, displacement (a) d = 40µm (b) d = 100µm.

As shown in the signal output of Fig. 6, a signal can be immediately detected (B) from lifting up the structure, then the probe-tip undergoes a vibration which indicates either the micro-structure is broken, or the structure is giving a reactive force to the probe-tip, or the system is vibrating due to a sudden stop. After the vibration of the sensor due to the reaction of the structure, we observed that the signal returned to the original value after a certain time (0.25sec). This phenomenon is due to the piezoelectric nature of the PVDF sensor – the change cannot be stored in the sensor under static deflection. We have observed experimentally, as predicted by Eq. (2.1), that the amplitude of the measured signal depends on the speed of movement of the probe-tip, which is coupled to the PVDF sensing elements.

3. 2-D PVDF-BASED SENSING SYSTEM

Our current focus is on producing a customized PVDF-based 2-D sensing system for commercially available micro-manipulation systems. We intend to demonstrate force-reflective commercial micro-manipulation tips using our PVDF sensors. We are also improving the low frequency response of our sensing system by using a charge amplifier to convert the high impedance output to a usable low-impedance voltage signal. The 2-D system constructed is shown in Fig. 7.

The 2-D sensing system was tested to investigate the performance of the probe-tip touching the substrate similar to the 1-D system. The controller was again used to command the manipulator to move the probe-tip downward to the substrate at different velocities. As shown in Fig. 8, signals from both sensors can be detected, hence the impact force information in 2-D can be obtained.

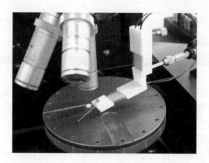

Fig. 7. Photograph of the actual 2-D sensing system integrated onto a micro-manipulator.

The probe-tip was also used to lift up micro-structures similar to the experiments performed for the 1-D system. Two-dimensional sensor signals were detectable, as shown in Fig. 9. Currently, we are able to detect distinctive signals when an excessive force applied by the probe-tip breaks a micro-structures (see Fig. 9). Our ongoing effort is to calibrate the PVDF sensing elements and to decipher the frequency contents of various signal errors, such as environmental vibration and vibration due to sudden stoppage of the sensing system. Eventually, we could extract only the interaction force between the probe-tip and the microstructures if these signal errors can be filtered.

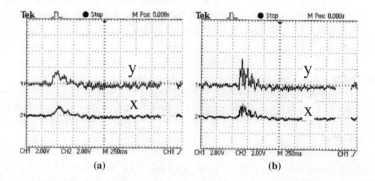

Fig. 8. 2-D sensor signals of the probe-tip impacting a substrate at (a) v =1000μm/sec, and (b) v =6000μm/sec.

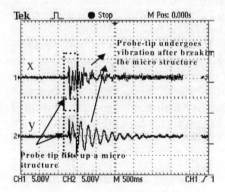

Fig. 9. 2-D sensor signal from the probe-tip lifting a micro-plate at v =3000μm/sec.

4. CONCLUSION

We have demonstrated that PVDF polymeric sensors can be used to sense micro-Newton forces when integrated with a commercial micro-manipulator with a probe-tip. Furthermore, we have demonstrated that a 2-D sensing system can be designed to perform force and impact detection – lifting of a micro-structure has been demonstrated. We now work to improve the PVDF sensing system into a force-feedback micro-manipulator to demonstrate automated micro-assembly of MUMPs structures with force-feedback control. We are also working on improving the low frequency response of our sensing elements and developing a 3-D sensing system that can be integrated with commercial micro-manipulation equipment. The goal for this project is to demonstrate a force-sensing micro-assembly system, including hardware and software, that is able to be integrated to existing commercial micro-manipulators, and capable of operating in automated assembling mode or tele-operated mode. Ultimately, this technology can be used to achieve micro-automation in batch assembling of MEMS devices such as micro-mirrors, micro-optical lenses, and general micro-sensing and actuation devices. In addition, this technology can potentially be used in bio-manipulation, including embryo injection and cell separation, and to understand the force interactions of micro-biological systems.

Acknowledgment
The authors would like to thank Mr. King W. C. Lai for his valuable contributions to this project. This work was funded by The Chinese University of Hong Kong and the NSF Grants IIS-9796300 and IIS-9796287 of Michigan State University.

References
1. Winston Sun, Ho, A.W.-T., Li, W.J., Mai, J.D. and Tao Mei, "A Foundry Fabricated High-speed Rotation Sensor Using Off-chip RF Wireless Signal Transmission", Micro Electro Mechanical Systems, 2000. MEMS 2000. The Thirteenth Annual International Conference on , January 2000, pp. 358-363.
2. P. Benech, E. Chamberod and C. Monllor, "Acceleration measurement using PVDF", Ultrasonics, Ferroelectrics and Frequency Control, IEEE Transactions on Vol.: 43, Issue: 5, Sept. 1996, pp. 838-843.
3. M. F. Barsky, D. K. Lindner and R. O. Claus, "Robot gripper control system using PVDF piezoelectric sensors", Ultrasonics, Ferroelectrics and Frequency Control, IEEE Transactions on , Volume: 36 Issue: 1, Jan. 1989, pp. 129-134.
4. C.K. M. Fung, W. J. Li, I. Elhaji and N. Xi, "Internet-based Remote Sensing and Manipulation in Micro Enviornment", 2001 IEEE/ASME International Conference on Advanced Intelligent Mechantronics Proceedings, 2001, Vol. 2, pp. 695-700.

Design of Manipulators and MEMS Assembly

This section is the largest in the book and contains nine papers covering a number of different subjects which have been broadly classified as the title suggests. They range from actual manipulator design to robot arm control, through simulation for educational purposes to MEMS applications.

The first paper is concerned with analysing the forces in a six-degree-of-freedom manipulator, with the aim of designing in sufficient motive power to overcome self weight and load inertia. By extending the Denavit-Hartenberg paradigm to eight parameters, the computational transformations of the simulation can deal with forces as well as displacements.

The next paper looks at a relatively new mechatronics design approach, i.e., the Design For Control (DFC). This is used for the design and control of parallel robots. The underlying idea of the DFC approach is to design the mechanical structure of a parallel robot judiciously such that it can result in a simple dynamic model; a simple control algorithm is then good enough for a satisfactory control performance. As such, complicated controller design can be avoided, on-line computation load can be reduced and better control performance can be achieved.

The third paper describes an integrated virtual construction, visualisation and control tool for complex mechatronic devices. ROCON allows the building of virtual robots from geometric elements connected by rotational and linear actuators. It also includes the facility to define sequences of motion patterns for exploring the complex mechanical constructions. Sensor information from a real robot, fed back to the visualisation system, supports the presentation of a realistic view of the robot, particularly concerning orientation in space which cannot be derived from the visualization alone.

The fourth paper presents the design and development of a multi-module deployable manipulator system. The system is designed for experimental investigations into dynamics and control of a variable geometry manipulator, particularly concerning its performance under various control schemes. The design process involves the selection and sizing of actuators, the design of mounting and connecting components, and the selection of hardware as well as software for real-time control

The fifth paper proposes a 'memoized' function to speed up on-line evolution of robot programs. On-line evolution is performed on a physical robot. It has an advantage over an off-line method as being robust and does not require the robot model. The 'memoized' function receives joint angles as its input. It outputs the positions of all joints.

The next paper presents the design of a gravity compensation system for flexible structure mounted manipulators. The paper describes the construction of a gravity compensation system that ideally eliminates the effects of gravity on a robot manipulator while using a minimum of power. It applies the concept to a manipulator that is mounted on a flexible structure.

The seventh paper considers the application of a relay feedback approach towards modeling of frictional effects in servo-mechanisms. The friction model consists of Coulomb and viscous friction components, both of which can be automatically extracted from the oscillation signals induced by suitably designed relay experiments. With the models, friction compensators can be designed in addition to the feedback motion controller.

Due to the minute scale of micro-engineered mechanical systems (MEMS), inertia forces are often neglected. The eighth paper proves that these forces can be significant even if a micro-structure's mass is less than a milligram. The authors have demonstrated that at this scale, mass inertia force can overcome surface forces and be used for non-contact self-assembly of MEMS structures.

The final paper in this section describes a novel distributive tactile sensor for determining the shape of a contacting object. The distributive approach uses the coupling information between elements that sense changes in the properties of a common tactile surface to infer the type of contact. It suggests that a smaller number of sensing elements will achieve the effectiveness of 'mainstream' discrete sensors.

Resolving the Tasks of the Dynamics for the Control of a Single-Planimetric Multimobile Manipulator

Prof. Korgan S. Sholanov,
The Kazakh National Technical University (K. I. Satpayev),
Republic of Kazakhstan, Almaty,
E-mail: shol@nursat.kz

Abstract
In this paper, the power demand for driving a six-mobile manipulator with reference to single-planimetric multimobile manipulators is obtained with the help of an simulation model, worked out especially for this purpose. The analytical approach of the model has been devised using the Newton-Euler's method for recurrent shapes for a local system. Numerical results were obtained by defining the structure, geometrical, kinematic and mass characteristics of the system.

Keywords: *Six-mobile, single-planimetric manipulator, simulated programme, modeling.*

1. INTRODUCTION

Research carried out [1,2] shows that a set of single-planimetric multimobile manipulators (SMM) consists of three constituents, namely: two basic functional groups (BFG) and a connective link (CL). Methods for the analysis, synthesis of constitution and methods for the kinematic analysis of these manipulators also are given.

The design of new manipulators requires the proper selection of the drive, thus ensuring that the operational requirements are met, for example, when dealing with the weight of the manipulated object. To implement the given programme, it is also necessary to know the required forces for the drives to perform adequately. However, problems in the dynamic analysis occur; in particular, where the efforts necessary for the drives and the design and controls of SMM remain unsolved. In designing new manipulators, an important aspect is the selection of the drive, to satisfy given operational requirements, for example, when the weight of the rigged object is in question. For the implementation of a programmed movement, during control, it is also necessary to know the required forces for the drives, with the purposes of defining the effects on the drives.

In this regard, there is a problem in elaborating the simulation model to determine the drives' efforts. The main requirement of a dynamic analysis method, used as a simulation model, is that it should be able to control robots in real time.

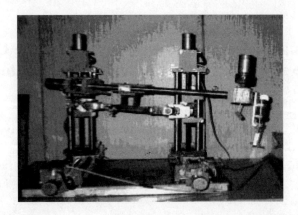

Fig. 1. A pilot model

On the other hand, the methods should be computer oriented, executable within a control program and effective in the computing field. The Newton-Euler's method in the recurrent form [3], earlier used in open-ended manipulators, appears to respond better in these particular conditions.

In practice, method [3] is applied to manipulators with a self-contained kinematics chain. On this basis, a simulation model is created and the dynamic research of a successful SMM is made. The pilot model is shown on Fig. 1.

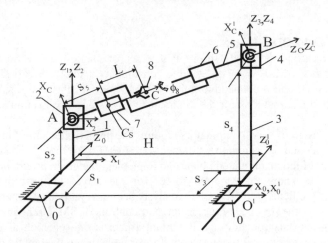

Fig. 2. The six-mobile manipulator

As shown in Fig. 2, the manipulator is synthesised into two inexact beam functional groups (BFG) consisting of links 1,2 and 3,4. The BFGs are paired at points A and B by a connective link from links 5,6,7 to a working body 8. In the six-mobile manipulator, the drives execute relative movements to links 0-1, 1-2, 0-3, 3-4, 6-7, 7-8. Thus the generalised co-ordinates will change accordingly: S_1, S_2,

S_3, S_4, S_5, ϕ_8. A base distance between BFG-H, and working body leave L, are assigned to the manipulator geometrical characteristic.

2. DYNAMIC CL

The dynamic analysis of the SMM starts with an output link. Therefore, we consider problems for the power analysis of a CL by applying the kinetostatics method. Fig. 3 shows a typical scheme for a CL, consisting of links 5-7 and working body (gripper) 8. All forces, including the force of inertia, are applied to the CL. So, for example, in the connections between the CL with links with the BFG (at a point A, with the help of a three-mobile hinge, and point B with the help of a two-mobile hinge), the forces and moments from the forces of reactions act. Therefore, at point A, the forces of reaction are applied as $\overline{A}^c = (X_a^c, Y_a^c, Z_a^c)^T$, and in a point B, it is a force $\overline{B}^c = (X_b^c, Y_b^c, Z_b^c)^T$. The moment of the forces of reactions is given by $\overline{M}_B^c = (0, 0, M_{Bc})^T$. (Here and hereinafter the superscript denotes an accessory to the co-ordinate system. For example, the index (c) means, that the indicated values are determined in a system $BX_cY_cZ_c$.) The availability of only one amounting moment of reaction forces is conditioned by the design features of a connective link and special selection of a direction of an axis BZ_c. In a centre of mass CL, without a gripper, to the point C_s, the gravity is $\overline{G}_c = m_c \overline{g}$, inertias $\overline{P}_c^J = -m_c \overline{a}_c$ are applied, and in the centre of mass C_s the gravity of a gripper and load are equal to $\overline{G}_g = (m_g + m_m)\overline{g}$. The force of inertia of their mass is $\overline{P}_g^J = -(m_g + m_m)\overline{a}_g$.

All of the above reaction forces and moments from forces have unknown values. Thus, it would appears that the problem is statistically indefinable.

To achieve a working equation, the relation of deformation will be used. It is accepted that the movement of the cross-section link 7, conducted perpendicularly to axes BZ_c, caused by deformation in a direction of the axis AB, is equal to zero.

For a system of bodies consisting of a gripper and a CL, it is possible to obtain the following kinetostatics equations in a local co-ordinate system:

$$\overline{G}_c^c + \overline{G}_g^c + (\overline{P}_c^J)^c + (\overline{P}_g^J)^c + \overline{A}^c + \overline{B}^c = 0$$
$$R_0^c(\overline{r}_c \times \overline{G}_c + \overline{BC} \times \overline{G}_g) + (\overline{M}_c^J)^c + (\overline{M}_g^J)^c + \overline{M}_B^J + \overline{BA} \times \overline{A}^c + \overline{M}_B^c = 0. \qquad (2.1)$$

In (2.1), $\overline{r}_c = (0\ 0\ -BC_s)^T$, $\overline{BA} = (0\ 0\ -l_{AB})^T$; a component BC_s means spacing interval from a point B up to a centre of a connective link mass without acquisition; $\overline{M}_c^J, \overline{M}_g^J$ - is the resultant moment from forces of inertia for CL and gripper with load; $\overline{M}_B^J$ is a moment obtained as a result of all forces of inertia reduction at point B.

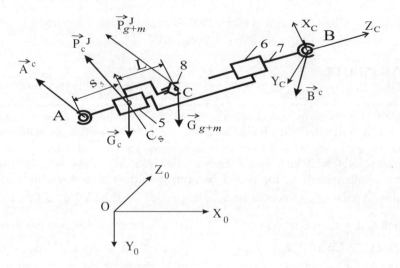

Fig. 3. A connecting link

The solution of the set of equations (2.1), together with a deformation consistency relation equation, allows to find the required forces of reaction. The forces thus obtained, and their moments, are untilised for the power analysis of the BFGs and the definition of the forces necessary to drive the SMM.

3. POWER ANALYSIS OF BFG

The dynamic analysis of the SMM requires the power analysis of the BFG using method [3]. The generalised computational scheme of Fig. 4, presenting the connection from links $i-1,i,i+1$, is used. As the scheme shows, position of points O_i, O_{i+1}, being the beginning of the co-ordinate systems, bound according to $i, i+1$, are links concerning the fixed co-ordinate system determined by position vectors $\overline{p}_i, \overline{p}_{i+1}$. Position vector $\overline{p}_{si}$ will determine the position S_i which is a centre of mass of the i of a link in the kernel system of the readout. The vectors $\overline{r}_i, \overline{O_i S_i}$ characterise the position of a beginning of the $i+1$ co-ordinate systems, regarding the i, as well as a centre of mass S_i about the $i+1$ co-ordinate system, accordingly. At point O_i the forces of reaction $\overline{F}_i$ are applied from the movement of the $i-1$ link to the i link. Throughout, $\overline{M}_i$, the main vector of the moments from forces of reaction of the i-1 link operating on the i- link, is indicated. The force developed by the drive P_i^d acts at point K. It is necessary to note that in a considered case, as opposed to that accepted in [3], efforts resulting from the translational motion drives are secured separately from all the set of forces which are operating on the i link. The drives for the majority of manipulators are not built in articulations. In that case, when the motion in an articulated piece is imparted by a rotary motion, the moment transmitted by the drive is regarded as a component from the moment of forces of reactions $\overline{M}_i$ in a hinge. As the *i* link is regarded as a system of bodies, in a centre of mass of a system S_i, gravity $\overline{G}_i$ is applied (including the drive's weights). Let the *i+1* link from the side of the *i* link and the general force $\overline{F}_{i+1}$ and moment of force $\overline{M}_{i+1}$ act. Then, accordingly to Newton's 3rd law, the force equal to (-$\overline{F}_{i+1}$) and the moment ($-\overline{M}_{i+1}$) acts on the *i* link from the side of *i+1* link.

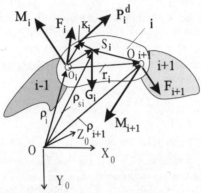

Fig. 4. The generalised computational scheme

Similarly, the recurrent relations for a connective link are obtained in a local co-ordinate system, it is possible to obtain recurrent relations for the power analysis of BFGs, recorded in a co-ordinate system, bound with the *i* link.

A further simplification of these expressions is possible if, instead of the Denavit-Hartenberg's transformation, we use the transformation on 8 parameters, as described in [4]. One of features of this method is that the co-ordinate system bound with a mobile link can be selected from the beginning from a centre of mass, having directed axes of co-ordinates along the principal axes of inertia. In this case the recurrent relations become

$$R_0^i \overline{F}_i = R_{i+1}^i (R_0^{i+1} \overline{F}_{i+1}) + m_i R_0^i \overline{a}_i - R_0^i (\overline{G}_i + \overline{P}_i^d) \quad (3.1)$$

$$R_0^i \overline{M}_i = R_{i+1}^i [R_0^{i+1} \overline{M}_{i+1} + (R_0^{i+1} \overline{r}_i) \times (R_0^{i+1} \overline{F}_{i+1})] + (R_0^i \overline{r}_i) \times (R_0^i m_i a_i -$$

$$- R_0^i \overline{G}_i) + R_0^i P_i^d \times R_0^i \overline{O_i K_i} + (R_0^i I_i R_i^0)(R_0^i \overline{\varepsilon}_i) + (R_0^i \overline{\omega}_i) \times [(R_0^i I_i R_i^0)(R_0^i \overline{\omega}_i)].$$

Here $\overline{a}_i$ is acceleration of the *i* co-ordinate system, which is expressed by a more simple relation. All members, except for diagonal members, are peer-to-zero point in the matrix ($R_o^i I_i R_i^o$), due to the special selection of the co-ordinates axes. The same quantity of additions and multiples is reduced here, important for increasing the calculation speed.

Power calculation starts from the last link calculation. Defines $\overline{F}_{i+1}, \overline{M}_{i+1}$ as equal loads which are operating on a gripper or on the part of a coupling link. A set of equations (3.1) allow to compute forces of reaction in kinematic pairs and efforts indispensable for drives.

4. DYNAMIC ANALYSIS OF THE SIX-MOBILE MANIPULATORS

The ratios obtained above are applied to the dynamic analysis of the SMM, as shown on Fig. 2.

The dynamic analysis of the manipulator starts with the CL and is made by applying the set of equations (2.1). The padding equation, which follows from a consistency relation of deformation, depends on a constitution of the actual manipulator.

Taking into account projections of all forces on an axis BZ_c of a connective link obtained with the help of a transformation matrix, we shall obtain consistency relations of deformation as:

$$m_c g \cdot l_{AS} \cdot Cos\phi_2 / EA - Z_B^c \cdot l_{AB} / EA + (m_g + m_m)g \cdot l_{AC} \cdot Cos\phi_2 / EA = 0. \quad (4.1)$$

Here ϕ_1 is a turn angle CL in a plane OXZ; ϕ_2 is an angle, padding to an angle between an axis BZ_c and axis OX_0 [2]; l_{AS}, l_{AB}, l_{AC} are variable spacing intervals from point A, accordingly up to points A, B, C (Fig. 3).

If we accept that the rigidity (EA) of a connective link on all length is identical, it is easy to receive the missing solution of equations (4.1), relating to a force of reaction component Z_B^c from a recorded consistency relation of deformation (2.1).

The value of the moment of rotation, indispensable for implementation, of a gripper 8, if friction is neglected, is derived from a differential equation for rotary motion

$$\tau_8 = J_{z8} \ddot{\varphi}_8, \quad (4.2)$$

The effort for the drive CL follows from

$$\tau_5 = (m_5 + m_g + m_m)(\ddot{s}_5 + gCos\phi_2) \quad (4.3)$$

Let's calculate the first BFG with application of recurrent relations (3.1). The forces of reactions X_A^c, Y_A^c, Z_A^c are the input data for calculations and kinematic parameters of BFG links. For a link 2, we have

$$\tau_2 = X_A^c C\varphi_1 S\varphi_2 + Y_A^c S\varphi_1 S\varphi_2 - Z_A^c C\varphi_2 + m_2(\ddot{s}_2 - g). \quad (4.4)$$

Here, and hereinafter, it is accepted that the factor of tenacious friction is equal to zero point.

For the power analysis of a link 1, we shall receive an effort of drive 1 in the form:

$$\tau_1 = X_A^c S\varphi_1 - Y_A^c C\varphi_1 + m_2 \ddot{s}_1 + m_1 \ddot{s}_1. \tag{4.5}$$

The power analysis of links of the second BFG is obtained similarly and the efforts (τ_4) (indispensable for the drive, executing motion on a link 4 concerning a link 3 and effort (τ_3) for the drive of a link 3), are determined.

Now the forces of reaction and moment of reacting in a hinge B - ($\overline{B}^c, \overline{M}_B^c$) are initially prescribed values, as well as kinematic parameters. Thus, the analytical relations for efforts of drives (τ_3, τ_4) will look like the following

$$\tau_4 = X_B^c C\varphi_3 C\varphi_4 + Y_B^c S\varphi_3 S\varphi_4 - Z_B^c C\varphi_4 + m_4(\ddot{s}_2 - g). \tag{4.6}$$

$$\tau_3 = X_B^c S\varphi_3 - Y_B^c C\varphi_3 + m_4 \ddot{s}_3 + m_3 \ddot{s}_3. \tag{4.7}$$

Expressions (4.3-4.7) determine the dynamic analysis of SMM and establish the influence of different geometrical, kinematic parameters and mass characteristics on efforts of drives.

5. SIMULATION MODELLING

The simulation model is a program developed in Turbo Pascal. The *simulation* of motion is made by changing the generalized co-ordinates. Thus the drive's cycle of activity is broken into three intervals conforming to a mode of boost, settled motion and inhibition. For each *k*-drive *(k=1,...,5)* there are a set initial *(q_{kb})* and final values of generalized co-ordinates *(q_{ke})* and time of improvement of the drive *(t_k)*. Thereby relative movement speed is set at the trapezoid law of motion. For calculation purposes, mass of load *(m_q)* and quantity of positions *(N)*, are the input data.

During the calculations, the efforts indispensable for each of drives *(τ_k)* are determined, depending on the numbers and position of the manipulator's links. The simulation model allows the study influencing the efforts of the drives in the following parameters: the geometrical characteristics *(H,L)*, speed of drives', motion sequence, load mass. Some of dynamic calculations outcomes are shown by the way of charts. Fig. 5 shows the change of efforts on drives at simultaneous translational motion on all of the 5 links 1,2,3,4,7. Thus $H=1m.$, $L=0.2m$, $m_q=20$ kg, $q_{1b}= q_{2b}= q_{3b}= q_{4b}= q_{5b}=0$, $q_{1e}= q_{2e}= q_{3e}= q_{4e}= q_{5e}=1m.t=1c.$. Fig. 6 shows change of efforts on drives at turn of links. In this case, the input data have the following values: $H=1m.$, $L=0.2m$, $m_q=20$ kg, $q_{1b}= q_{2b}= q_{5b}= 0$; $q_{3b}= q_{4b}= 1m.$; $q_{1e}= 1m. q_{3e}= q_{4e}= 0$; $t=1c$.

It follows from the charts: gravity's influence is more fundamental than the force of inertias, as seen from the efforts-change curves in the vertical drives. The maximum efforts arise in the vertical motion drives, and these efforts are 10 times more on the load mass. Also, the effects of load mass is investigated and the change of efforts on drives is reviewed at spacing interval H. The increase of the load mass in two occasions results in an increase of effort τ_2 of less than two times. Simulations of the mechanism are carried out provided that the gripper is structurally arranged and goes outside the reference points.

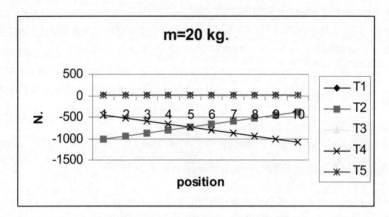

Fig. 5. Change of efforts at translational motion of links

It appears that the location in each moment of a gripper essentially influences the efforts value and nature of their change. The results obtained testify that in case of the arrangement of a gripper outside of the reference points (with leave) efforts τ_2, τ_4, are increased by almost 2 times.

The program allows to imitate the mechanism with an open-ended kinematic chain. In fact, having accepted value $H \approx 0$, it is possible to receive 3 mobile mechanisms ($W=3$) with an open-ended kinematic chain. The simulation modelling program, based on application of the Newton-Euler's recurrent equations can be utilised for the control of the mechatronic system in real time. It is necessary to note that the program can be included by the way of sub-programs in a general program of the robot's drives control.

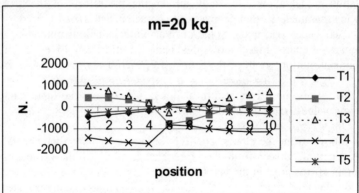

Fig. 6. Change of efforts at turn CL

6. CONCLUSIONS

A model for simulating a programmed motion of a six-mobile manipulator, with a given mass-inertial characteristics, is discussed. The model operates under an external loading, on the basis of obtained analytical relations. The simulation model allows the calculation of the forces necessary for the drives to carry out a prescribed motion and the software is included in the robot. The outcomes of the dynamic analysis, obtained with the help of an imitative program, can be utilised for the selection of drives in the design of single-planimetric multimobile manipulators.

References

1. Z. Baigunchekov, S. Joldasbekov, K. Sholanov, R. Gill: New Controlled Mechanisms with Parallel Structures. Tenth World Congress on the Theory of Machines and Mechanisms, Oulu, Finland. 1999. pp 240-246.
2. K. Sholanov: Synthesis of multimobile controlled mechanisms with closed kinematical chains. Vestnik, MGTU. Priborostroenie. 2000. no. 1. pp 111-119.
3. J.Y.S. Luh, M. W. Walker, R. P. C. Paul: On-Line computational scheme for mechanical manipulators. Trans. Asme, J. Dyn. Systems, Measurements and Control. 1980. 102, no. 2, pp 69-76.
4. Байгунчеков Ж.Ж., Шоланов К.С. Математическое описание взаимосвязанных тел пространственной механической системы // Доклады МН-АН РК. Алматы, 1999.№3. С. 75-80. (Zh. Zh. Baigunchekov, K. S. Sholanov: Mathematical description of the space mechanical system interdependent bodies. Reports of the Ministry of Sciences of the Academy of Sciences of the Republic of Kazakhnstan. Almaty, 1999. no. 3. pp 75-80.)

Design and Control of a Parallel Robot Based on the Design For Control Approach

Qing Li,
School of Mechanical and Production Engineering,
Nanyang Technological University, Singapore
Fang Xiang Wu,
Department of Mechanical Engineering,
University of Saskatchewan, Canada

Abstract

This paper applies an effective mechatronics design approach, i.e., the Design For Control (DFC) approach, for the design and control of parallel robots. The underlying idea of the DFC approach is to design the mechanical structure of a parallel robot judiciously such that it can result in a simple dynamic model; a simple control algorithm is then good enough for a satisfactory control performance. As such, complicated controller design can be avoided, on-line computation load can be reduced and better control performance can be achieved. Throughout the discussion in the paper, the design and control of a two-degree of freedom (DOF) parallel robot is studied as an illustrated example. The resulting control performances of two different mechanical designs, derived from the same robot structure topology, demonstrate the effectiveness of the DFC approach.

Keywords: Parallel robots, mechatronic systems design, Design For Control approach.

1. INTRODUCTION

The dynamic models of parallel robots, in general, are highly nonlinear and highly coupled. To achieve a satisfactory control performance, it normally requires an advanced control algorithm. However, the design of such a control algorithm can be difficult and the control performance is also hard to predict due to modelling uncertainties and unmodelled dynamics. In this paper, an effective mechatronic systems design approach, namely, the Design For Control (DFC) approach [1], is adopted in the design and control of a parallel robot. In most of the existing works for mechatronic systems development, control design is usually considered at the completion of the design and construction of the mechanical structure. However, DFC suggests that the facilitation of control design, as well as the execution of control action with the least hardware restriction, can actually be taken into consideration at the stage of the mechanical structure design. An intuitive way to implement this idea is to design an appropriate mechanical structure which leads to a simple dynamic model with simple dynamic response characteristics. To design and implement a controller for this system is thus a less difficult task.

Through out the discussion in the paper, a two DOF parallel robot is studied as an illustrated example. At first, the mechanical structure of the parallel robot is designed following the conventional design procedure. As a general robot's design, the resultant dynamic model consists of: the nonlinear and coupling terms such as the inertia term, the Coriolis plus centrifugal term and the gravitational term. In the subsequent study, the mass-distribution of the robot structure is carefully synthesised following the DFC idea. The resultant dynamic model is greatly simplified. Satisfactory control performance can be achieved via a simple PD control.

2. DYNAMIC MODEL OF THE PARALLEL ROBOT

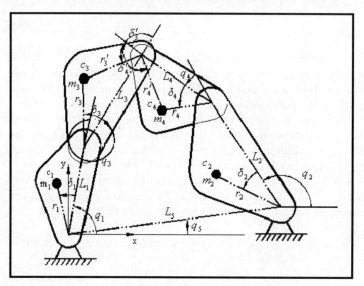

Fig. 1. The Structure of the two-DOF Parallel Robot

Fig. 1 depicts the structure of the two-DOF parallel robot, where the joint angles q_1 and q_2 of links 1 and 2 are driven by their respective servo motors; m_i and L_i are used to denote the mass and the length of the link respectively; and r_i, r_i' and δ_i are used to denote the location of the mass centre, indicated by a darkened circle c_i. The dynamic model of the robot designed based on the conventional mechanical design procedures can be obtained as follows [2]:

$$\mathbf{D}(\mathbf{q}')\ddot{\mathbf{q}}' + \mathbf{C}(\dot{\mathbf{q}},\dot{\mathbf{q}}')\dot{\mathbf{q}}' + \mathbf{g}(\mathbf{q}') = \mathbf{B}\tau^{*} \qquad (1)$$

---

[*]Actually, a reduced dynamic model is used in this work. Due to the limited paper length, Eqn. (1) is used in the presentation because it has a compact form.

where **q'** is the joint vector of the coordinates of the robot; $\mathbf{D}(\mathbf{q}')$ is the inertia matrix, defined as follows:

$$\mathbf{D}(\mathbf{q}') = \begin{bmatrix} d'_{11} & 0 & d'_{13} & 0 \\ 0 & d'_{22} & 0 & d'_{24} \\ d'_{31} & 0 & d'_{33} & 0 \\ 0 & d'_{42} & 0 & d'_{44} \end{bmatrix}, \text{ with}$$

$d'_{11} = m_1 r_1^2 + m_3 (L_1^2 + r_3^2 + 2L_1 r_3 \cos(q_3 + \delta_3)) + J_1 + J_3$

$d'_{13} = d'_{31} = m_3 (r_3^2 + L_1 r_3 \cos(q_3 + \delta_3)) + J_3$,

$d'_{22} = m_2 r_2^2 + m_4 (L_2^2 + r_4^2 + 2L_2 r_4 \cos(q_4 + \delta_4)) + J_2 + J_4$,

$d'_{24} = d'_{42} = m_4 (r_4^2 + L_2 r_4 \cos(q_4 + \delta_4)) + J_4$, and

$d'_{33} = m_3 r_3^2 + J_3$, and $d'_{44} = m_4 r_4^2 + J_4$;

$\mathbf{C}(\mathbf{q}', \dot{\mathbf{q}}')\dot{\mathbf{q}}'$ is the centrifugal and Coriolis term defined as :

$$\mathbf{C}(\mathbf{q}', \dot{\mathbf{q}}') = \begin{bmatrix} h_1 \dot{q}_3 & 0 & h_1(\dot{q}_1 + \dot{q}_3) & 0 \\ 0 & h_2 \dot{q}_4 & 0 & h_2(\dot{q}_2 + \dot{q}_4) \\ -h_1 \dot{q}_1 & 0 & 0 & 0 \\ 0 & -h_2 \dot{q}_2 & 0 & 0 \end{bmatrix}, \text{ with}$$

$h_1 = -m_3 L_1 r_3 \sin(q_3 + \delta_3)$ and $h_2 = -m_4 L_2 r_4 \sin(q_4 + \delta_4)$;

$\mathbf{g}(\mathbf{q}')$ is the gravitational term, and $\mathbf{B}\tau$ is the input torque. It can be seen that the dynamic model is very complicated.

3. REDESIGN OF THE PARALLEL ROBOT

To facilitate the control design, the robot structure will be redesigned. In the dynamic model given in (1), the inertia matrix, the Coriolis and centrifugal matrix and the gravitational vector are all highly nonlinear and coupled. The purpose of the redesign is to linearise and decouple these three terms to their full extent by using a mass-distribution scheme [3].

The robot dynamic in (1) is derived from the Lagrangian, which is defined as the difference between the kinetic and potential energy. If the robot structure is designed in such a way that the potential energy can be maintained as a constant during motion, the gravitational term can thus be eliminated. To meet this requirement, the global mass centre of the structure must be kept stationary during motion. Based on this design criterion, the following mass-distribution conditions are derived:

$$m_1 r_1 L_3 = m_3 L_1 r_3', \quad \delta_1 = \delta_3' \tag{2}$$
$$m_2 r_2 L_4 = m_4 L_2 r_4', \quad \delta_2 = \delta_4' \tag{3}$$
$$m_4 r_4 L_3 = m_3 L_4 r_3, \quad \delta_4 = \delta_3 + \pi \tag{4}$$

When the above conditions are held, it can be derived that $\mathbf{g}(\mathbf{q}') = \mathbf{0}$, and hence it leads to the following simpler dynamic model:

$$\mathbf{D}(\mathbf{q}')\ddot{\mathbf{q}}' + \mathbf{C}(\mathbf{q}',\dot{\mathbf{q}}')\dot{\mathbf{q}}' = \mathbf{B}\tau \tag{5}$$

Now, the possibility to linearise and decouple the inertia matrix and the Coriolis plus centrifugal matrix in the dynamic model is explored. Let the mass centres of links 3 and 4 be placed at the ends of links 1 and 2 respectively, thus, we have:

$$r_3 = r_4 = 0 \tag{6}$$

Once (6) is held, (4) is always true, (2) and (3) are reduced to:

$$m_1 r_1 = m_3 L_1, \quad \delta_1 = \pi \tag{7}$$
$$m_2 r_2 = m_4 L_2, \quad \delta_2 = \pi \tag{8}$$

Refer to (1), when (6), (7) and (8) are held, all elements associated with r_3 and r_4 in the dynamic model are vanished. $\mathbf{D}(\mathbf{q}')$ and $\mathbf{C}(\mathbf{q}',\dot{\mathbf{q}}')$ are partially linearised and decoupled. The resultant dynamic model of the robot can be denoted as:

$$\overline{\mathbf{D}}(\mathbf{q}')\ddot{\mathbf{q}}' + \overline{\mathbf{C}}(\mathbf{q}',\dot{\mathbf{q}}')\dot{\mathbf{q}}' = \mathbf{B}\tau \tag{9}$$

with hat "-" indicating the simplified matrices.

4. CONTROLLER DESIGN

As mentioned earlier, the basic idea of the DFC approach is to spare control design effort and improve control performance by providing a simple dynamic model through judicious mechanical design. Hence, in the first control design step, a simple PD controller is implemented to control the robot with and without redesign. A trajectory tracking control with a time span of 4 seconds is considered in the simulation study. The PD controller is given as:

$$\tau = \mathbf{K}_p \mathbf{e} + \mathbf{K}_d \dot{\mathbf{e}} \tag{10}$$

where vector $\mathbf{e} = \mathbf{q}_d - \mathbf{q}$ is the joint displacement tracking error, and $\mathbf{K}_p$ and $\mathbf{K}_d$ are the positive definite gain matrices. In this simulation study, the gain matrices are selected based on the optimisation of the following performance index:

$$I_R = E_R + W \tag{11}$$

where E_R and W represents tracking errors and control energy respectively:

$$E_R \quad \min(\ \alpha_1 \int e^2(t)dt + \alpha_2 \int_0^4 \dot{e}^2(t)dt) \tag{12}$$

$$W \quad \min(\ \beta \int_0^4 |\tau(t)| dt) \tag{13}$$

with α_1, α_2 and β as the weighting factors.

The simulation results for link 1 are shown in Fig. 2 (at the end of this paper). The results of link 2 are similar to link 1 and are therefore omitted. Fig. 2(a) shows that the angular displacement tracking error of the redesigned robot has been reduced significantly. A similar result is also observed for the angular velocity tracking error which is not presented here. From the control torque profiles shown in Fig. 2(b), it is indicated that less control energy is consumed in the redesigned robot control. We can conclude that, through redesign, not only can the motion tracking performance be improved, but also the control energy is significantly reduced.

In the second control design step, a more complex computed torque control (CTC) algorithm is employed to control the original robot with the more complicated dynamic model. The motivation behind this simulation study is to compare the performances between two situations arisen from this particular mechatronic system: Situation 1) a complex controller + a complex dynamic model; Situation 2) a simple controller + a simple dynamic model.

The CTC algorithm is chosen as follows

$$\tau = \mathbf{D}(\mathbf{q}')(\ddot{\mathbf{q}}_d + \mathbf{K}_p \mathbf{e} + \mathbf{K}_d \dot{\mathbf{e}}) + \mathbf{C}(\mathbf{q}',\dot{\mathbf{q}}')\dot{\mathbf{q}} + \mathbf{g}(\mathbf{q}') \tag{14}$$

with $\mathbf{K}_p$ and $\mathbf{K}_d$ as the feedback PD loop gains.

The simulation results for link 1 are shown in Fig. 3 (at the end of this paper). Fig. 3(a) depicts profiles of the joint displacement errors under these two situations. Although it seems that the tracking error for Situation 2) is much smaller than that for Situation 1), by taking into consideration the scales of the figures, the actual difference between the two results is only 0.00025rad, which is negligible. A similar phenomenon also occurs to the angular velocity errors, which is not shown here. These observations reveal that the difference of the tracking performance between the two design situations is not significant. However, for Situation 2), to produce a similar tracking performance as Situation 1), a much higher control torques, as shown in Fig. 3(b), is required. It can thus be concluded that, for this mechatronic system, the tracking performance of the original parallel robot controlled by the complex CTC algorithm does not surpass that of the redesigned robot structure controlled by the simple PD algorithm. However, the former design requires much more control energy.

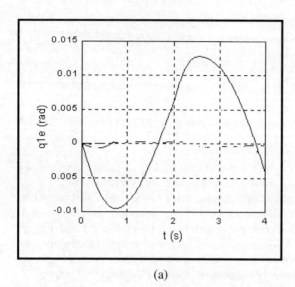

(a)

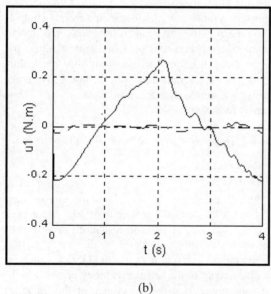

(b)

Fig. 2. Tracking Performance with PD Control
(Solid line: for original design; Dashed line: for redesign)
(a) Profiles of the angular displacement errors of the first input link.
(b) Profiles of the torques of the first input link.

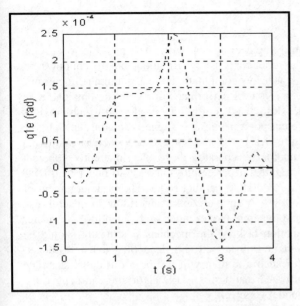

(a)

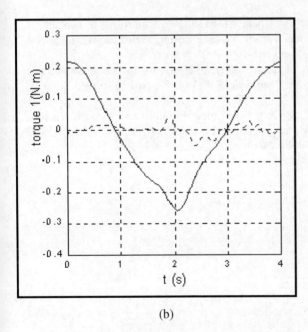

(b)

Fig. 3. Tracking Performances with CTC Control
(Dotted line: for Situation 1; Solid line: for Situation 2)
(a) Profiles of the angular displacement errors of the first input link.
(b) Profiles of the torques of the first input link.

5. CONCLUSIONS

Following the traditional control engineering approach, it is possible to develop advanced algorithms for parallel robot control. This traditional approach, however, may encounter difficulties, such as heavy computational load and modelling errors, to name it a few. The satisfactory control performance may not be obtainable. In this paper, a novel mechatronics design and control idea is adopted which provides a feasible solution from a different point of view. It suggests to consider both the control and mechanical structure designs at the same time, whereas traditional mechatronic systems design approach separates these two inherently coherent designs. Following this suggestion, this paper employs an easy-to-understand DFC approach to design and control a two-DOF parallel robot. The basic argument of DFC can be summarised as follows. A simple control algorithm, in general, can control a structure with a simple dynamic model quite well. Therefore, no matter how sophisticated a desired motion task is, the mechanical structure should be designed such that it can result in a linearised and decoupled dynamic model. Then, to design a controller for such a system will not be a difficult issue. The simulation results have demonstrated that a simple PD algorithm can control the robot with a simpler dynamic model very well.

6. REFERENCES

1. Q. Li, W.J. Zhang and L. Chen, "Design For Control - a concurrent engineering approach for mechatronic systems design", *IEEE/ASME Trans. on Mechatronics*, vol. 6, No. 2, pp.161-169, 2001.
2. A. Codourey, "Dynamic modeling of parallel robots for computed-torque control implementation", *Int. J. of Rob. Res.* Vol. 17, No. 12, pp. 1325-1336, 1998.
3. W.J. Zhang, Q. Li, and L.S. Guo, "Integrated design of mechanical structure and control algorithm for a programmable four-bar-linkage", *IEEE/ASME Trans. on Mechatronics*, Vol.4, No.4, pp. 345-362, 1999.

ROCON – A Virtual Construction Kit, Visualization Tool and Remote Control System for Mechatronic Devices

Jörg Kaiser, Thomas Fries
Department of Computer Structures,
University of Ulm,
Germany

Abstract

ROCON (<u>RO</u>bot visualization and <u>CON</u>trol system) is an integrated virtual construction, visualization and control tool for complex mechatronic devices. ROCON allows to built virtual robots from geometric elements connected by rotational and linear actuators. It also includes the facility to define sequences of motion patterns and to explore the complex mechanical constructions. In addition, ROCON enables the control of a physical equivalent by generating the necessary control signals derived from the simulation. This can be exploited for remote robot control. Sensor information from a real robot, which is fed back to the visualization system, supports the presentation of a realistic view of the robot, particularly concerning orientation in space which cannot be derived from the visualization only.

Keywords: 3-D visualization, augmented reality, remote robot control.

1. INTRODUCTION

Complex mechatronic devices like legged robots, robot arms and sophisticated gripper devices are usually composed from multiple geometric elements connected via flexible joints [1], [2]. These joints may be implemented as mechanical actuators like electric, pneumatic or hydraulic servos. Rapid prototyping and testing of the general movement of these devices is difficult because of the complex mechanical design. In addition, due to the many degrees of freedom, the movement of these devices is difficult to control. The high mechanical forces generated by the servos may lead to a certain probability of mechanical destruction and, therefore, it may be desirable to test the complex control algorithms and motion patterns in virtual reality, in which it is possible to observe the behaviour and detect collisions of the moving parts. Once tested, these sequences could later be used to control the real actuator.

When controlling the actuators of a robot, a visualization of the actuation is highly beneficial. Firstly, it can be used to support the operator which can directly manipulate the 3-D image by an adequate input device and derive the control

signals from the animation. This is, of course, particularly important for remote control where the robot cannot be observed directly. In such an application, a 3-D animation can show the status and the actions of the remote robot where a video camera would cause problems, e.g. if the bandwidth of the control channel is very low or the robot operates in opaque fluids or dark environments. Also it is not always possible to position an camera adequately outside the robot. Finally, in a 3-D representation, the perspective can be chosen freely and any angle of view can be generated which is also not possible for a camera. However, a visualization needs feedback from the real robot to match the calculated and the real conditions. Particularly, orientation in the real world cannot be derived reliably from a simulation. Therefore, to obtain the real orientation and actuator positions, appropriate sensor feedback from the physical device is needed, like inclination and direction in some system of coordinates. The ROCON system addresses these needs. It provides:
- a construction and visualization tool to design a virtual robot,
- the management of an extensible construction kit of reusable elements,
- the facility to define motion sequences and detecting collisions,
- the ability to control a physical robot remotely by exploiting sensor feedback.

ROCON is completely written in Java except the low level parts of the communication and the instrumentation interface.

2. ROCON ARCHITECTURE

Fig. 1 depicts the components of the ROCON architecture.

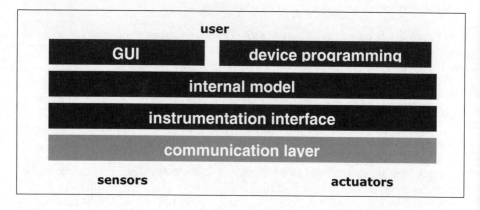

Fig. 1 - The components of the ROCON architecture

The core of ROCON is the internal object model which represents all elements and their relationships. The internal model will be further described in chapter 3.

The graphical user interface (GUI) visualizes the internal model. There is a rich set of functions to construct and manipulate the geometric representation of the robot device. Fig. 2 shows a screenshot of ROCON in the manipulation mode, where it is possible to pick geometric blocks of the virtual device and move them (currently with an ordinary mouse). As is described in the next chapter, all blocks which are connected to a selected one are also moved accordingly. It is also possible to point to joints, in which case the characteristics of the joints are displayed, e.g. whether it is a rotational or a linear joint, the constraints of movement and the current position.

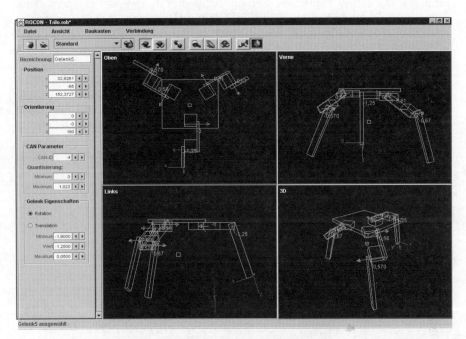

Fig. 2 - Operator Screen of ROCON

The device programming interface allows to define motion sequences for the internally represented mechanical device. Instead of using a special script language, this can be done in Java. The programmer can define arbitrary movements within the constraints of geometric rules and the joints which connect the blocks. When the movements are executed, the internal model could be used for collision detection.

From the geometric representation of the internal model, the instrumentation interface layer calculates the respective control signal for the real robot. Some of the basic parameters, like the granularity of movement, are also adjustable on-the-fly from the GUI. In case the real device is not directly attached to the operator host, a communication layer will pack the control signals in respective messages and forward it to the real device. This is detailed in chapter 5.

3. THE VIRTUAL CONSTRUCTION KIT

When considering how to model a complex robot, we have to distinguish two aspects. Firstly, we have to visualize the robot. This requires the model to provide information about the shape of the robot which, at least to some extent, should reflect the real appearance, although details can be omitted. Examples for such models are the Java 3-D Scene Graph [3] and VRML [4]. Secondly, we have to describe the robot in terms of geometric elements related by joints which have positions and certain degrees of freedom of motion in a system of coordinates. These models are often expressed in Denavit-Hartenberg [5] coordinates or equivalent representations. The model which we present to visualize the robot has to integrate both aspects of modelling. The construction tool allows to built a virtual device from rectangular 3-D geometric elements, as shown in Fig. 3. We distinguish three basic types of elements: (geometric) *element*, *joint* and *anchor*. An element abstracts the physical building blocks and stores attributes necessary for the visualization, like points, which describe the surfaces and lists these surfaces to make up a 3-D geometric body. At the moment we only use rectangular building blocks. A point, where a connection to another block is possible, is marked as an anchor. These anchors provide means to build up a set of reusable parts.

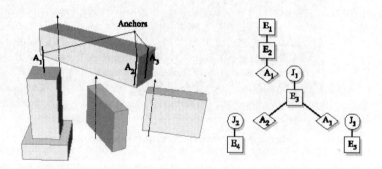

Fig. 3 - Elements, joints and anchors: building blocks in ROCON

Anchor points represent the relative positions of a part, where other parts can be plugged in. Two parts of the construction kit can be combined by connecting a joint of one part to an anchor of the parent part. The relation between the building blocks can be specified as trees (cf. Fig. 3.). A joint has one degree of freedom and may have a circular (rotation) or linear (translation) characteristic. In Fig. 3, only rotational joints are used and their axes depicted. The relations between the construction elements are described in a graph, shown at the right hand side in Fig. 3. The static pillar on which the robot arm will be placed is made of two elements E_1 and E_2 and an anchor A_1. Similarly, the building block representing the arm comprises the joint J_1 which will be connected to A_1, an element E_3 and two anchors to connect the gripper elements. Fig. 2 displays the respective robot arm and the related graph.

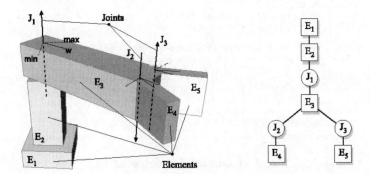

Fig. 4 - An example of a virtual robot arm and the related connection graph

As can be seen from the figures, the anchors only define the coordinates where a joint can be attached and are obsolete when the connection is defined. Elements, anchors and joints are represented as Java classes.

There are two systems of coordinates used: world coordinates which refer to an absolute reference in the universe, while the object coordinates describe the positions of the building blocks relative to each other. The first is necessary because a device may have an arbitrary position and orientation in space and there may be many independent objects for which it is necessary to display their absolute positions. Object coordinates are used to calculate the position of a geometric block relative to its predecessor in the graph. They also support a hierarchy of movements. When a component down the hierarchy is moved, its movement does not affect components at a higher level, however, vice versa, a motion of a higher level component will obviously affect all components down in the hierarchy.

4. ANIMATING THE VIRTUAL DEVICE

In addition to a direct manipulation by appropriate input devices, the virtual model of the mechatronic device can be animated. It is possible to define sequences of movements for each block according to the specifications of the joints. The model includes for each joint the specification of angle or linear ranges. Each joint can be manipulated within their ranges by an arbitrary Java program that accesses the internal ROCON model by an API (application programming interface). A simple Java class that implements a specific interface can be loaded dynamically to control the joints of the actual construction. Fig. 5 shows an example of a Java program defining the synchronous movement of the joints J2 and J3 of the example in Fig. 4, which results in a symmetric gripper property. As can be seen from the code, we use an event-based programming model which is particularly suited for control systems. As an extension, the geometric information of the model can be used to detect collisions along the movement.

```
1    ...
2    Public class MyControl extends Control {
3
4      Class MyListener implements ValueChangedListener {
5        Joint otherJoint;
6        Public MyListener(Joint otherJoint) {
7          This.otherJoint = otherJoint;
8        }
9        public void valueChanged(ValueChangedEvent thisJointEvent) {
10         otherJoint.setValue(thisJointEvent.getValue());
11       }
12     }                     // init() overrides a method of class Control
13                           // It is called by ROCON after loading.
14     public void init()    // getJoint(x) retrieves the Joint object with name x
15       getJoint("J2").addValueChangedListener(new MyListener(getJoint("J3")));
16     }
17   }
```

Fig. 5 - Example of a motion script

ROCON does not provide a complete physical simulation of the mechanical devices, taking into consideration masses, acceleration, etc. However, when the physical device is available, ROCON includes the possibility to use sensor feedback from the real device to reflect the physical conditions.

5. CONTROLLING A ROBOT REMOTELY

One of the most important features is the remote control capability provided by ROCON. Fig. 6 depicts the overall system.

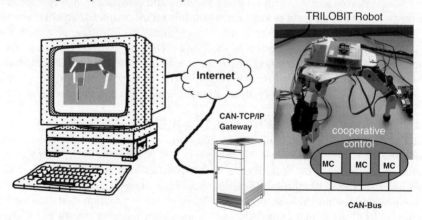

Fig. 6 - Remote Control Scheme

On the local control system the virtual representation of the real device is visualized. In this example, we modelled an active tripot. The operator can, for example, move the table by just pointing to a specific joint and drag it to a certain position. The new values for the joint will be calculated and sent via the internet to the remote host. Here, a gateway will feed it into a fieldbus network (CAN-Bus [6]) to which the micro-controllers for the real device are connected. Because the orientation of the real device in space (i.e. the orientation in the world coordination system), and particularly its orientation relative to gravitation cannot be derived from the simulation, we use a smart acceleration sensor on the real robot and sensor feed back to visualize its actual inclination in a plane. Fig. 2 depicts the operator interface screen with four selected views on the active tripot. The left side includes fields for the position, orientation, joint characteristics and network parameters.

6. DISCUSSION AND RELATED WORK

ROCON combines a simple 3-D visualization and virtual construction kit with the ability to derive control signals for a real equivalent of the virtual device directly from manipulating the virtual image and to integrate sensor feedback into the visualization. For each of the specific functions of ROCON there exist similar or better solutions in the CAD, simulation and control area. However, the combination is hardly found in other systems. Driven by the internet, during the last years there appeared a large number of systems which allow the 3-D visualization of complex mechanical devices. VRML97 [4], Java 3D [3] or X3D [12] encourage the modelling and animation of virtual robots. However, neither allow these systems to control a physical robot nor do they support the on-the-fly construction of virtual robots similar to a LEGO construction kit.

There are also systems which allow to test data developed for a real robot by a simulator before they are used to control the physical device, e.g. [7] [8]. Other systems allow the control of robots via the internet by special control panels [9]. However, ROCON takes another way, it directly derives the control information from the motion of the visualization, not from the motion of the input device or some control panel. Additionally, it is also possible to specify algorithms for robot control and motion patterns in Java to be visualized.

7. FUTURE WORK

Visualization combined with the appropriate sensor feedback can replace a camera controlled remote operation, in many cases with a fraction of the required bandwidth. In addition, a virtual representation of a robot, a complex tool or gripper device gives the advantage to take any view point and perspective to observe its operation. ROCON was designed specifically with applications in mind, which have to cope with low bandwidth communication channels where the control information and the feedback from the sensors do not require large data volume. At the moment, our experimental system has an inclination sensor for its absolute orientation in space and position sensors in the joints. We use smart sensors connected to the CAN-Bus, as shown in Fig. 6, and will further integrate tactile and distance sensors to allow a better perception of the environment. The low communication requirements enable a remote control over narrow, loaded and unreliable channels. However, there are applications in which latency may become the major problem. Long latencies can be encountered either when multi-hop connections are required with long delays in the routing nodes, the robot is far away e.g. in space missions, or the communication medium is slow, e.g. for AUVs (Autonomous Underwater Vehicles) where we have acoustic networks with propagation delays of 0.67 sec/km, signal ranges of 10-90 km and a bandwidth of 8-15 Khz [10], [11]. In these applications we firstly need local autonomy of the controlled entities, i.e. they must incorporate some goals, planning and adaptive algorithms. Secondly, we need learning or predictive filters to match the visualization with the (predicted) remote situation to mask the latency of the channel. Future work will include these properties in ROCON.

Acknowledgment
This work was partly funded by the EU in the CORTEX Project under the contract No. IST-2000-26031. (CORTEX: **CO**operating **R**eal-**T**ime s**E**ntient objects: architecture and e**X**perimental evaluation.)

References
1. Paul R.P.: Robot manipulators: Mathematics, Programming and Control, MIT Series in Artificial Intelligence, The MIT Press, 1981.
2. Featherstone R.: Robotic dynamics algorithms", The Kluwer International Series in Engineering and Computer Science, SECS 22, Robotics : Vision, Manipulation and Sensors, Kluwer, 1987.
3. Deering, M. and Sowizra H.: Java3D Specification, Version 1.1., Sun Microsystems, 2550 Garcia Av. Mountain View, CA, USA, 1998.
4. VRML97 International Standard, ISO/IEC 14772-1:1997.
5. Denavit J. and Hartenberg R.S.: A Kinematic Notation for Lower-Pair Mechanisms Based on Matrices, Journal on Applied Mechanics, June 1955, 22:215-221.
6. Robert Bosch GmbH: CAN Specification Version 2.0, 1991.
7. Speck A. and Klaeren H.: RoboSiM: Java 3D Robot Visualization", Proceedings of the IECON'99, The 26th Annual Conference of the IEEE Industrial Electronics Society, IES., San Jose, CA, 821 – 826, 1999.
8. EASY-ROB: http://www.easy-rob.de/, May 2002.
9. Internet Robotics: http://www.keldysh.ru/i-robotics/home.html.
10. A networking protocol for underwater acoustic networks, Technical report, TR-CS-00-02, Department of Computer Science, Naval Postgraduate School, December 2000.
11. Xie G.G. and Gibson J. : A network layer protocol for UANs to address propagation delay Induced performance limitations, Proceedings of MTS/IEEE Oceans 2001 Conference, Honolulu, HI, November 2001.
12 X3D: web 3D consortium: Extensible 3D, www.web3d.org.

Development of a Novel Multi-module Manipulator System: Dynamic Model and Prototype Design

C. W. de Silva, K. H. Wong, and V. J. Modi
Department of Mechanical Engineering
University of British Columbia
Vancouver, BC, Canada V6T 1Z4

Abstract
This paper presents the design and development of a Multi-module Deployable Manipulator System (MDMS) and a dynamical formulation of the manipulator. The system is designed for experimental investigations on dynamics and control of this variable geometry manipulator, particularly its performance under various control schemes. The planar manipulator that is developed here is somewhat unique in that it comprises four modules, each of which has one revolute joint and one prismatic joint, connected in a chain topology. The design process involves the selection and sizing of actuators, the design of mounting and connecting components, and the selection of hardware as well as software for real-time control. The dynamical model is formulated using an $O(N)$ algorithm, based on the Lagrangian approach and velocity transformations. The algorithm is computationally efficient permitting real-time control of the system.

Keywords: *Manipulator design, manipulator dynamic model, manipulator control.*

1. INTRODUCTION

Robotic manipulators play an important role in space exploration because of the harsh environment in which they have to operate and the challenges associated with it. Their tasks include capture and release of spacecraft, maneuvering of payload, and support of extra-vehicular activities (EVA). One example is the Mobile Servicing System (MSS), Canada's contribution to the International Space Station project. Manipulators with a combination of revolute and prismatic joints offer several useful characteristics with respect to the dynamics and control. A Multi-module Deployable Manipulator System (MDMS) is evolved from several modules, each having both revolute and prismatic degrees of freedom, connected in series. The MDMS offers several advantages over the manipulator designs that involve only revolute joints:
- Simpler decision making;
- Reduced inertial coupling, for the same number of joints;
- Better capability to overcome obstacles;
- Reduced number of singular positions for a given number of joints.

Although the particular manipulator design has these desirable features, it has not received widespread attention. This is particularly the case concerning experimental investigations using a satisfactory ground-based manipulator. A prototype manipulator can be used to assess, through real-time experiments, the effectiveness of a variety of control procedures for their possible application to space-based systems. This paper describes the development of a prototype MDMS with experimental results and presents the validation of an analytical model of the manipulator that is formulated using an $O(N)$ Lagrangian approach.

2. DESIGN OF THE MDMS PROTOTYPE

This section presents the design, construction and integration of the prototype Multi-module Deployable Manipulator System (MDMS). The objective is to develop a working system for real-time tests and control implementation. Fig. 1 shows the integrated and operational manipulator, presently located in the Robotics Laboratory of the Institute for Computing, Information and Cognitive Systems (ICICS), at the University of British Columbia. The configuration of the prototype manipulator has the following main features:

- It is a planar, eight-axis robotic manipulator with four modules, each consisting of one slewing link and one deployable link;
- It uses rolling supports on a flat surface to compensate for gravity;
- Maximum extension of each deploying link is 15cm ($\approx$ 6in).

The availability rolling supports is an important feature because the length of the manipulator induces a high bending load at each unit, as well as at the shoulder joint. The 15cm extension provides sufficient change in length for demonstrating the characteristically improved performance due to prismatic joints.

Fig. 1. The MDMS prototype

The workspace of the prototype manipulator is a circle of approximately 4.5m in diameter. This manipulator was designed employing four harmonic-drive gear actuators [1] from HD Systems Inc. (see Table 1) and four Pulse Power I (PPI) linear actuators from Dynact Inc. Once the actuators were selected, the remaining components, mounting brackets and connectors for the MDMS were designed with the aid of a CAD software package. The assembled CAD model is shown in Fig. 2. The frame plate is a wooden board, for mounting onto a support frame. The modules 1 to 4 are the PPI actuators.

Table 1. Selected actuators and their specifications for revolute joints

	Joint 1	Joint 2	Joint 3	Joint 4
Model	RFS-20-3007	RH-14C-3002	RH-11C-3001	RH-8C-3006
Max torque (Nm)	84	20	7.8	3.5
Max speed (rpm)	40	50	50	50
Diameter (m)	0.085	0.05	0.04	0.033
Length (m)	0.216	0.148	0.125	0.107
Mass (kg)	3.6	0.78	0.51	0.32
Inertia (kgm^2)	1.2	0.082	0.043	0.015

A detailed drawing of one of the revolute joints is given in Fig. 3. The module connector integrates the joint bracket to the end of the deploying shaft of the previous module through a taper pin. Two deep groove ball bearings are installed, one each at the top and bottom plates, to support the slewing link. The slewing link consists of the PPI linear actuator, which is based on the actuator mount, and top and bottom connectors between the mount and the bearings. The top connector is attached to the harmonic drive gear actuator and is secured in place with two setscrews. Two screws are required to prevent any possible slack during operation. Rolling support, with ball transfers, is attached to the bottom plate of the joint.

3. OPERATION OF THE MDMS

Controlled motion of the MDMS is achieved using optical encoders as feedback devices, which come integral with the actuators, a data acquisition board, and an IBM-PC compatible computer. Components of the control system are schematically shown in Fig. 4. To implement different control algorithms on this robotic manipulator system, an open architecture real-time control system has been established using an 8-axis ISA bus servo I/O card from Servo To Go, Inc. The actuators are driven by brush type PWM servo amplifiers and power supplies from Advanced Motion Controls. For real-time control, the 8-axis card is operated under a QNX real-time operating system. The necessary drivers and sample control programs are available in C language from Quality Real-Time Systems (QRTS).

As a reference, the operation of the MDMS is carried out with a proportional-integral-derivative (PID) controller written in C language. The control program also serves as the manager for data exchange and the coordinator of the following functions: setting the sampling rate, acquiring encoder signals [1], and sending out command signals to the amplifiers through digital-to-analogue conversion (DAC)

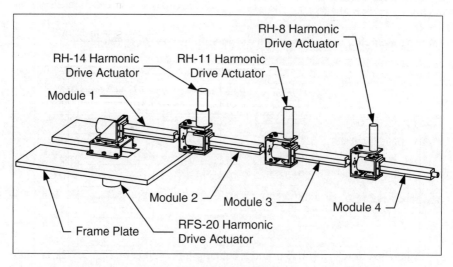

Fig. 2. Overall CAD model for the MDMS prototype

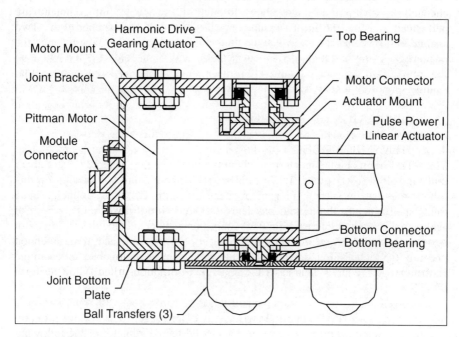

Fig. 3. Details of a revolute joint

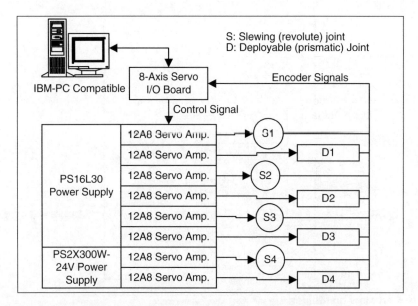

Fig. 4. Schematic diagram of the overall control system

channels. A variety of manoeuvres through independent joint command are performed to assess the ability of the controller to follow a prescribed path. Two sample results are shown in Fig. 5 and Fig. 6. In these experiments, the manipulator carries out tracking of a slew manoeuvre of 45° and a 5cm deployment, using sine-on-ramp trajectories. Although further tuning of the PID controller is required to obtain improved performance, the prototype MDMS promises to fulfil its role in evaluating different control strategies during execution of manoeuvres and tracking of trajectories.

4. DYNAMIC MODEL

The MDMS is modelled as an open chain of rigid bodies. Generally, there are N units and a flexible payload attached at the end of the Nth unit [2]. Each body can both rotate and translate with respect to its neighbours. Flexibility in the revolute joints is also included in the model. The $O(N)$ algorithm [3], where the number of arithmetic operations increases linearly with the number of modules in the robot system, is employed in deriving the equations of motion, and the simulation program is written in C language. The equations of motion are obtained using the Lagrangian procedure. This casts the equations of motion in the following form [4],

$$\ddot{\mathbf{q}} = \mathbf{M}^{-1}\mathbf{Q} - \mathbf{M}^{-1}\left(\dot{\mathbf{M}}\dot{\mathbf{q}} - \frac{1}{2}\frac{\partial(\dot{\mathbf{q}}^T\mathbf{M}\dot{\mathbf{q}})}{\partial \mathbf{q}} + \frac{\partial V_e}{\partial \mathbf{q}} + \frac{\partial R_d}{\partial \dot{\mathbf{q}}}\right) \quad (4.1)$$

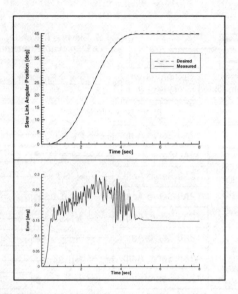

Fig. 5. A slew maneuver

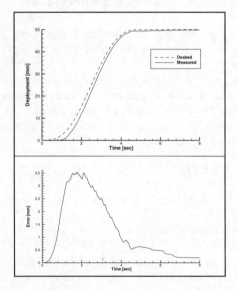

Fig. 6. A deployment maneuver

where M is the inertia matrix; Q is the generalized force vector; q is the generalized coordinate vector; V_e is the system potential energy; and R_d is the energy dissipation. In particular,

$$\mathbf{M}^{-1} = \mathbf{R}^{-1}\left[\mathbf{I}-\mathbf{R}^c\right]\tilde{\mathbf{M}}^{-1}\left[\mathbf{I}-\mathbf{R}^c\right]^T \mathbf{R}^{-T} \tag{4.2}$$

$$\mathbf{M} = \mathbf{R}^{v^T}\tilde{\mathbf{M}}\mathbf{R}^v \tag{4.3}$$

where $\mathbf{R}$, $\mathbf{R}^c$ and $\mathbf{R}^v$ are velocity transformation matrices between two sets of generalized coordinates: one based on $\mathbf{M}$ and the other on $\tilde{\mathbf{M}}$. In (4.2), $\mathbf{R}$ and $\tilde{\mathbf{M}}$, the matrices to be inverted, are both block diagonal, thus their inversion is an $O(N)$ process. Furthermore, the structure of the remaining matrices allows their multiplication to be $O(N)$ as well. Thus, inversion of the system mass matrix is now an $O(N)$ process. This formulation reduces the computational time and memory requirements considerably, making real-time applications possible.

The size of the governing equations, apart from the large number of operations required to derive them, can easily lead to errors in formulation and programming. A convenient verification procedure is to check the conservation of energy in the absence of dissipation. One set of sample results from a 2-module manipulator is shown in Fig. 7. The manipulator was set to execute a double pendulum-like motion. Initially, the two modules were aligned with each other and at 45° with the vertical axis. α_1 and α_2 are the shoulder and elbow joint angles, respectively, measured downward from horizontal axis. With the high stiffness values of the prototype MDMS, the vibration at both shoulder and elbow joints was found to be less than 0.06°, as shown in the top-right plot. The bottom plots show the time response of the system energy terms and the relative change in total energy. With the latter term being in the order of 10^{-5}, it is clear that the energy is conserved in the absence of dissipation. After carrying out a variety of further simulation runs, the dynamic model was considered to be validated.

5. CONCLUSION

A new robotic manipulator design, called the Multi-module Deployable Manipulator System (MDMS), which includes both deployable links and revolute joints was presented, and its advantages were pointed out. The mechanical design and the control system hardware and software were outlined. Results from preliminary operation of the MDMS with a PID controller were given. The MDMS will serve as an experimental platform for evaluating various control strategies during execution of manoeuvres and tracking of trajectories. Formulation of the dynamic model of the MDMS was summarized. The $O(N)$ Lagrangian approach has reduced the computational and memory requirements considerably. The model and the simulation program were verified through conservation of energy. Representative experimental results were given.

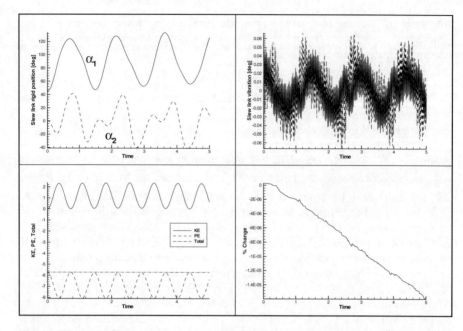

Fig. 7. Free response of a 2-module deployable manipulator in vertical plane

Acknowledgements

This work has been funded by the Natural Sciences and Engineering Research Council (NSERC) of Canada.

References

1. De Silva, C.W., *"Control Sensors and Actuators"*, Prentice-Hall, Englewood Cliffs, NJ, 1989.

2. Wong, K., "Design, Integration, and Dynamical Model of a Multi-module Deployable Manipulator System (MDMS)", M.A.Sc. Thesis, University of British Columbia, Vancouver, Canada, 2000, pp. 26-28.

3. Pradhan, S., Modi, V.J., and Misra, A.K., "Order N Formulation for Flexible Multibody Systems in Tree Topology – The Lagrangian Approach", Proceedings of the AIAA/AAS Astrodynamics Conference, San Diego, California, U.S.A., July 1996, AIAA Publisher, Paper no. AIAA-96-3624-CP, pp. 480-490; also Journal of Guidance, Control, and Dyanmics, Vol. 20, No. 4, July-August 1997, pp. 665-672.

4. Wong, K.H., Modi, V.J., de Silva, C.W., and Misra, A.K., "Design of a Novel Multi-module Manipulator", *Proceedings of the Second IASTED International Conference on Control and Applications*, Banff, Canada, July 1999, Editor: M.H. Hamza, Acta Press, Calgary, Canada, pp. 539-544.

On-line Evolution of a Robot Program using a Memoized Function

Worasait Suwannik and Prabhas Chongstitvatana
Department of Computer Engineering
Chulalongkorn University
Phayatai Road, Bangkok, Thailand 10330
Email: prabhas@chula.ac.th

Abstract
This work proposes a memoized function to speed up on-line evolution of robot programs. On-line evolution is performed on a physical robot. It has an advantage over an off-line method as being robust and does not require the robot model. However, on-line evolution is very time consuming. To validate our proposal, an experiment with visual-reaching tasks is carried out. The result shows that the memoized function can speed up on-line evolution by 23 times and the resulting control program performs robustly.

Keywords: *On-line evolution, genetic programming, robot program, memoized function.*

1. INTRODUCTION

Classical approaches to robotics require mathematical models in order to plan the robot operations. Because of this requirement, robots that adopt those approaches have to be well engineered and are normally found in a structured environment. In the cases where proper models are difficult to be obtained, the traditional approaches fail to work. To generate a robot controller in such situation, two approaches can be used. The first approach required human intuition to program the robot. A successful example of this approach is Subsumption Architecture proposed by Brooks [1]. The second approach used learning algorithms to automatically synthesize a robot controller. Evolutionary algorithms such as Genetic Algorithms [2] or Genetic Programming [3] allow robots to learn.

Natural evolution is a very slow process. Progress has been made over thousands of generations, which take billions of years. However, with the speed of today's computer, thousands of generations in off-line artificial evolution might take only few hours. Off-line evolution of robot controllers can be done very fast because it occurs in simulation. However, the controller obtained from simulation is not likely to work robustly when transferred to the real robot. This is due to the inaccurate model of the physical world. To overcome this shortcoming, the evolution can be performed with a real robot in the physical world. This is called on-line evolution. However, it is very time consuming. For example, Floreano and Mondada [4] could evolve an obstacle avoidance behavior on a physical mobile robot in about thirty hours. Chongstitvatana and Polvichai [5] estimates that genetic learning of a robot arm control program will take about two thousands hours.

Artificial evolution can be sped up by using some techniques such as: modifying evolutionary algorithms, changing the genetics operators, or reducing the time used in evaluating an individual. Our approach falls into the last category. We reduce the evaluation time by letting the robot learn the effect of its action and use it while evolving a controller for the task.

This paper is organized as follows. Evolutionary robotics is explained in Section 2. Section 3 describes the experiment. Section 4 discusses the result. Section 5 concludes this paper.

2. EVOLUTIONARY ROBOTICS

Natural evolution has created various types of highly fit biological creatures that can survive well in their environments. A group of robotics researchers borrowed this idea in order to create an artificial creature that can successfully perform a task with little human intervention. Evolutionary algorithms have been used successfully to evolve robot programs. Genetic Algorithms (GA) were introduced by Holland [2]. Each individual in a GA population is a fixed-length binary string. Genetic Programming (GP) is a variation of GA introduced by Koza [3]. The major difference between GA and GP is the representation of an individual. Instead of being a fixed-length binary string, an individual in GP is a computer program whose length is varied.

Simulation is useful for study some fundamental problems in evolutionary robotics [6-9]. However, Brooks [10] pointed that when using simulation, we might solve a problem that does not exist in the real world or the solution of the problem cannot be used in the real world. This is because a simulated robot is usually oversimplified. Many aspects of the real world such as noise, uncertainty, mass, friction, inertial forces are ignored.

For these reasons, the evolved controller from simulation does not work robustly when transferred to the real world [11]. Several approaches were proposed to improve the robustness. Some researchers added noise to the simulation to make more accurate simulation [12-14]. Some researchers performed experiments using simulation built from real robot's sensory and actuator data to model the real world more accurately [5,15].

Crossing the simulation-reality gap is not trivial because of two major reasons [13]. First, it is difficult to model any aspect of the real world accurately. Second, it is very difficult to include every aspects of the reality in simulation. Brooks recommended discarding simulation model and using a real robot as its model [1].

On-line evolution of a robot controller is similar to natural evolution in the sense that it occurs in the real world. The tremendous amount of time in on-line evolution leads to several problems [16]. This might be a reason that there are a few attempts to evolve a robot controller on-line. Floreano and Mondada evolved an obstacle avoidance behavior for a mobile robot [4]. The behavior of the neuron network controller converges in about thirty hours. Dittrich and his colleagues evolved a controller for a robot with arbitrary structures [17].

3. MEMOIZED FUNCTION

To evaluate an individual on-line, each robot motion is performed in the physical world. However, many of these motions are repetitive hence their effects are already known. If these effects are stored and reused, then a large number of actual motions can be eliminated. We propose a memoized function to store the effects of robot motions.

The memoized function receives joint angles as its input. It outputs the positions of all joints. Our implementation of the memoized function can hold every possible combination in the joint space. Since each joint of our robot arm has 60 discrete steps, there are a 60^3 or 216,000 entries. However, for larger size of configurations, we do not have to allocate all entries to implement the function. The technique of virtual memory can be used. The memoized function can be implemented with a much smaller physical memory.

4. EXPERIMENT

The aim of this experiment is to evolve a robot program without using any simulation model. The robot learned the visual reaching task on-line using GP. Before the on-line evolution took place, we used simulation to estimate the time that the robot learned the task without using a memoized function. Different sets of genetic parameter were used in the simulation. The best parameter set was used in on-line learning. We compared the estimated time with the exact time spent in on-line evolution using a memoized function. Finally, the robustness of a resulting program from each run is measured.

4.1 The Robot Arm and Its Task

We constructed a three DOF robot arm as our experimental platform, as shown in Figure 1. Each joint of the arm is made of servos normally used in a hobby radio-controlled airplane or car. The arm can move only in a plane. A CCD camera located above the arm provides a robot vision. The vision system monitors the distance between the tip and the target and checks whether the robot hit any obstacles.

Although the robot looks simple, building an accurate model is not obvious. The real robot as its model gives the following properties:
- Lens distortion. An inexpensive surveillance CCD camera used in this experiment has pretty high distortion. The length of each links varied from angle to angle.
- Camera calibration. By not depending on any mathematical model, there is no need to calibrate the camera to fit with the model. The camera can be placed at any height from the robot-moving plane as long as it can see the arm movement.
- Motor effect. The robot can learn the effect of its motor command such as joint positions during the evolution of control program.

As shown in Figure 2, five instances of the visual-reaching problem similar to those in [14] were created as the representatives of the problem. They are varied in degree of difficulty. A circle in each picture is a target. Black rectangles are obstacles. During the learning period, the robot can freely move from one configuration to another without hitting any obstacles. However, in the testing period, it has to reach the target while avoiding the obstacles.

4.2 The Control Program

GP used primitives in a terminal and a function set to construct a robot program. The terminals set contains robot motion command and sensing primitives. A motion command primitive moves a specific joint one step. A sensing primitive senses if any of the robot's links hit an obstacle or whether the tip is closer to the target. The function set contains basic control flow primitives, which are IF, IF_AND, IF_OR, NOT. The structure of the robot program makes the robot reactive.

Figure 1. A robot arm seen from its vision

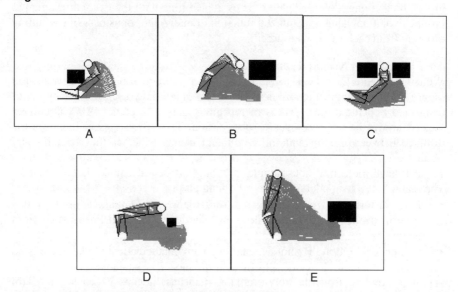

Figure 2. Instances of a target-reaching problem

4.3 Fitness Function

Each robot program is given a limited amount of steps to be executed before its fitness is evaluated. The further the distance from the tip of the arm to the target, the lower the fitness. A fitness function f is defined as follows.

$$f(v) = 1000; \qquad d(v, tip) \leq 2$$
$$ = 1000 - d(v, tip); \quad \text{otherwise}$$

where $d(v1, v2)$ is the distance between point $v1$ and $v2$.

4.4 Genetic Parameters

Standard GP is used in our experiment. As the population size affects the on-line evolution time, we conducted an experiment in simulation to determine the appropriate population size. We found that the size of 200 individuals gives the best estimation result in evaluation time.

We made no efforts in optimizing other genetic parameters. The parameters are similar to those in [14], except that the evolution is continued until no progress has been made during 10 consecutive generations. If the evolution stops without a solution, it will be rerun with the memoized function that has already been filled.

5. RESULTS

As shown in Table 1, the robot can learn the task in less than one hour on average when using a memoized function. We estimated the time that the robot can learn the task on-line without using a memoized function. The estimation is calculated by assuming that each movement of the robot took one second to complete. When compared the estimated time with the real evolution time, the average speed-up is about 23 times.

The recall rate is very high. Each recall means the required data is found in memory, which means that the robot does not have to move in the physical world. Less than 3% of the memory allocated for the memoized function were used. The usage of the memory depends on the time it takes to learn the task. The first generation filled more entries compared to later generations. This is because the offspring is likely to be in the same configurations as its parents is. The new configurations implies the exploration of the offspring.

Table 1. Evolution time

Problem	With Memoize (hours)	Without Memoize (hours)	Speed-up (times)	Recall (%)	Memory used (%)
A	0.51	5.00	9.64	89.52	0.90
B	0.64	6.33	7.80	85.65	1.13
C	1.19	25.32	20.93	94.82	2.20
D	1.28	132.68	59.09	93.45	2.31
E	0.93	34.81	17.18	81.91	1.79
Average	0.91	40.83	22.93	89.07	1.67

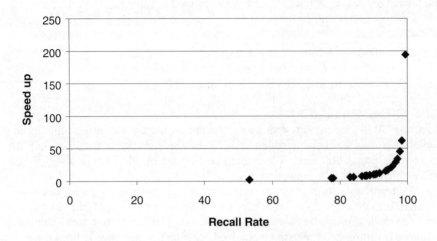

Figure 3. Recall rate versus speed-up

As show in Figure 3, the percentage of recall and speed-up are exponentially related. Starting from zero recall, the speed-up is equal to one (i.e. no speed-up). The speed-up grows much larger when the recall rate is above 95%. A controller for a more complex task might be able to evolve on-line in a reasonable amount of time if the recall rate is very high.

The robustness of a robot program is the percentage of times the robot can successfully perform the task. The robustness of each control program is shown in Table 2. We found that the failed program moved to the configuration that is not found in memory many times more than the successful program does. In other words, the failed program moved to the configuration it was not learned during the evolution. We hypothesize that this is due to noise in the system that leads the robot to move to unknown configurations. One way to improve the robustness is to add the same level of noise to the memoized function when evolving a control program.

Table 2. Robustness of an evolve program

Problem	Robustness of a program evolved on-line with memoized function
A	98
B	76
C	82
D	60
E	92
Average	82

6. CONCLUSION

Inaccurate models resulted in the fragile behavior of a robot. On-line evolution eliminates the use of any mathematical models of the robot. However, on-line evolution is very time consuming. By using a memoized function, on-line evolution of a visual-reaching task can be speeded up by about 23 times. The evolved robot program works robustly in the testing environment. The size of required memory for implementing memoized function grows exponentially with the degree of freedom of the robot. However, a memoized function can be implemented with the technique of virtual memory, as it is found that the percent of use is low.

To improve the robustness, noise can be added to a memoized function. The memoized function can be replaced by another learning method, which will learn the effect of robot movement. The remaining question is that the speed-up is enough to learn a more difficult task or use a high DOF robot.

References

1. Brooks R., "New Approaches to Robotics", Science 254, pp. 1227-1232, Sep. 1991.
2. Holland J., Adaptation in Natural and Artificial Systems, University of Michigan Press, 1975.
3. Koza J., Genetic Programming, volume (1), MIT Press, 1992.
4. Floreano D., Mondada F., "Automatic Creation of an Autonomous Agent: Genetic Evolution of a Neural-Network Driven Robot", Proceeding of the 3rd Inter. Conf. on Simulation of Adaptive Behavior, 1994.
5. Chongstitvatana P., Polvichai J., "Learning a visual task by genetic programming", Proceedings of IEEE/RSJ Inter. Conf. on Intelligent Robots and Systems, 1996.
6. Koza J., "Evolution of Subsumption using Genetic Programming", Proceedings of the First European Conference on Artificial Life, pp. 110-119, 1992.
7. Nolfi S., Floreano D., "Learning and Evolution", Autonomous Robots, 1999.
8. Chongstitvatana P., "Using Perturbation to Improve Robustness of Solutions Generated by Genetic Programming for Robot Learning", Journal of Circuits, Systems and Computer, vol. 9, no. 1 & 2, pp. 133-143, 1999.
9. Suwannik, W., Chongstitvatana, P., "Improving the Robustness of Evolved Robot Arm Control Programs Generated by Genetic Programming", Proceedings of Inter. Conf. on Intelligent Technologies, pp. 149-153, December 2000.
10. Brooks R., "Artificial Life and Real Robots", Proceedings of the First European Conference on Artificial Life, pp. 3-10, 1992.
11. Polvichai J., Chongstitvatana, P., "Visually-guided reaching by genetic programming", Proceedings of 2nd Asian Conf. on Computer Vision, pp. 329-333, December 1995.

12. Miglino O., Lund H., Nolfi S., "Evolving Mobile Robots in Simulated and Real Environments", Artificial Life, pp. 417-434, 1995.
13. Jakobi N., Minimal Simulations for Evolutionary Robotics, PhD. thesis, University of Sussex, 1998.
14. Suwannik, W., Chongstitvatana, P., "Improving the Robustness of Evolved Robot Arm Control Programs with Multiple Configurations", Proceedings of Asian Symposium on Industrial Automation and Robotics, pp. 87-90, May 2001.
15. Lund H., Hallam J., "Evolving Sufficient Robot Controllers", Proceeding of IEEE Inter. Conf. on Evolutionary Computation, 1996.
16. Mataric M., Cliff D., "Challenges in Evolving Controllers for Physical Robots", Special Issues of Robotics and Autonomous System, Vol. 19, No. 1, pp. 67-83, October 1996.
17. Dittrich P., Bürgel A., Banzhaf W., "Learning to Move a Robot with Random Morphology", Proceedings of the First European Workshop on Evolutionary Robotics, pp. 168-178, 1998.

Design of Gravity Compensation System for Flexible Structure Mounted Manipulators

Theeraphong Wongratanaphisan[1], Meng-Sang Chew[2], and Thongchai Fongsamootr[1]

[1] Department of Mechanical Engineering, Chiang Mai University, Chiang Mai, THAILAND

[2] Department of Mechanical Engineering and Applied Mechanics, Lehgih University, Pennsylvania, U.S.A.

Abstract
This paper presents the design of a gravity compensation system for flexible structure mounted manipulators (FSMM). The purpose of the study is twofold: 1) to construct a gravity compensation system that ideally eliminates the effects of gravity on a robot manipulator while using the least amount of power; 2) to apply the gravity compensation concept to a manipulator that is mounted on a flexible structure. First, a passive gravity compensation system was designed to counteract gravity forces in any posture of the manipulator. Second, the gravity compensation system was applied to a FSMM. A prototype of gravity compensated FSMM was constructed and tested. The experiments on vibration control of the designed system were performed and the results are provided. The advantage of employing the gravity compensation is discussed.

Keywords: *Gravity compensation, spring suspension, vibration control, flexible structure mounted manipulator, micro-macro manipulator.*

1. INTRODUCTION

Two common assumptions in the study of robot manipulators are that the robot structure is rigid, and that the robot's base is stationary. Contemporary robot applications have gone in the direction of larger and faster systems of which the structure can no longer be considered rigid. This has lead to many studies in the area of flexible robot manipulators on a stationary base.

Recently, a relatively new system which consists of a small rigid manipulator mounted on a large flexible structure has been studied. This system is broadly called the "Flexible Structure Mounted Manipulator (FSMM)" [1], [2]. Its applications are in the area of space robotic missions. Unlike conventional robots, such as the industrial ones, the base of the FSMM is not stationary. Instead, it is mounted on a large structure that is considered flexible. The dynamics of the

FSMM may simply be illustrated by a manipulator sitting on a spring-mass-damper system (see Fig. 1).

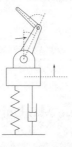

Fig. 1 Flexible structure mounted manipulator (FSMM)

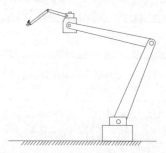

Fig. 2 Macro-Micro Manipulator

The "Macro-Micro Manipulator (MMM)" [3]-[5] is another system that operates similarly to the FSMM. It consists of a small rigid manipulator mounted on the tip of a large flexible manipulator (see Fig. 2). Its potential applications are in the area of nuclear waste remediation and other operations that require large working space. The macro manipulator has long arms and hence provides large working space. However, due to structural flexibility, its end-point cannot be positioned accurately. In the operation, the macro manipulator is employed for coarse positioning of the micro manipulator which then operates in a smaller region with higher accuracy. In each coarse positioning, the macro manipulator holds its posture until the micro manipulator finishes its tasks in its micro workspace. Then, the macro manipulator changes it posture to relocate the micro manipulator, and the operation continues. For each coarse positioning, therefore, the whole system can be regarded as a FSMM. It has been shown that the dynamic structure of the MMM system is more feasible to control than that of the flexible manipulator alone [3].

2. KINEMATICS AND DYNAMICS OF A FSMM SYSTEM

The analysis of kinematics and dynamics of rigid robot manipulators have very well been established, as can be found in many textbooks such as [6] and [9]. These analyses can be extended to take into account the dynamics of the flexible structure as briefly presented in the following.

2.1 Kinematics

Let $q \in \Re^{n \times 1}$ denote a vector of generalized coordinates that correspond to the joint angle of the rigid manipulator, $\varsigma \in \Re^{r \times 1}$ a vector of generalized coordinates that correspond to the deformation of the flexible structure, $p \in \Re^{m \times 1}$ an $m (\leq 6)$-dimensional task space vector of the end-effector of the manipulator. $\Re^{n \times 1}$ represents n-dimensional Euclidean space. Vector p is called the "manipulation vector". The vector ς results from a finite dimension approximation of the flexible modes in the structure.

In general, the manipulation vector can be related to the joint space vector and the deformation vector by nonlinear functions

$$\mathbf{p} = f(\mathbf{q}, \varsigma, \phi) \tag{2.1}$$

where ϕ denotes a set of system inertial and geometric parameters of the links. The first and second derivatives of $\mathbf{p}$ are

$$\dot{\mathbf{p}} = J_q \dot{\mathbf{q}} + J_\varsigma \dot{\varsigma} \quad \text{and} \quad \ddot{\mathbf{p}} = J_q \ddot{\mathbf{q}} + J_\varsigma \ddot{\varsigma} + \dot{J}_q \dot{\mathbf{q}} + \dot{J}_\varsigma \dot{\varsigma} \tag{2.2}$$

where J_q and J_ς are the Jacobian matrices of appropriate dimension of which the elements are functions of $\mathbf{q}$, ς, and ϕ.

2.2 Dynamics

Using the Lagrangian method, the dynamic equation of a FSMM system can be modeled and expressed in the following mathematical form

$$\begin{bmatrix} M_m & M_{mb} \\ M_{mb}^T & M_b \end{bmatrix} \begin{Bmatrix} \ddot{\mathbf{q}} \\ \ddot{\varsigma} \end{Bmatrix} + \begin{Bmatrix} 0 \\ K_b \end{Bmatrix} + \begin{Bmatrix} c_m + g_m \\ c_b \end{Bmatrix} = \begin{Bmatrix} \tau \\ 0 \end{Bmatrix} \tag{2.3}$$

where M_b and K_b $\Re^{m \times m}$ are inertia and stiffness matrices of the flexible base, respectively, M_m $\Re^{m \times m}$ the inertia matrix of the manipulator, M_{mb} $\Re^{m \times n}$ the inertia coupling matrix between the manipulator and the base, c_m and c_b vectors of nonlinear velocity-dependent terms, g_m a vector of gravity induced torque of the manipulator, and τ a vector of applied joint torques. The degree of dynamic coupling between the flexible structure and the manipulator depends on M_{mb}.

3. CONTROL OF VIBRATION

Vibration in the flexible structure is the main disturbance that affects the performance of the manipulator in the FSMM system. As the manipulator arms move, part of their kinetic energy is transferred to the flexible structure, thereby inducing vibration which in turn reduces positioning accuracy of the manipulator end-effector. It may be concluded that there are generally two schemes in dealing with vibration in the FSMM system: 1) vibration avoidance and 2) vibration control.

3.1 Vibration Avoidance

In this category, there are mainly two techniques: path planning and input shaping. In the path planning technique, the goal is to plan the motion of the manipulator for a given task such that the forces transferred from the manipulator through its base to the flexible structure is minimum. To completely avoid such induced vibration, this technique requires a redundant manipulator. In the input shaping technique, the reference input command is filtered before being sent to the manipulator. This means that the real command sent to the manipulator is shaped. A number of studies have employed this technique because of its simplicity [8],[9].

3.2 Vibration Control

In this category, the purpose is to deal with vibration that has occurred or cannot be avoided. There are two main control schemes: vibration suppression and vibration compensation. In the vibration suppression, the main concern is to eliminate or reduce the vibration that has been induced into the flexible structure by utilizing reaction forces from the manipulator. Once the vibration is damped out, the accuracy of the overall system is achieved. In vibration compensation scheme, the main objective is to achieve global positioning accuracy of the end-effector as fast as possible. The vibration may not completely die out before global positioning is achieved. Such a technique is not suitable for systems where residual vibrations are crucial to its operation, such as space robots. A variety of techniques have been developed for vibration compensation controls of FSMM and MMM systems. See [10], [11].

Most of the studies on the FSMM system have focused only on the vibration control aspect on the existing system. The design of a FSMM that best suits the task has not been the subject of interest. As a result, many systems have to cope with inherently structurally difficult control problems. This study proposes the design of a FSMM that helps to solve some difficulties in control.

4. GRAVITY COMPENSATION DESIGN FOR FSMM

Gravity causes many negative effects on the robot system. It necessitates the use of large actuators and employment of gear units which exhibit nonlinear characteristic such as backlash and friction. In this study, a gravity compensation system has been designed for a FSMM with a serial manipulator that employs electro-mechanical drives. The primary goals of the gravity compensation system are as follows.
    a.   To allow the use of smaller actuators leading to low current operation.
    b.   To eliminate the bias due to the gravity induced torques at the actuators, hence allowing the actuator to work more efficiently.
    c.   To eliminate the use of gearing units leading to the direct-drive system.
    d.   To make the actuator relocation possible, thus allowing the system to be configured so that the control problem is structurally easier.

Generally, there are two methods of compensating for gravity: mass balancing and spring suspension. Since the mass balancing technique adds substantial mass into the system causing the system to be heavy which is not suitable for the flexible structure, in this study, the spring suspension was employed. There have been a number of studies on gravity compensation by spring suspension of manipulators which achieve perfect compensation [12]-[14]. By perfect compensation, it means the manipulator stays in balance for all its postures without any supplied force from the actuators. Consider Fig. 3 which shows a one-

DOF manipulator operating on the horizontal plane and being suspended by a linear spring. It has been found that the compensation will be perfect if the following conditions are satisfied;

$$kab = mgp \quad \text{and} \quad L_0 = 0 \tag{4.1}$$

where P is the center of mass of the link, g is the gravity acceleration, m is the mass of the link, L_0 is the free-length of the spring. Note in Eq. 4.1 that $L_0 = 0$ means that the free-length of the spring is zero, which is not physically realizable. However, it has been found that the spring can be set up such that it behaves as if its free-length is effectively zero [12]. This type of setup can be also be applied for a two-DOF and a multi-DOF manipulator through a parallelogram structure [12]. In this study, this kind of setup is proposed for gravity compensation of a FSMM as explained in the following section.

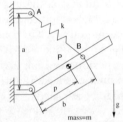

Fig. 3 One-DOF manipulator suspended by linear spring

5. THE TWO-DOF GRAVITY COMPENSATED FSMM PROTOTYPE

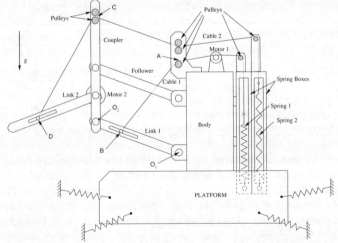

Fig. 4 Drawing of a two-DOF gravity compensated FSMM prototype

Fig. 4 shows the drawing of a two-DOF gravity compensated FSMM prototype developed at the Mechatronic Lab, Lehigh University. The prototype consists of a two-DOF gravity compensated manipulator sitting on a flexible platform. The platform is suspended by a number of springs around the side allowing it to move in all six possible motions (three translations and three rotations). The suspended platform is used to simulate the vibrational motion of the flexible structure. The manipulator consists of five rigid linkages: main body, Link 1, Link 2, the follower and the coupler. Link 1, the follower, the coupler and the main body are connected together to form a parallelogram structure that provides the first degree of freedom. Link 2 is connected to the coupler by a revolute joint to provide the second degree of freedom. Link 2 and the parallelogram structure are assembled together such that their motions are in the same plane.

The manipulator is mounted on the platform such that the plane of motion of its arms is maintained as close to the vertical plane as possible. As part of the gravity compensation system, two linear springs are employed in the zero-free-length-spring setup. The two springs reside in the spring boxes that are attachment to the main body. The first spring, Spring 1, is tied to Cable 1 which is routed through two pulleys to connect to Link 1 at the attachment point B. On the other hand, Spring 2 is tied to Cable 2 which is routed through a series of five pulleys to connect to Link 2 at the attachment point D. The routing of Cable 2 is set up such that the portion between coupler and the main body is parallel to Link 1 and the follower. In this way, the static position of Link 2 will not be disturbed when the static position of the parallelogram is changed. Link 1 and Link 2 each has a slot to allow the cable attachment point to be adjusted.

Table 1 Parameters of the prototype

Link Parameters	Link 1	Link 2	Follower	Coupler
Length (mm.)	178^*	178^{**}	178^*	-
Center of Gravity^1 (mm.)	60	114	63	-
Mass^2 (g.)	213	310	286	436
Spring Parameters	Spring 1		Spring 2	
Stiffness (N/mm.)	0.070		0.029	
Free Length (mm.)	127		127	
Extended Length (mm.)	443		535	

* measured between two joint axes that are associated with the link
** measured from the joint axes to the tip
^{1} measured from the axis of rotation
^{2} including mass of the actuators and/or sensors attached to the link

Two d.c. motors are employed for driving the manipulator arms. Motor 1 remotely drives Link 1 through Cable 1. Attached to the shaft of Motor 1 is a driving pulley that Cable 1 is routed past before reaching point A. The friction between the driving pulley and Cable 1 is sufficiently high for Motor 1 to drive the manipulator arms without slippage. Motor 2 is installed to drive Link 2 directly at its joint axis. No gearing units are employed in the system. Both motors are very small - without the gravity compensation system, they would not be able to bear the gravity loads of the manipulator arms. Motor 1 is less powerful than Motor 2 although its load is larger. This is due to the mechanical advantage that is gained from remotely driving Link 1. The reason that Motor 1 is placed at the assigned position is to illustrate that with the gravity compensation, the actuator can be relocated. The ability to relocate the actuators can be beneficial to the control system in terms of reducing the degree of non-collocation. Consider a typical system that uses joint actuators to control the tip. Because the input and the output are far away from each other, this system represents a non-collocation system which normally results in a non-minimum phase system known to be difficult to control. By driving the arm through cables, the force is then directly applied to the attached point which is closer to the tip. Therefore, the degree of non-collocation is reduced.

Table 1 provides the parameters of the manipulator arms and the springs. Fig. 5 shows several photographs of the FSMM prototype taken from various views

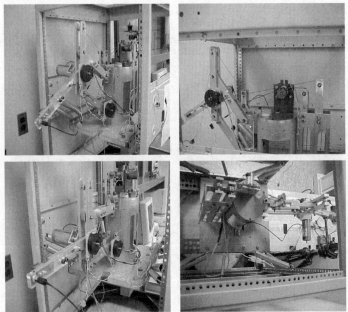

Fig. 5 Photographs from various views of the FSMM prototype

6. EXPERIMENT ON VIBRATION CONTROL

The experiment was conducted to illustrate the effectiveness of the proposed system. Fig. 6 shows the arrangement of the hardware equipment used in the experiment. In the setup, the main computation is carried out by the dSPACE's DS1003 Processor Board which is connected to a host personal computer. This board is equipped with DSP which allows for fast computation. Two additional I/O boards are employed: 1) dSPACE's DS2201 multi-I/O board and 2) dSPACE's DS3002 incremental encoder board. The control signals are sent from the DS2201 board to Coupley's 4113 d.c. Servo Amplifier which drives two Pittman's servo d.c. motors. The angular positions of Link 1 and Link 2 are detected by incremental encoders which send a series of pulsing signals back to the DS3002 board. Not shown in the figure is a d.c. unregulated power supply which provides the energy source for the servo amplifier.

To detect the position of the tip of Link 2, a laser diode is attached to the tip of Link 2. The laser beam is projected on a screen creating a light spot. The On-Trak's PSM2-10 position sensing detector (PSD) module was used to detect the position of the light spot by collecting the light that is reflected from the screen. A plano-convex lens was mounted in front of the PSD sensor in order to increase the detectable area on the screen. The signals from the PSD sensor are sent to the On-Trak's OT301-DL 2-axis Position Sensing Amplifier which translates the input signals from the PSD to voltage signals. These voltage signals are then received by the DS2201 board - hence closing the loop.

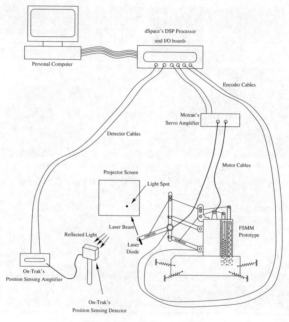

Fig. 6 Experiment setup

Since the manipulator has only two degrees of freedom, it is not possible for the tip of the manipulator arm to access all the surfaces in the task space. Then the goal was set to control only the projected position of the tip of Link 2 on the vertical plane. The manipulator was instructed to travel from one location to another in a step manner through a PD controller. The predictive control law developed by Phan et al. [15] was adopted on top of the PD controller for the control of residual vibration. In Fig. 7 the dash-dot line represents the open-loop response of the x- and y-coordinate of the light spot on the screen when the manipulator was commanded to move at time 1 second. As can be seen, the residual vibration lasted for more than 20 seconds. The solid line represents the response after the predictive controller was turned on at time 1.1 second. From the figure, it can be seen that the amplitude of the vibration was less than the opened loop response and vibration settled at about time 4 second. The result shows that the proposed concept works promisingly for vibration control in a FSMM system.

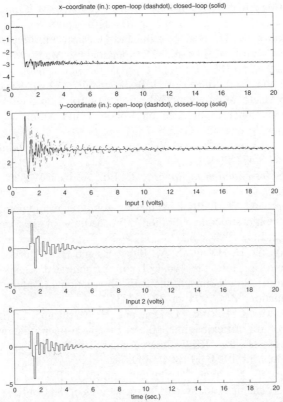

Fig. 7 Time histories of the coordinate of the light spot on the screen: dash-dot --- opened-loop response; solid --- closed loop response

7. CONCLUSION

In this study, the design of gravity compensated FSMM system has been illustrated. The prototype was constructed and the experiment on vibration control has been conducted. The experimental result shows that the application of gravity compensation system to a FSMM is effective. Not only does it help to reduce the size of the actuators and eliminate the gear units, but it also allows for actuator relocation which lessen the degree of non-collocation in control problem.

References

1. K .Yashida, D. N. Nenchev, and M. Uchiyama, "Vibration Suppression and Zero Reaction Maneuvers of Flexible Space Structure Mounted manipulators," *Smart Materials and Structures*, Vol. 8, 1999, pp 847-856.
2. M. A. Torres, S. Dubowsky, and A. C. Pisoni, "Vibration Control of Development Structures' Long-Reach Space Manipulators: The P-PED

Method," Proceedings of the 1996 IEEE International Conference on Robotics and Automation, April 1996, Minneapolis, Minnesota, pp 2498-2504.
3. A. Sharon, N. Hogan, and D. E. Hardt, "The Macro/Micro Manipulator: An Improved Architecture for Robot Control," *Robotics & Computer Integrated Manufacturing*, V. 10, n. 3, 1993, pp 209-222.
4. T. Yoshikawa, K. Hosoda, T. Doi, and H Marukani, "Dynamic Trajectory Tracking Control of Flexible Manipulator by Macro-Micro Manipulator System," 1994, pp 1804-1809.
5. D. W. Cannon, D. P. Magee, W. J. Book, and J. Y. Lew, "Experimental Study on Micro/Macro Manipulator Vibration Control," Proceedings of the 1996 IEEE International Conference on Robotics and Automation, April 1996, Minneapolis, Minnesota, pp 2549-2554.
6. H. Asada and J.-J.E. Slotine, *Robot Analysis and Control*, John Wiley & Sons 1986, ISBN 0-471-86029-1.
7. J. J. Craig, *Introduction to Robotic*, 2nd Ed., Addison-Wesley 1989, ISBN 0-201-09528-9.
8. D.P. Magee and W. J. Book, "Filtering schilling manipulator commands to prevent flexible structure vibration," Proceedings of the American Control Conference, June 1994, pp 2538-2542.
9. D. S. Kwon, D. H. Hwang, S. M. Babcock, and B. L. Burks, "Input shaping filter methods for the control of structurally flexible, long-reach manipulators," Proceedings of the IEEE International Conference on Robotics and Automation, May 1994, pp 2227-2233.
10. T. Yoshikawa, K. Hosoda, T. Doi, and H. Marukani, "Dynamic trajectory tracking control of flexible manipulator by macro-micro manipulator system," Proceedings of the IEEE International Conference on Robotics and Automation, May 1994, pp 1804-1809.
11. W. Yim and S. N. Singh, "Trajectory control of flexible manipulator using macro-micro manipulator system," Proceedings of the Conference on Decision & Control, New Orleans, 1995, pp 2841-2846.
12. R. H. Nathan, "A constant force generation mechanism," Journal of Mechanism, Transmissions, and Automation in Design, ASME, 1985, V. 107, pp 508-512.
13. D. A. Streit and B. J. Gilmore, "Perfect spring equilibrators for rotatable bodies," *Journal of Mechanisms, Transmissions, and Automation in Design*, 1989, V. 111, pp 451-458.
14. T. Wongratanaphisan and M. Chew, "Gravity compensation of spatial 2-DOF serial manipulators," *Journal of Robotic Systems*, 2002, V. 9, n. 7.
15. M. Q. Phan, R. K. Lim, and R.W. Longman, "Unifying input-output and state-space perspectives of predictive control," Report 3044, Princeton University, NJ, September 1998.

Relay-based Friction Modeling Technique for Servo-Mechanical Systems

K.K. Tan and X. Jiang,
Department of Electrical and Computer Engineering
National University of Singapore

Abstract
This paper considers the application of a relay feedback approach towards modeling frictional effects in servo-mechanisms. The friction model consists of Coulomb and viscous friction components, both of which can be automatically extracted from the oscillation signals induced by suitably-designed relay experiments. At the same time, the dynamic model of the servo-mechanical system can be obtained from the experiments. With the models, friction compensators can be designed in addition to the feedback motion controller.

Keywords: Friction modeling, servo-mechanical systems, friction compensation, relay identification techniques.

1. INTRODUCTION

Relay feedback has enjoyed tremendous success in the automatic tuning of controllers [1]. Today, automatic tuning features, in one form or another, can be readily found in standard industrial controllers. In these applications, a relay feedback apparatus is used to excite a sustained oscillation, from which information of the system can be inferred and used to tune the controller. The application domain of relay automatic tuning has been expanded since its inauguration, and today, the technology is applicable to many industrial processes [1] and servo-mechanisms [2]. However, thus far, the system models (implicit or explicit) derived from the relay experiments for tuning purposes are mainly linear ones.

In this paper, we explore an application of relay feedback in the identification of a friction model for servo-mechanisms. The frictional characteristics associated with servo-mechanisms are highly nonlinear in nature, and a good friction model is especially important for applications involving high precision motion control of servo-mechanisms, where the frictional force needs to be adequately compensated in order to improve the transient performance and to reduce steady state tracking errors. Even in adaptive control of servo-mechanisms, as will be illustrated in the paper, an initial friction model is also crucial to ensure smooth control signals and rapid parameter convergence [3]. Friction modeling has always been a difficult and challenging problem [4],[5]. Models of varying complexity have been used to approximate the dynamics of friction [4]. However, in practical applications, models of the Tustin type are usually used which yield good consistent results, in many cases without undue structural complexity.

In [6], the idea of using relay feedback to identify a simple Coulomb friction model was first mooted. Either an iterative procedure or a noise-sensitive analysis of the limit cycles of a servo-mechanical system under relay feedback can be invoked to yield an approximation of the Coulomb friction coefficient. In [7], a relay method using only the amplitude and frequency of the limit cycle oscillations induced is presented. In this paper, under the same configuration as [7], we explore the use of a nonlinear least squares estimation method to identify the key parameters from the input and output signals of the system. This technique is reasonably robust to the effects of noise measurement.

2. MODEL OF SERVO-MECHANICAL SYSTEMS

Following [7], the dynamics of a servo-mechanical system can be described using a nonlinear mathematical model:

$$\ddot{x} = (a\dot{x} - u - f_{fric} - f_{load} - f_{res})/b, \tag{2.1}$$

where u(t) is the time-varying motor terminal voltage; x(t) is the motor position; f_{load} is the applied load force respectively. f_{fric} denotes the frictional force and f_{res} represents any remaining small and uncertain unaccounted dynamics; a and b are constants relating to the dynamics of the motor.

The frictional force affecting the movement of the translator can be modeled as a combination of Coulomb and viscous friction as:

$$f_{fric} = [f_c + f_v |\dot{x}|]\operatorname{sgn}(\dot{x}) \tag{2.2}$$

where f_c is the minimum level of Coulomb friction and f_v is associated with the viscosity constant. For loading effects which are independent of the direction of motion, f_{load} can be described as:

$$f_{load} = f_l \operatorname{sgn}(\dot{x}). \tag{2.3}$$

Cumulatively, the frictional and load force can be described as one external disturbance F given by:

$$F = [f_1 + f_2|\dot{x}|]\operatorname{sgn}(\dot{x}), \tag{2.4}$$

where $f_1=f_l+f_c$ and $f_2=f_v$. Fig. 1 graphically illustrates the characteristics of F. It is an objective of this paper to estimate the key characteristics of F as well as the model parameters of the motor using a relay feedback experiment.

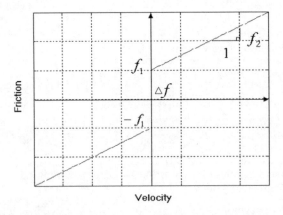

Fig. 1. F- $\dot{x}$ characteristics

3. PROPOSED IDENTIFICATION METHOD

As discussed in [7], a dual relay feedback for servo-mechanical systems may be used to automatically generate excitation signals. The proposed system is shown in Fig. 2.

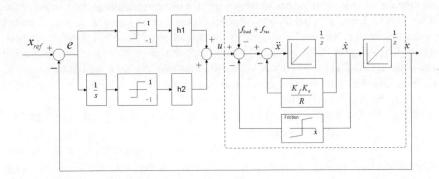

Fig. 2. Proposed dual relay setup

Mathematical model (2.1) is in the linear regression form and it is straightforward to apply a least squares estimation method to obtain the unknown parameters. h_1 and h_2 can be manipulated so as to obtain a sufficiently exciting signal. In [7], an equivalent setup, as shown in Fig. 3, is considered where an approximation technique based on a describing function approach can be applied. The parallel relay construct (henceforth called the equivalent relay ER) consists of feedback relays FR1 and FR2, as well as the inherent system relay SR due to frictional and load forces. The describing function (DF) approximation is thus directly applicable towards the analysis of the feedback system.

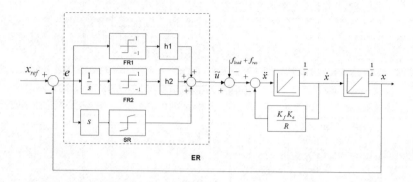

Fig. 3. Equivalent system

It may appear that the method will be sensitive to noise, since the derivatives of x are needed in the identification. However, it should be pointed out that with the relay setup, the limit cycle oscillation with the typical low-pass property of any practical system will yield a sinusoidal output from the system. Once the key parameters of this sinusoid are determined, it will be straightforward to derive the derivatives. Thus, a reliable and accurate identification of the key parameters associated with relay oscillations under the influence of noise is important. A non-linear least-squares (LS) method can be applied in a two-stage identification experiment [7]. The basic idea is to find the optimal true sine parameters (i.e., A, ω, θ in (3.1)) in the least squares sense, assuming the oscillations are stationary and periodic. This is given by

$$x(t) = A\sin(\omega t + \theta). \quad (3.1)$$

Mathematically, this is equivalent to minimizing a performance index J(A,ω,θ) given by

$$J(A,\omega,\theta) = \sum_{j=0}^{N_p-1}\left[\bar{x}(t_0 + jT_s) - x(t_0 + jT_s)\right]^2 \quad (3.2)$$

where $\{\bar{x}(t)\,|\,t=t_0, t_0+T_s,...,t_0+(N_p-1)T_s\}$ is a data series of a sampled noisy sinusoidal signal where N_p is the total number of points, T_s the sampling period and t_0 is the initial time.

Stage 1: Fixed ωω

When ω is fixed, (3.2) can be converted to a linear LS problem. Defining $\alpha_1 = A\sin(\theta)$ and $\alpha_2 = A\cos(\theta)$ for a given ω, the optimization problem is to locate A and θ so that J_ω is minimized where

$$J_\omega(A,\theta) = \sum_{j=0}^{N_p-1} \left[\bar{x}(t_0 + jT_s) - \alpha_1 \sin(\omega(t_0 + jT_s)) - \alpha_2 \cos(\omega(t_0 + jT_s))\right]^2 . \quad (3.3)$$

This is clearly a linear LS problem which can be directly solved.

Stage 2: Varying ωω

The parameter optimization can be repeated for a range of frequency ω in the neighbourhood of the estimated value used in the earlier stage. It can be defined that

$$\min_{A,\omega,\theta} J(A,\omega,\theta) = \min_\omega \left\{ \min_{A,\theta} J_\omega(A,\theta) \right\}. \quad (3.4)$$

The complete optimal set (A, ω, θ) can thus be identified. Fig. 4 shows the extraction of a sinusoidal profile from the noisy oscillation signal of a relay feedback experiment.

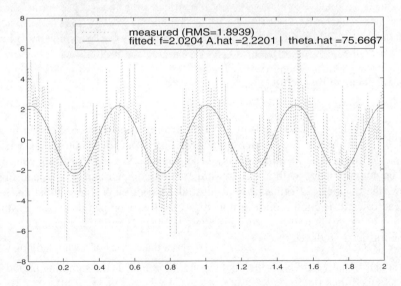

Fig. 4. Feature extraction from a noisy sinusoidal signal (solid-extracted sinusoid, dotted-actual sinusoid)

4. EXPERIMENT

In this section, experimental results are provided to illustrate the effectiveness of the proposed method. Fig. 5 shows the experimental setup. The linear motor used is a direct thrust tubular servo motor manufactured by Linear Drives Ltd (LDL)(LD 2504), which has a travel length of 500mm and it is equipped with a Renishaw optical encoder with an effective resolution of 1μm. The dSPACE control development and rapid prototyping system, in particular the DS1102 board, is used. dSPACE integrates the entire development cycle seamlessly into a single environment, so that individual development stages between simulation and test can be run and rerun, without frequent re-adjustment. MATLAB and SIMULINK can be directly used in the development of the final dSPACE real-time system. The proposed algorithm is written in C and embedded in one S-function block.

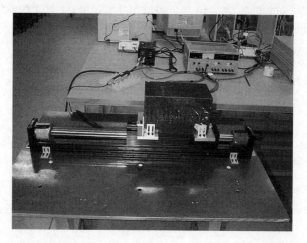

Fig. 5. Linear motor setup

The limit cycle oscillations arising from the two experiments are shown in Fig. 6 and 7, from which the model parameters for both the motor and friction characteristics can be identified. With these parameters, a PID feedback controller can be commissioned with a feedforward friction compensator. Fig. 8 and Fig. 9 shows the tracking performance to a sinusoidal trajectory profile without and with a feedforward friction compensator respectivelly. Compared to one without friction compensation, the root-mean-square (RMS) value of the tracking error has been drastically reduced by 50%.

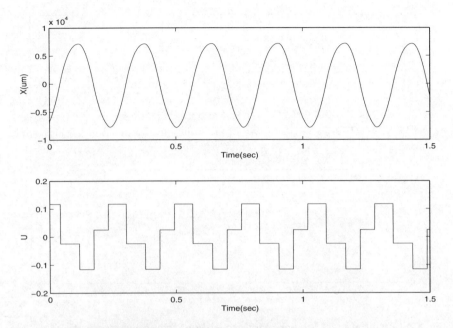

Fig. 6. Limit cycle oscillations- first experiment

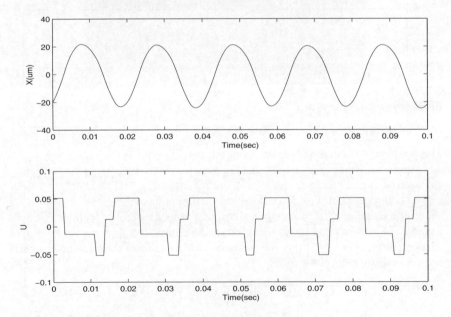

Fig. 7. Limit cycle oscillations- second experiment

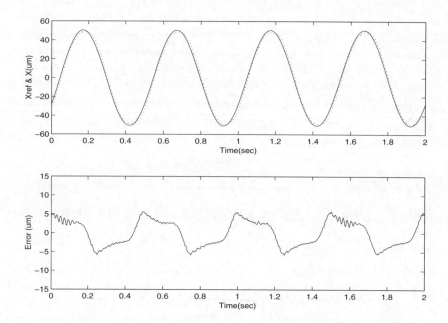

Fig. 8. Sinusoidal tracking performance without friction compensation

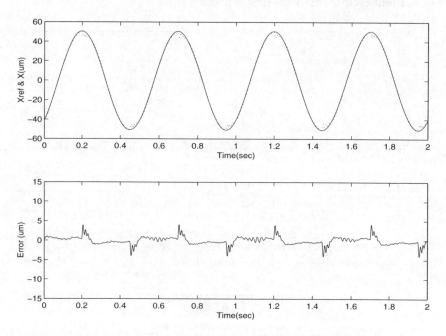

Fig. 9. Sinusoidal tracking performance with friction compensation

5. CONCLUSION

A new approach towards modeling of friction in servo-mechanisms using relay feedback has been developed in this paper. The friction model consists of Coulomb and viscous friction components, both of which can be automatically extracted from suitably-designed relay experiments. At the same time, the dynamical model of the servo-mechanical system can be obtained from the experiments. In this way, both a feedback motion controller and a feedforward friction compensator can be automatically tuned.

REFERENCES

1. Astrom, K.J. and T. Hagglund (1995), *PID Controllers: Theory, Design, and Tuning,* 2nd Edition. Instrument Society of America.

2. Tan, K.K., T.H. Lee and F.M. Leu (2000), Automatic tuning of 2DOF control for D.C.Servo Motor Systems, *Intelligent automation and soft computing,* vol. 6, no.4, pp.281-290.

3. Krstic,M., I. Kanellakopoulos and P. Kokotovic (1995). *Nonlinear and Adaptive Control Design,* John Wiley & Sons.

4. Armstrong-Helouvry, B., P. Dupont and C.C. de Wit (1994), A survey of models, analysis tools and compensation methods for the control of machines with friction, *Automatica,* vol. 30, no.7, pp. 1083-1138.

5. Taghirad, H.D. and P.R.Belanger (1998), Robust friction compensator for harmonic drive transmission, *Proc. of 1998 IEEE Int'l Coni on Control Applications,* pp. 547-551.

6. Besancon- Voda and G. Besancon (1999), Analysis of a class of two-relay systems, with application to Coulomb friction identification, *Automatica,* vol. 35, no.8, pp. 1391-1399.

7. Tan, K.K., T.H. Lee, S. Huang and X. Jiang (2001), Friction modeling and adaptive compensation using a relay feedback approach, *IEEE Transactions on Industrial Electronics,* vol. 48, no.1, pp. 169-176.

8. Tan, K.K., T.H. Lee, H. Dou and S. Huang (2001), Precision motion control-design and implementation, Springer-verlag (London).

Automated Micro-Assembly of MEMS by Centrifugal Force

King W. C. Lai and Wen J. Li
Dept. of Automation and Computer Aided Eng.,
The Chinese University of Hong Kong

Abstract

Due to the minute scale of MEMS, inertia forces are often neglected. However, we have proved that these forces can be significant even if a microstructure's mass is <1mg (a 250μmx100μm mass with MUMPs poly1, poly2, and Au layers). We have demonstrated that at this scale, mass inertia force can overcome surface forces and be used for non-contact self-assembly of MEMS structures. Centrifugal force was applied to hinged MUMPs#43 structures, causing these structures to self-assemble by rotating themselves $90°$ out of the substrate plane and automatically locked themselves to designed latches. This batch-assembly technique is very fast, low-cost, non-contact, and non-destructive Moreover, we have successfully characterized the centrifugal forces needed to assemble these microstructures by integrating sensors on the same MUMPs chips to provide wireless signal that relate to the dynamic behaviour of the microstructures. This is a very important outcome in terms of making feasible quantitative analyses of surface forces acting on surface micromachined MEMS devices Our current results will be reported in this paper.

Keywords: *micro-assembly, non-contact micro-assembly, batch micro-assembly, centrifugal micro-assembly, MUMPs assembly.*

1. INTRODUCTION

New designs in surface-micromachining that requires micro-assembly to form 3D devices have now focus on the ease of assembly. An example of a simple technique was reported as a single-step assembly system in [1], where movement of a single plate or structure assembles the entire structure constrained by hinges. Automatic latches have also been used to engage and lock these structures into position. Numerous techniques to move this plate or assemble other hinged structures are available and have been summarized in [2]. These methods include using on-substrate actuators, manual probing, or tensile films to assemble. Other techniques include magnetic actuation and tribo-electricity. These methods all have inherent disadvantages such as yield, cost, and difficulty in implementation. We propose a fast, reliable, and low-cost batch assembly technique using centrifugal force to lift MEMS structures.

It is generally accepted that, for MEMS devices, surface forces dominate over volume forces since, by isometric scaling argument, surface area reduction is proportional to only 2/3 power of the reduction in volume. Consequently, it is often assumed that inertia force is negligible in the presence of other forces, such as friction or surface forces. However, our research proved that mass does matter to a certain scale in MEMS. Our work demonstrated that for many MEMS structures, their inertia force could overcome friction and surface forces. From this, we present a novel non-contact method of batch auto-assembly of surface-micromachined devices that has many advantages over conventional micro-assembling methods.

2. MICRO-ASSEMBLY BY CENTRIFUGAL FORCE

Theoretically, gravitational force can become significant compared to other micro-scale related forces if the sizes of objects fall within a certain dimensional range. By replacing the gravitational force with centrifugal force, the surface forces, e.g., friction, electrostatic, etc., may be overcome, thus allowing microstructures to be lifted. However, some engineering considerations must be given, such as a hinge that undergo significant stress from a pulling mass may break.

As the suitable force to manipulate microstructures should be in the range of micro-Newtons to avoid damaging the microstructures under manipulation, centrifugal force is appropriate to perform the task because the force can be applied to the structures uniformly, and the force can be made in the order of micro-Newtons. As an example of application, micro-mirrors can be rotated from horizontal to a desired position (see Fig. 1), where the final position of the mirror can be decided by the geometry of a pre-designed locking system. The applied centrifugal force should be large enough to overcome the surface forces, otherwise, the microstructures would not be released from the substrate. On the other hand, the upper limit for applied force should not cause a stress greater than the allowable stress on the hinge structure.

The centrifugal force acting on a microstructure depends on the its mass and angular velocity as,

$$F = mr\omega^2 \qquad (2.1)$$

where m is the mass of a structure to be assembled, r is the radius of a spinning disc which holds the substrate containing the structure, and ω is the angular velocity of the disc. Hence, the angular velocity and size of a spinning disc can be used to assemble a microstructure with a given mass. Based on MUMPs fabrication data, the thickness of the various structural layers are known, thus, along with given material density data, the mass of a MUMPs structure can be estimated. In general, the mass of poly1 and poly2 layers can be neglected if a gold layer is used to build a structure, due to their significant difference in density [3]. Hence, the centrifugal force acting on a MUMPs mass can be related to the angular velocity in rpm by:

$$F = \frac{V\rho r R^2 \pi^2}{900} \tag{2.2}$$

where V is the volume of the gold layer, ρ is the density of gold, r is the distance for the rotation and R is the angular velocity in rpm.

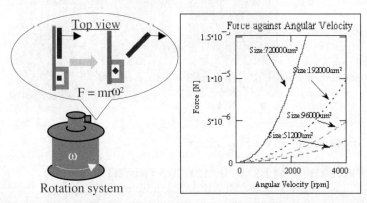

Fig. 1. Conceptual drawing for centrifugal force applied on micro mirrors.

Theoretically, centrifugal force can be generated on different sizes of MUMPs structures (poly1, poly2, and gold) by changing the angular velocity of a spinning disc.

3. STRESS ON POLYSILICON HINGES

In order not to destroy the hinges used to clamp the microstructures during the centrifugal assembly process, stress on the polysilicon should be estimated. For a typical Pister hinge (shown in Fig. 2), 3 areas on the hinges are the weakest part in the whole structures.

The tensile stress acting on these 3 areas can be estimated by the following equation,

$$\sigma = \frac{F_{applied}}{A_{total}} \tag{2.3}$$

where $F_{applied}$ is the minimum centrifugal force to overcome the surface force, A_{total} is the total area on the specific region.

We have estimated that if we applied 1μN centrifugal force to one hinge, the stress on the weakest point would be 3.629×10^4Pa, which does not exceed the maximum tensile strength of polysilicon, given by Sharpe et al. [4] for MUMPs polysilicon as 1.20±0.15GPa. In general, a sufficient number of hinges should be used to hold the microstructures to be assembled to ensure that the structures will not *fly* off the substrate during the centrifugal assembly process because centrifugal force would be about 22µN for a 600x300µm² mass platform at 6250rpm.

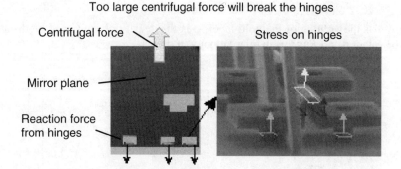

Fig. 2. Stress acting on the polysilicon.

4. CENTRIFUGALLY ASSEMBLED STRUCTURES

We have assembled various MUMPs structures using centrifugal force. Some detailed experiments were performed by applying a centrifugal force on micro mass (mirror) structures. A rotation system was setup to realize our non-contact batch micro-assembly process, which is able to provide steady angular velocity up to 6250rpm. In order to reduce the airflow from the surroundings during high speed rotation, a cover is used over the chip. (MUMPs structures have been observed to break and fly off the substrate by a small amount of air flow.)

A number of MUMPs micro masses mirrors have been fabricated for centrifugal assembly tests. The masses are classified into four different groups based on their size and locker system. The different mass designs are given in Table 1.

Table 1. Sizes of mirrors tested.

Type	Size of mass	Locker system
I	600x300um²	Traditional latches
II	300x200um²	Traditional latches
III	250x100um²	Traditional latches
IV	300x200um²	V-shape latches
[Image deleted from web version for lack of space]	[Image deleted from web version for lack of space]	
Traditional Latch	V-shape Latch	

The batch micro-assembly process was tested to investigate the performance and repeatability. Two identical MUMPs chips were tested and the results are very consistent. By increasing the angular velocity, different sizes of mirrors rotated up from horizontal to vertical position. When the rotation system reaches the maximum angular velocity of about 6250rpm (a limitation of our current rotation system), all 600x300μm² micro-mirrors were lifted successfully into the vertical position. Applied centrifugal forces are greater for larger mass size at the same angular velocity. So, Type I mirrors can be assembled at a smaller angular velocity than Type II and III mirrors. When the angular velocity is at 6250rpm, the centrifugal forces applied on Type I, Type II, and Type III mirrors are 2.23×10^{-5}N, 7.45×10^{-6}N, and 3.10×10^{-6}N, respectively.

In Fig. 3, arrays of assembled mirrors are shown after a MUMPs chip is rotated under a certain angular velocity. A plot of the number of mirrors assembled versus rotational speed is given in Fig. 4. An SEM picture showing an array of assembled mass platforms is shown in Fig. 5.

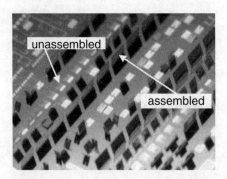

Fig. 3. Different sizes of mass assembled by centrifugal force. The white structures represent the unassembled structures (reflection from gold layer), and the gray are the assembled structures with black shadows.

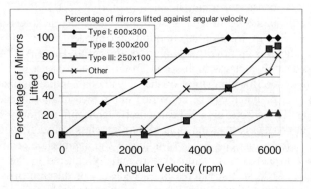

Fig. 4. Percentage of micro mirrors assembled versus angular velocity, with mirror size as a parameter.

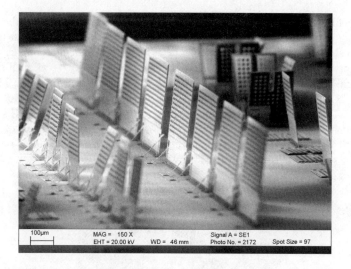

Fig. 5. The batch micro-assembled mirror arrays.

This self-assembly process has been verified to be very successful. None of the micro mirrors were destroyed during the assembling process because the applied centrifugal force was small and did not exceed the maximum tensile strength of polysilicon.

5. SURFACE FORCE MEASUREMENT

To determine the limitations of the centrifugal-force-assisted assembling method, we have performed systematic experiments to ascertain the multi-force interactions of structures that consist of a simple mass platform supported by 2 cantilevers. These platform-cantilever structures were designed as piezoresistive sensors capable of wirelessly transmitting motion information under rotation up to 6250rpm (similar to our prior work reported in [5]). These MUMPs surface-micromachined non-contact high-speed rotation sensors will convert the mechanical deflection or elongation of the polysilicon cantilevers into change of electrical resistance, which can be converted into a measurable change of voltage by connecting the sensors in a Wheatstone-bridge configuration. The voltage output form the bridge is then transmitted by a wireless transceiver after voltage-to-frequency conversion. This allows finding of the appropriate centrifugal force needed to *free* a particular structure that is initially adhered to the substrate by surface forces. At least 2 structures of each design of different mass size were tested. Typical dynamic motion of such a structure as a function of the angular velocity of a spinning disc is shown in Fig. 6. The structures consistently will be freed when sufficient centrifugal force (spin speed) is applied on the mass platform. However, when the speed is decreased, the platform will *snap down* to the substrate at a lower angular speed than the *freed-state*. This hysteresis characteristic is repeatable and may be attributed to surface-force effects. The force value measuring the freed-state rpm would be the force required to overcome the surface force for a particular mass size.

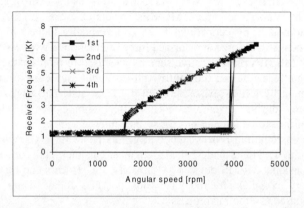

Fig. 6. Typical motion of a platform suspended by two cantilevers beams under rotation (frequency is the wireless signal received and is proportional to the position of a platforms).

The relationship between the *freed-state* and the *snap-down-state* of the platform was also analysed for various platform geometries. In general, larger mass platforms will need lower rpm to free them. Although platforms as small as 320μm x160μm were all freed successfully, the range of required spin speed to free them is more sporadic than the large platforms (greater experimental error bars as shown in Fig. 7). This may be an indication of surface forces, which depends on many factors, including humidity and temperature, becoming more dominate as a mass becomes smaller.

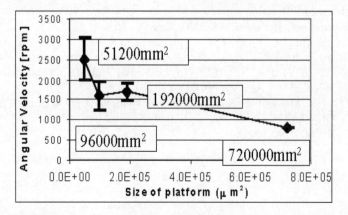

Fig. 7. The "freed-state" for 4 different sizes of MUMPs mass platforms. The angular velocity at which each platform type is freed from the substrate can be related to the centrifugal force acting on the micro platforms.

6. CONCLUSION

Batch micro-assembly of MUMPs microstructures using centrifugal force was demonstrated. Various complex MUMPs structures were rotated about micro hinges autonomously by centrifugal force and vertically locked by latches. To quantify the centrifugal force and surface force interaction, a wireless force sensing system using MUMPs piezoresistive sensors was used to monitor the dynamics of micro mass platforms during the centrifugal-assembling process. Results indicate that a consistent hysteresis exists between the freed-state and snap-down-state of microstructures during assembly. This effect should be studied further to understand surface force interactions for surface micromachined structures. The non-contact, batch-assembling process reported in this paper is very low-cost and non-destructive, thus it will provide MEMS engineers a quick and convenient way to make 3-dimensional MEMS devices. Ultimately, this process can be used to batch assemble very complex microdevices not possible today using conventional manipulator-based assembling process.

Acknowledgement

We would like to thank Dr. Winston Sun for his help in setting up the wireless sensing data acquisition system for analysing the micro-structural dynamics during rotation assembly.

References

1. E. E. Hui, R. T. Howe and M. S. Rodgers, "Single-Step Assembly of Complex 3-D Microstructures", IEEE MEMS 2000, pp. 602-607.
2. V. Kaajakari and A. Lal, "Electrostatic Batch Assembly of Surface MEMS Using Ultrasonic Triboelectrocity", IEEE MEMS 2001, pp. 10-13.
3. W. Sun, A.W.-T. Ho, W. J. Li, J. D. Mai and T. Mei, "A Foundry Fabricated High-speed Rotation Sensor Using Off-chip RF Wireless Signal Transmission", IEEE MEMS 2000, pp. 358–363.
4. W. N. Sharpe, Bin Yuan, R. Vaidyanathan and R. L. Edwards, "Measurements of Young's modulus, Poisson's ratio, and Tensile Strength of Polysilicon", Proceedings of the 10th IEEE International Workshop on Micro Electro Mechanical Systems, 1997, pp. 424-429.
5. W. J. Li, T. Mei and W. Sun, "A Micro Polysilicon High-Angular-Rate Sensor with Off-Chip Wireless Transmission", Sensors and Actuators A: Physical, 89, 1-2, 2001, pp. 56-63.

An Efficient Distributive Tactile Sensor for Recognising Contacting Objects

P. Tongpadungrod[*] and P.N. Brett[+]
*Department of Production Engineering,
King Mongkut Institute of Technology, North Bangkok (KMITNB),
Phibulsongkram Rd,
Bangkok,
10800 Thailand
E-mail: pensiri@kmitnb.ac.th
[+]Professor of Biomedical Engineering systems,
University of Aston,
Birmingham, UK
E-mail: p.n.brett@aston.ac.uk

Abstract
This paper describes a novel distributive tactile sensor for the discrimination of object shape. The distributive approach uses the coupling information between sensing elements that captures changes in properties of a common tactile surface to infer contact types. It offers a reduction in the number of sensing elements compared to the mainstream discrete type sensors.

Keywords: *Tactile sensing, distributive sensing, object recognition, neural network.*

1. INTRODUCTION
Types of tactile sensors range from a simple construction with few sensing points to complex construction with arrays of many sensing elements. Simple tactile sensors are appropriate for giving basic properties of the contacting object, for example a binary system for identifying the presence/absence of contact or a sensor for measuring the resultant force of a contacting object [1]. To identify more complex contacting properties, an array of sensors may be required [2], [3]. Discrete sensors can be formed into an array that not only detects changes in the environment with respect to load application but that also produces an image of contacting shapes. Although array sensors may be used to provide information concerning contacting properties, they are often of a complex construction, and are computationally intensive.

An alternative to the conventional array (discrete) sensor is the distributive approach which has some similarities to the human system. The technique relies on strong coupling between sensor outputs that is provided by the response of the surface to the contacting load [4]. This opposes the approach to sensing of most discrete tactile devices that attempt to ensure normal properties of the measurement between sensing elements are decoupled. Utilising the continuum of a process, such as the deformation of a surface in response to an applied load, forms the distinguishing characteristic of the distributive approach [5]: Additional examples can be found in temperature sensing [6]. The coupled information between sensing elements enables a distributive tactile sensor to use previous knowledge of the surface behaviour under contact to identify the contact type. Like the human system, distributive sensors are appropriate for discriminating between different contact types rather than deriving exact measurements.

The most obvious advantage of distributive tactile sensors against the conventional array type is the reduction in the number of sensing elements to achieve similar spatial resolution, as areas between sensing elements are also sensitive to an applied load [7]. This leads to a reduction in cost, construction complexity, and computational load. Unlike most array tactile sensors, the sensing elements are often not an integral part of the contacting surface [4]. Medical devices can benefit from this feature as risks to contamination between contact and sensing elements are reduced [8]. However, distributive tactile sensors are suited as a discrimination devices, providing for the recognition or estimation of the state of contact, rather than as a device to measure the exact shape/profile of an object. In practice, in most manufacturing processes and other application sectors, it is often the case that objects presented to a system are of a previously known type. By making the broad assumption of consistent groups of product types it is possible to offer deterministic solutions to identify categories and to determine the relevant properties within those categories.

This paper describes a distributive tactile sensing system to discriminate objects of known shapes. The experimental rig was constructed using a surface that deforms principally in bending and proximity sensors to detect changes in surface deflection as a result of an applied object. A computational algorithm is used to discriminate object shapes.

2. EXPERIMENTAL RIG

An experimental rig was constructed from a mild steel surface of dimension 250×340 mm and thickness 0.5 mm, supported along the edges on a rigid frame. Underneath the surface was a platform on which infrared proximity sensors were mounted to measure the surface deflections of an applied load. Sixteen sensing elements were placed at an equal pitch as shown in Fig. 1. The clearance of the sensors from the surface were adjusted individually to achieve a similar sensitivities over the operating range.

Two circular objects of diameters 30 (C1) and 50 (C2) mm and three rectangular objects of dimensions 5×16 (R1), 16×45 (R2) and 27×72 (R) mm were used in the experiment to applied load on the surface. The functional area was defined such that the objects were concentrated near the centre of the surface (within an envelope of 115×185 mm). This was the working range of the surface, having acceptable responses.

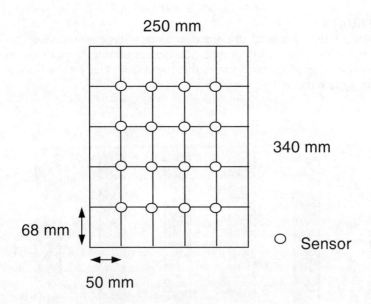

Fig. 1. Sensing positions for the 2-D system

3. NEURAL NETWORK TRAINING

In this example, back-propagation neural networks were implemented as the interpretation algorithm to discriminate contact shape from the sensor outputs. They are commonly used and have also been applied by other tactile sensing applications [3]. In back-propagation neural networks, the weighted values that mathematically relate inputs to outputs are established off-line in a training process. During training, weighted connections are adjusted through a combination of feedforward output generation and backward error calculation [9]. A network is successfully trained when the error between the target and network outputs converges to within a specified threshold. The process of training is a significant factor in deciding its suitability for the application. It is often the case that network parameters, such as the number of hidden layers and PEs, have to be adjusted to achieve satisfactory performance.

In this study, neural network inputs were derived from sensor responses due to an applied load (shape) and the output vector contained information of contacting parameters. The network was trained to discriminate between rectangle and circular shapes and to determine the load centroid position. An input string consisted of 16 entries, equal to the number of sensing elements employed and the output string consisted of 4 entries, two of which corresponded to each contact shape and the others were related to load position in each direction. The sensor responses of the rectangular and the circular objects of the largest dimensions for applied positions at an interval of 50 mm were used to train the network.

4. RESULTS

Performance of the system to discriminate contact shape was defined by the success rate that was referred to as the percentage of correct discriminated shape over the total number of samples presented to the system. Fig. 2 shows the plot of success rate against the applied shapes.

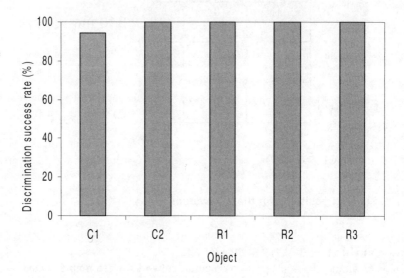

Fig. 2. Shape and size discrimination results

The system was able to discriminate object shapes with 100% success rate in most cases, despite presented with objects of different sizes (compared to the trained objects). It can be deduced from such results that changes in object size had little effect on shape discrimination. However, exceptionally low success rate was achieved when the small circular object was presented because the system incorrectly perceived as if it was a rectangular object. There was evidence that the system was biased towards the rectangular object as all were correctly discriminated, whereas some circular objects were discriminated incorrectly. An increase in success rate tended to increase with the diameter of the circular objects towards the trained size (50 mm). Note that although the network was trained with 2 objects each of a single size, it was able to discriminate correctly objects of the same shape but different sizes. Such results suggested that the system was more sensitive to shapes than sizes.

Load position was determined with an average accuracy within 3.1% and 1.9% of the dimensions in the x and y directions respectively. This relates to 7.8 mm and 5.9 mm respectively. With optimal placement of as little as 4 sensing elements, the system successfully discriminated object shape with 100% success with reduced accuracy in determining load position to approximately 7.5% and 5.4% in the x and y directions respectively. However, due to the small number of sensing elements it was not able to discriminate shape and position of untrained objects satisfactorily.

5. CONCLUSIONS

This work has shown an example of a distributive tactile sensing system that can be used to discriminate object shapes. The spatial resolution of the system has shown to be higher than the number of the sensing elements as this is the characteristic of a distributive system as opposed to array (discrete) sensors. The method was shown to correctly identify object size and shape to within 94% for a range of sizes and shapes. However, the system performs a recognition process in a mechanically and computationally efficient way. It is not suited to describe the exact profile of the contacting shapes, as only a small number of sensing elements are involved and due the nature of the deformation of the contacting surface.

6. REFERENCES

1. Harrison, A.J. and Hillard, P.J., "A moment-based technique for the automatic spatial alignment of plantar pressure data", *Proceedings of the Institution of Mechanical Engineers Part H. Journal of Engineering in Medicine*, 214(3), 257-264, 2000.

2. Benhadj, R. and Dawson, B., "Air jets imaging tactile sensing device for automation applications", *Robotica*, 13(5), 521-529, 1995.

3. Holweg, E.G.M. and Jongkind, W., "Object recognition using a tactile matrix sensor", *EURISCON' 94 Advanced Manufacturing and Automation RES 1994*, Malaga, Spain, 1379-1383, 1994.

4. Brett, P.N. and Li, Z., "A tactile sensing surface for artificial neural network based automatic recognition of the contact force position", *Proceedings of the Institution of Mechcanical Engineers*, 214(I), 207-215, 2000.

5. Ma, X. and Brett, P.N., "A distributive 1 dimensional tactile surface for determining the position, width and intensity of a distributive load", *IEEE Transactions on Instrumentation and Measurement*, 51(2), 331-336, 2002.

6. Banim, R.S., Tierney, M.J. and Brett, P.N., "The Estimation of Fluid Temperatures through an Inverse Heat Conduction Technique", *J. Num. Heat Transfer*, 37(5), 465-476, 2000.

7. Dargahi, J., Parameswaran, M. and Payandeh, S., "A micromachined piezoelectric tactile sensor for an endoscopic grasper : Theory, fabrication and experiments", *Journal of Microelectromechanical Systems*, 9(3), 329-335, 2000.

8. Brett, P.N. and Stone, R.W.S., "A technique for measuring contact force distribution in minimally invasive surgical procedures", *Proc.Instn Mech Engrs*, 211(H part 4), 309-316, 1997.

9. Eberhart, R.C., Simpson, P.K. and Dubbins R.C., *Computational Intelligence PC Tools*, Academic Press Inc. USA, 1996.

Robots and Machine Vision in Agriculture and Food Processing

Agriculture is an interesting hybrid of mundane and earthy operations, mixed with engineering ingenuity. Even when I was a child, I remember the readiness of my farmer cousin to weld together any new device and add a hydraulic cylinder for its actuation. We now see new breeds of tractors entering the market with steering that is not only hydraulically driven, but commanded from an electronic interface.

The use of GPS satellites for yield mapping and guidance is becoming well accepted, although the price tag often puts the system out of reach. With the economic pressure on the farmer to make his time and efforts go even further, agriculture is likely to see ever more dramatic robotic advances in the next few years.

Vision based colour sorting is becoming almost commonplace in the selection and grading of a wide variety of produce, such as tomatoes, oranges, macadamia nuts, sugar cane, dates, pepper berries, etc. In the first a paper, from Saudi Arabia, dates are the produce being graded; reference is made to a history of over fifteen years of vision sorting. In the present work, shape also plays an important part in selection. Features are extracted and neural net methods are used to train the decision process to arrive at a practical system.

The second paper has a more limited objective, to count macadamia nuts as they are harvested. Vision appears to offer the best chance of attributing an accurate count to each individual tree when the harvest is to be a factor in selective breeding. Some low cost 'tricks of the trade' are described that allow a low-cost webcam to be used as the primary image source.

Vision grading is again the focus of the third paper, in this case of white pepper berries. With a combination of back and front lighting, a single-hidden-layer model can be trained to achieve an accuracy of a few percent.

The fourth paper has a more robotic flavour. The object is a small robot that can perform agricultural tasks such as weed-removal and spraying without the need for human supervision. The vehicle acts as a test bed for some low-cost GPS techniques, reinforced with aspects of vision guidance designed a few years ago. If the paradigm proves successful, it could have a substantial impact on future agricultural methods.

The next paper deals with a control problem, that of chilling meat. It leans on a mathematical model of the partial differential equations to optimise a strategy that can minimise chilling time while avoiding freezing of the outside surface.

For the final paper we return to the paddock. The aim is to reduce the control burden on the driver of a sugar-cane harvester by providing automatic 'topper' control. As the stem is harvested, the leafy tops are cut and discarded, since they would add needlessly to the vegetable matter to be processed at the mill. If cut too low, however, valuable sugar will go to waste. Practical results are presented of the use of a novel and robust sensor.

A Prototype Mechatronic System for Inspection of Date Fruits

Abdulrahman Al-Janobi
Department of Agricultural Engineering, King Saud University
P.O. Box 2460, Riyadh 11451, Saudi Arabia

Abstract
A prototype mechatronic system based on machine vision has been developed for inspection and grading of date fruits. The system consisted of four integrated units, namely, the feeder, lighting system, imaging system, and grading mechanism. The feeder was a belt conveyor, which carried dates through a specially designed illumination chamber. The imaging system consisted of a personal computer equipped with a frame grabber and a colour camera. The system captured the images of the dates moving on the belt conveyor. The acquired digital images were sent to the computer for processing and the grade of the date was determined after analysing a set of features extracted from the date images. A feed forward multilayer perceptron network trained with the back-propagation algorithm was used for classification of the dates. The sorting mechanism installed at one end of the belt conveyor was operated by a TTL signal from the computer to push the graded date into the corresponding grade box. The system successfully graded samples at a rate of 2 dates/s, giving a throughput of approximately 108 kg/hour.

Keywords: *Machine vision, colour features, neural networks, classification.*

1. INTRODUCTION

Manual inspection, grading and sorting of fresh produce are all very subjective, monotonous, inconsistent and labour intensive. Also they are highly variable and difficult to evaluate. For example, inspectors position themselves along a packing line to scan the fruit as it travels on a belt conveyor and remove the defective dates from the flow. Usually the inspectors make a judgment about the quality of the product without any special tooling or other instruments during sorting. The current grading is shown in Fig. 1. In the time-constrained and fatiguing atmosphere associated with the various processes of sorting and packing, accuracy and consistency are bound to suffer. Date quality is determined by many factors such as: visual appearance, moisture and sugar content, surface defects, etc. In general, dates are graded based on size, shape, surface colour, surface defects and texture. Surface colour is an important factor considered to distinguish between acceptable dates and damaged or immature dates. The colour of acceptable dates is relatively uniform and predominantly light amber in colour. Size is affected by the variety and the physical condition of the producing trees. Dates are rejected if they are significantly larger or smaller than the subjective average size of the dates. Texture is a useful factor to identify over-dried, hard dates. Shape is also an important factor in identifying over-dried dates and dates with defects.

Fig. 1. Date fruits manual inspectors positioned along the belt conveyor sort of date fruits

The newly developed field of computer vision shows a trend to spread through every part of engineering research. Its potential as a powerful sensing technique is accepted and the range of possible application is growing steadily. There are several areas of application of computer vision in agriculture. Some examples are: guidance of equipment, inspection of products, and sorting and packing of products. Real-time computer vision for sorting and grading of fruits and vegetables has the potential for improving product quality and the speed of operation [5]. A number of vision-based sorting systems to grade different fruits and vegetables have been reported, but to a much lesser extent for the grading of dates. Therefore, the need for an automated date grading system is increasing significantly. The first work on sorting dates with computer vision was done in 1986 by VARTEC Co. in California [3]. The use of image processing in separating good dates from the defective dates using thresholding technique was discussed in [6]. A machine vision system to grade date fruits (Deglet noor variety) into quality classes based on colour and texture analysis was reported in [1]. The classification accuracies were 98.4 and 77% for colour and monochrome models respectively. In a further work, a computer vision system based on colour thresholding technique for grading Sifri variety of date fruits was developed [2]. The system showed an average classification error of 1.8% using features from the red colour band.

The overall objective of this project was to implement an automated grading system for some of the most common varieties of Saudi dates using imaging techniques. Specific objectives were:
1) Develop image processing techniques to grade dates into quality classes based on colour, and
2) Develop a prototype date inspection system for grading date fruits by implementing the developed imaging techniques.

2. MATERIALS AND METHODS

A prototype mechatronic date inspection system based on machine vision was designed and fabricated for grading and inspection of dates. Figure 2 shows the setup of the system. The system consisted of several parts and integrated as a built-in system for the specific task.

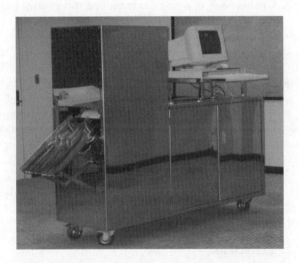

Fig. 2. Setup of the prototype mechatronic system for inspection of date fruits

2.1 Feeder
A constant speed DC motor-driven conveyor belt, provided with two channels, was used to carry dates under inspection. The speed of the motor could be adjusted and controlled through a variable DC power supply.

2.2 Lighting Chamber
The lighting chamber is rectangular in shape and mounted over the belt conveyor. It consisted of eight 100W tungsten halogen lamps. The inner surface of it was coated with black paint to reduce spectral reflectance. It also accommodated the video camera at the top centre (Fig. 3).

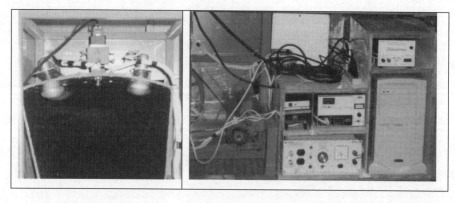

Fig. 3. The lighting, imaging systems and the control unit of the system

2.3 Imaging System

The heart of the system was an imaging system used for capturing and processing digital images of date samples. The system consisted of a SONY XC711CCD colour video camera, an IBM compatible Pentium IV 1.5GHz computer, an image monitor, and μTech's MV1000 image frame grabber and MV1300 colour camera interface module. The video camera was positioned above the belt conveyor and fitted at the top centre of the lighting chamber (Fig. 3). The MV1000 digitized the analog video signals and stored the corresponding digital signals in an on-board VRAM of 1 MB capacity. The MV1300 plugs in as a daughter card to the MV1000 and provides the interface to cameras with an RGB video. The digitized video is in RGB 24 bits/pixel format. The system captured the images of the dates moving on the belt conveyor. All operations from the frame grabbing to grading dates were performed with a set of programs developed in Visual C++ environment (Fig. 4).

2.4 Feature Extraction

Feature extraction is the procedure of generating descriptions of an object in terms of measurable parameters. The extracted features represent the relevant properties of the object, and may be used with a classifier to assign the object to a class or grade. The objects are usually classified on the basis of some measurements or features such as shape, colour, size, texture, etc. Several feature extraction techniques were tested and from these colour features were found to be the more accurate and used for the classification of date samples into quality classes.

For automated image analysis, colour is a powerful descriptor that simplifies object identification and extraction from a scene. There are many colour models in use in various applications and the most commonly used in machine vision applications are the RGB and HSI models. In this work, the RGB colour model was used. Algorithms have been developed and incorporated with the developed software to extract a set of colour features of date samples from the three colour band images. From the R, G, and B values, the values of their mean, standard deviation, kurtosis, skewness, and wrinkle ratio as measure of features, were calculated.

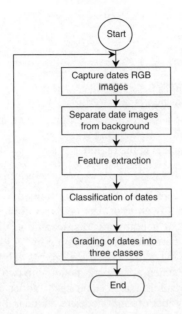

Fig. 4. Flowchart of the software developed for the system

These values are called amplitude histogram features and can be computed in terms of the histogram of image pixels within a neighbourhood as follows:

$$\text{Mean, } \bar{a} = \sum_{a=0}^{a=255} a\, p(a) \tag{1}$$

$$\text{Standard deviation, } \sigma_\rho = \sqrt{\frac{1}{N} \sum_{a=0}^{a=255} (a-\bar{a})^2} \tag{2}$$

$$\text{Skewness, } = \frac{1}{\sigma_x^3} \sum_{a=0}^{a=255} (a-\bar{a})^3\, p(a) \tag{3}$$

$$\text{Kurtosis} = \frac{1}{\sigma_x^4} \sum_{a=0}^{a=255} (a-\bar{a})^4\, p(a) - 3 \tag{4}$$

$$\text{Wrinkle Ratio} = \frac{\sum_{a=b+1}^{a=255} a\, p(a)}{\sum_{a=0}^{a=b} a\, p(a)} \tag{5}$$

Where:

a = gray level ranges $0 \leq a \leq 255$.
b = threshold of wrinkle light.
$p(a)$ = ratio of the number of pixels of gray level 'a' in a region to the total number of pixels in the region.
N = Number of pixels in the region.

2.5 Classification

A multi layer feed forward neural network, trained using back propagation algorithm was used in classification of date cultivars. Several aspects of the state-of-the-art implementation of neural networks, leading to the architecture of multi layer feed forward neural networks and its training by back-propagation, are found in [4]. In order to validate the neural classifiers, a cross validation method was used for training and testing. The neural network model with two hidden layers with two sets of hidden nodes 22 and 15, one set on each hidden layer was used for the classification task. The number of input nodes to the neural network model was 15, corresponding to the number of features from the colour model. To train and test the classification performance of the neural network, a software to access the power of neural network environment while working in the Excel spreadsheet, was used.

2.6 Grading Mechanism

A sorting system built around a set of rotary solenoids operating on a 12 V DC supply, installed at the outgoing end of the belt conveyor, was used to grade the dates moving along the belt conveyor into quality classes. A circuit was designed and built mainly with a set of reed relays and electromechanical relays to activate the sorting system upon the proper TTL signals from the personal computer (Fig. 5). The proper binary pattern of the TTL signals were fed to the reed relay circuit to activate the sorting system when the dates whose grades determined were about to fall at the end of the belt conveyor.

3. PERFORMANCE AND ACCURACY

The trained feed forward multilayer perceptron neural network together with the classification algorithm of colour features was implemented in the prototype date inspection system. The system was tested for its performance by grading two most common varieties of Saudi date cultivars - Sukkari and Maneefi. The date samples of these varieties, obtained from a local date packing factory, were initially graded into three classes, A, B, and C corresponding to premium, good, and low quality respectively by manual graders. During the test run of the system, the imaging system captured the images of the dates moving on the belt conveyor, and with the developed program, the set of colour features from the date images was extracted. The trained feed forward multilayer perceptron network performed classification on the extracted colour features. The sorting mechanism fitted at the out going end of the belt conveyor activated to deflect the date fruits into the respective grade boxes. The two date cultivars, Sukkari and Maneefi, were classified with average classification accuracies of 88 and 93%, respectively. The test results are presented in Table 1. The system was able to successfully grade 200 date fruits in 100 seconds, i.e., at a rate of 2 dates/second, equivalent to about 108 kg dates/hour. This is a reasonable and highly acceptable rate as far as the date industry is concerned.

Table 1
Summary of the performance of the system for classification of the cultivars.

Cultivar	Percentage Classification Accuracy				No.of samples
	Class A	Class B	Class C	Average	
Sukkari	84	82	96	88	200
Maneefi	95	86	96	93	106

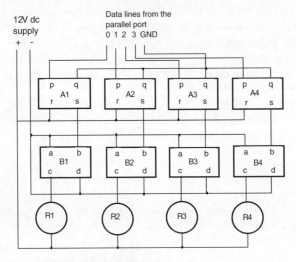

A1, A2, A3, A4 - Reed Relays
p q - relay coils energized by TTL signals
r s - relay contacts
B1, B2, B3, B4 - Electromechanical relays
a b - relay coils energized by 12V dc
c d - relay contacts
R1, R2, R3, R4 - Rotary solenoids

Fig. 5. Circuit diagram of the sorting system

4. CONCLUSIONS

A prototype date inspection system was designed and fabricated for grading Saudi date cultivars into three quality classes based on machine vision technology. A classification algorithm with colour features from the RGB colour model of the date images, together with a feed forward multi layer perceptron neural network trained with the back propagation algorithm, was implemented to make the system operating in real time. The system successfully graded 2 dates/second, equivalent to approximately 108 kg dates/hour. This attained rate is reasonable and highly acceptable, and it marks a milestone in the continuous scientific efforts towards achieving an advanced automated machine vision inspection system for date fruits with the highest possible accuracy and speed.

The system is expected to evaluate date quality fairly and impartially and assign the grade accurately. It would give economic benefits to the date industries and increase consumer confidence in the quality of the fruit. Also, it would significantly enhance the competitiveness of date industries.

References
1. A. Al-Janobi: Machine vision inspection of date fruits. Ph.D thesis, Oklahoma State University, 1993, Stillwater, Oklahoma, U.S.A.
2. A. Al-Janobi: Date inspection by colour machine vision. Journal of King Saud University, 2000, Vol. 12, Agricultural Science (1), pp 41-51.
3. G.K. Brown: Personal communication. Research leader, USDA ARS, 1991, Agricultural Engineering Department, East Lansing, MI.
4. R.C. Gonzalez, R.C. Woods: Digital image processing, 1992, Addison-Wesley Publishing Company, Inc.
5. G. Kranzler: Applying digital image processing in agriculture. Agricultural Engineering, 1985, 66(3):11-15.
6. D. Wulfson, Y. Sarig, R.V. Algazi: Preliminary investigation to identify parameters for sorting of dates by image processing, 1989, ASAE Paper No. 89-6610. ASAE, St. Joseph, MI.

The Counting of Macadamia Nuts

John Billingsley,
University of Southern Queensland, Toowoomba, Australia.

Abstract
As macadamia nuts are harvested, there is a requirement to monitor the yield. In one case, the individual yield of each tree must be determined and it is more appropriate to use counting rather than weighing. It is proposed to use machine vision to count the nuts as they are gathered. A software interface enables a digital 'webcam' style of camera to be used, an OCX control having been designed to provide image data at an adequate speed for analysis. Alternative measuring techniques are reviewed for routine harvesting.

Keywords: *Machine vision, webcam, counting, macadamia, harvesting.*

1. INTRODUCTION

There is a requirement for the monitoring of the yield of macadamia nuts as they are harvested, both under normal operations and for the varietal trials of trees for propagating stock. This latter case is particularly stringent, in that the yield must be assessed of each individual tree. Nuts are harvested after they have fallen to the ground, so the data requirement involves defining the boundary of the measurements within the 'territory' of each tree.

The nut and shell are surrounded by a substantial husk at least equal in mass, so if measurements are taken before de-husking it is more appropriate to use counting rather than weighing. For varietal testing, the gathering and de-husking process performed on the harvester will tend to 'blur' the boundaries between the produce of one tree and the next, so it has been proposed to use machine vision to count the nuts as they are gathered.

The machine employs a 'bristly roller' of plastic fingers to capture the nuts as they lie on the ground, carrying them to a stripper where they fall into a gathering auger. They are plainly visible as they travel some thirty centimetres, gripped in the roller. The problem becomes one of capturing an image of the roller, which can be up to four metres wide, at a resolution which will allow single kernels to be identified, even if several are touching. The image must be analysed at a forward speed of up to 2.8 metres per second.

A software interface has been developed to enable a digital 'webcam' style of camera to be used. An OCX control provides data at a rate of up to thirty frames per second. This constitutes a substantial data flow for analysis.

Conventional edge-finding is discussed and a pragmatic algorithm is presented which can achieve the necessary objectives. This is illustrated by image examples.

For routine harvesting, the 'blurring' effect of the harvesting operation is much less important. Methods are discussed for counting the shells as nuts enter the hopper after the de-husking process.

2. A COUNT FOR EACH TREE

For a tree-by-tree assessment it is necessary to define the boundary of the 'real estate' within which fallen nuts are attributed to each tree. The counting method must then be careful that only nuts (but all nuts) gathered within this boundary should join an individual count.

The normal harvesting process gathers the nuts on a roller, strips them into a trough and conveys them across the vehicle to be passed into the de-husking system. There is no 'time-slice' through this process which will map back into a rectangular plot on the ground from which the nuts have been gathered. Instead the shape is a severely sheared parallelogram.

To obtaining a simple mapping, we must therefore look before the nuts are stripped from the roller. At first, thoughts turned to a vision system which inspected the ground in front of the machine. However, looking at the gathering roller itself would have many advantages.

The ground offers an ill-defined background to any vision system. To achieve reasonable contrast between nuts and the ground, strong oblique lighting would be needed, suggesting a night-time operation. On the other hand, the roller gathers nuts in preference to leaves. These nuts can be seen against the background of the roller and a much simpler vision processing system is possible in normal light.

To make the task even easier, the roller can be coloured blue. The vision system makes a distinction based on colour and any non-blue item is examined.

2.1 Preliminary calculations

To examine the feasibility of this method, a number of simple initial calculations must be made. These are summarised in Table 1 as:
- The diameter of the roller is 510 mm or 720 mm. One rotation therefore represents progress of 1.6 or 2.26 metres.
- Forward velocity in normal harvesting is between 3.2 and 10 kph, which corresponds to 0.9 and 2.8 metres per second.
- A ball-park figure for the rotation of the roller might therefore be a half to one revolution per second in 'precision mode' and less than two revolutions per second at top speed.

From the view of the roller in Figure 1, it is plain that a full ninety degrees or more of the roller can be seen. The lower limit for processing speed might therefore be taken as four frames per revolution. That is two frames per second at the lowest speed, up to eight frames per second at top speed.

Speed	510mm diameter roller	720mm diameter roller
3.2 km/hr	0.56 rps	0.4 rps
10 km/hr	1.75 rps	1.24 rps

Table 1. Roller rotation speed range for measurements of individual trees

2.2 Experimental results.

Two experiments were performed, one to verify that images could be 'grabbed' sufficiently rapidly, the other to establish the speed with which the image data could be processed.

Figure 1. A view of the roller.

(a) A 'Smartcam' web camera was interfaced with the computer using the USB port and the corresponding driver software installed. This conventionally allows the image to be accessed for recording as an AVI file or allows images to be acquired using the TWAIN software system.

An Active-X control was written for the purpose of tapping the data source to make it available for analysis. Salient details of the interface to the operating system can be found via a web page [1]. Software to be found there is declared to be freely available without copyright restrictions. It was modified to provide a variety of functions covering a range of applications.

The simplest function is 'snap to clipboard', allowing one frame of an image stream to be snapped and pasted into a picture box. By using this feature in a loop, a moving picture can be displayed. At a resolution of 320x240 pixels, a frame rate of around thirty frames per second was comfortably achieved.

A further feature transfers image data into a byte array for rapid processing. When combined with a 'movie' display, this did not appreciably reduce the potential frame rate. Indeed, by suppressing the display of the image a substantially higher rate might be achieved.

(b) A single frame was captured and subjected to analysis. A property of the OCX control allows a frame of data to be transferred to a byte array defined in the calling program.

When a 'captive' image is used instead, the 'point' property returns a long integer containing all three colour components: red, green and blue. To separate these components and determine the hue they represent takes considerably longer than accessing the data from the byte array. The processing speed of the experimental algorithm therefore gives a very conservative estimate of the speed that can be achieved in practice.

To process every pixel of the image takes an unacceptably large part of one second. To locate a nut, however, it is only necessary to search on a sufficiently fine grid that the nut cannot 'fall between the cracks'. If every fifth column of pixels is searched, at intervals of five pixels, the routine is fundamentally twenty-five times as fast.

The complete method is not as crude as this, however. When a nut is detected by a non-blue pixel, a search is made up and down the image, at one pixel resolution, to find the top and bottom of the line segment which cuts the nut. From the midpoint of this segment, pixels are tested to the left and to the right to find the width of the nut. Knowing this dimension and the width of the segment, the location and diameter of the nut have been found. Any other search pixels which would hit the same nut can be skipped, so that the nut is only counted once. Only objects with a sufficient diameter are counted as nuts, while an upper limit is placed on the diameter of the region to be excluded, so that when two nuts touch, both will be counted.

Figure 1 shows an image of the pickup roller, as seen by the camera.

In Figure 2, all the pixels have been tested for blueness and those falling above the threshold have been made dark on this greyscale representation of the image. This is shown merely to illustrate the discrimination process.

Figure 3 shows a rectangular selection which will be analysed. The blue image pixels are all shown as black, but this time details from the original picture have been omitted.

Figure 4 shows an array of 'spots' representing those pixels which have been tested, with lines of closer spots demonstrating the precision search.

Figure 5 shows the result of using conventional image processing to find the edges of the nuts. It is a lengthy process which is not relevant to this study and has been included for interest only.

Figure 2. Blue pixels (dark area) have been identified.

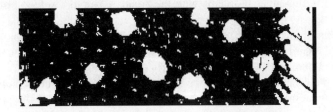

Figure 3. Blue pixels (dark area) without the original image.

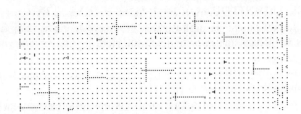

Figure 4. Points 'visited' in counting these nuts.

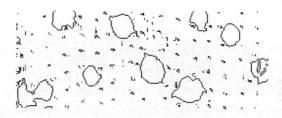

Figure 5. Edges found by conventional image processing - not relevant here.

2.3 Comments on the tests

These two tests have shown that the method gives certainty of success, although certain details must be adjusted for performance and economy, as follows.

Across a four metre roller there are some 160 disks of pickup bristles The diameter of a nut is slightly larger than the gap between these disks. A horizontal image resolution of 320 pixels would therefore only yield some two columns of 'hits' per nut. It would be necessary to process the image at full density.

There would be some saving, in that the roller would only occupy a small fraction of the image height, one quarter or less. Moreover, only half of this would need to be processed.

An immediate problem which can be foreseen is the vantage-point of the camera, to capture the entire four metre width of the roller in one image. Either the camera must be some two metres or more away, or a fish-eye lens must be used. In turn, this gives problems with the greatly reduced resolution of the ends of the roller.

The alternative is to use two cameras. This might well involve the use of two computers, too. However the price of the additional hardware might be kept to below AU$1000. Certainly this would make the problem much easier.

Another alternative, of course, is to use a harvester with a shorter roller. This remark is not meant facetiously. Since the counting of the yield is part of an experimental tree-selection process, harvesting efficiency is of lower importance than accuracy.

3. A YIELD MONITOR FOR ROUTINE HARVESTING

The second circumstance is monitoring of harvest yield rate for everyday harvesting. In this case, the requirement that yield be associated with individual trees does not apply.

It seems sensible to count the kernels after they have been de-husked, when leaves and detritus have also been eliminated.

A preliminary survey has shown that the maximum hopper rate which can be expected is 200 nuts per second. Several approaches have been considered.

1. If the nuts can be singulated into a single stream, a simple 'beam breaker' can drive an electronic counter to read the exact number. It is necessary to consider whether such singulation is practical.

Each nut has a diameter of some 1.3 centimetres. 200 nuts, end to end, will therefor occupy a line 2.6 metres in length. To singulate nuts at a rate of 200 per second into such a line implies that the line must move at a speed of 2.6 metres per second.

From the elementary dynamics formula $v^2 = 2\,g\,s$, we see that this velocity corresponds to a fall of 34 cm.

A v-channel would be mounted below the delivery chute, down which all the delivered nuts would roll. It is possible, however, that this fall would restrict the height to which the hopper could be filled.

If the delivery is split into two streams, the velocity is halved and the necessary fall is divided by four. If split into four, the reduction factor is sixteen. A fall of five or ten centimetres would almost certainly suffice. Since the 'beam-breaker' sensors are a low cost component, the need for four of them is insignificant.

2. A second approach, suggested by work on cane loss [2], is simply to allow the nuts to impact on a rigid plate. The nuts are hard and will cause a distinctive acoustic 'crack' as each hits the plate. A microphone sampled at 10 kHz will easily discriminate the sound from each single impact. The additional electronics would probably involve an amplifier, signal conditioner and counter chip. The alternative solution of inputting the acoustic signal direct to a computer would require considerably more processing than to preprocess the microphone signal into a stream of pulses.

3. Given the possibility of allowing the nuts to drop a small distance, their impact with a plate can be measured by a load-cell, in the manner of grain and other yield monitors. This will measure the mass of the nuts, rather than their number. However, acoustic counting is likely to be the more accurate.

4. CONCLUSIONS

For two apparently similar tasks it is appropriate to use totally dissimilar methods. Vision provides the precision of locating the original placement of the nuts for varietal tests, while a measurement made in the harvested and de-husked flow is more expedient for routine harvesting.

The methods investigated for extracting images from a video data stream have many more potential applications. Already a demand has been found for the optical measurement of the size of fossil dingo teeth, suggesting the need for novel structured light methods which will be reported elsewhere.

Acknowledgement
This work has been funded by a grant from Horticulture Australia Limited.

References
1   E.J Bantz, Programming Video for Windows, http://ej.bantz.com/video/detail/
2   S G McCarthy, J Billingsley, H Harris, Listening for Cane Loss, Mechatronics and Machine Vision, Research Studies Press, England, September 2000, ISBN 0-86380-261-3, pp 113-118

Machine Vision Application to Grading of White Pepper Berries

Mani Maran Ratnam, Weng Li Khor,
School of Mechanical Engineering,
Universiti Sains Malaysia.

Chee Peng Lim,
School of Electrical and Electronic Engineering,
Universiti Sains Malaysia.

Abstract
This paper describes the application of machine vision to the automatic grading of white pepper berries. The pepper berries were initially graded manually into three grades based on the amount, by weight, of black/dark gray berries in white pepper. The vision system was used to capture front and back illuminated images for each sample from which the mean and standard deviation of the pixel gray values of the berries were extracted. These parameters were used as input to a three-layer feed-forward back-propagation neural network. By using 300 training data samples and 60 unknown test samples, the prediction accuracy was found to be 96.7%. The grading carried out using discriminant analysis showed an accuracy of 95%.

Keywords: *Pepper grading, machine vision, neural network, discriminant analysis.*

1. INTRODUCTION

Machine vision is finding increasing applications in the agricultural industry, particularly for automatic quality inspection and grading of farm produce [1-4]. The implementation of vision system not only removes the tedium associated with manual inspection but also improves the efficiency and repeatability of the inspection. Extensive work has been reported in the literature on the application of statistical methods, such as discriminant analysis, and neural network in the grading of agricultural products [5-6]. One useful application of the technique is in the automatic grading of white pepper berries. The worldwide production of pepper in the year 2001 was about 300,000 tonnes and Malaysia is the fifth largest producer, contributing about 8.4% of the total world supply. Due to the high production volume, quality control requirements are becoming more stringent. In Malaysia, white pepper berries are graded into five grades based on four parameters calculated as a percentage by weight [7]. The four parameters are: (a) maximum moisture content, (b) amount of light berries, (c) extraneous matter and (d) amount of black/dark gray berries.

The grading procedure involves: (i) receiving and processing of exporters' applications, (ii) sampling of consignments (ii) laboratory testing of samples, (iii) grade certification, and (iv) labeling, sealing and releasing of consignments. The quality inspection of the pepper is carried out manually using chemical and physical analyses. Gas chromatographs and atomic absorption spectrophotometer are used for the identification of chemicals like pepper compounds, pesticides residues, heavy metals and trace elements. Estimation of the amount of black/dark gray berries in white pepper is also performed manually on random samples taken from the consignments. Since manual inspection to estimate the percentage of black/dark gray berries in white pepper is stressful and inaccurate, mainly due to the small percentage by weight of black/dark gray berries in white berries, it is highly desirable to develop and implement an automatic grading mechanism. Abdesselam et al. [8] proposed the use of machine vision and neural network for grading of pepper berries based on the mean and standard deviation of the red band in a RGB color images of the berries. Their optimum results showed an error of only 3% when using five hidden nodes in a feed-forward back-propagation neural network. The need to extract and store color information from the RGB images, however, is computationally expensive. Since parameter (d) is based on black/dark gray berries, the possibility of using grayscale images instead of color images for the grading purpose is investigated in the current work. Both neural network and discriminant analyses methods are compared with reference to the prediction accuracy.

2. SAMPLE PREPARATION AND IMAGE ACQUISITION

The white pepper berry samples were originally classified into three grades manually, i.e. grades A, B and C, based on the amount of black/dark gray berries by weight. Grade A contained 0% black/dark gray berries, grade B contained 10% black/dark gray berries and grade C contained 20% black/dark gray berries. For the purpose of investigating the possibility of using grayscale images for the grading and for comparing the neural network classification with discriminant analysis method, the percentage of black/dark gray berries in white pepper was selected arbitrarily. The samples of pepper berries were taken from each grade and laid out as a single layer on a white background. The reason for using a single layer was to avoid errors in pixel values caused by shadows or insufficient light reaching the inner berries if multiple layers are used. Two images, one back-illuminated and the other front-illuminated, both of dimensions 768 × 576 pixels, for each sample were recorded in sequence using a *Pulnix TMC6700* charged-couple-device (CCD) camera connected to a *Meteor-2* frame grabber board manufactured by Matrox Imaging Limited, Canada, in a personal computer as shown in Fig. 1. Back illumination was provided by a 25W round fluorescent tube in a dark-walled enclosure, whereas the front light was provided by a high-frequency ring illuminator. During recording of the two images the location of the samples were undisturbed and only the switching of the lights were changed.

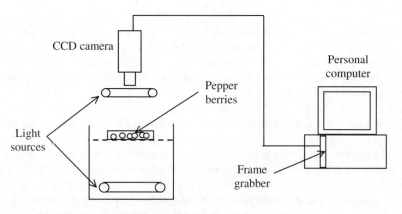

Fig. 1. Schematic of image capture system.

The captured images were cropped to 730 × 530 pixels to avoid edge pixels that may be affected by the diffused lighting. The back-illuminated image was binarized using a threshold of 180 gray value and used to identify the area of the pepper berries from the background in the front-illuminated image. The gray value for thresholding was selected based on the histogram of the back-illuminated image, where a clear separation was observed between the dark and bright pixels. In this work, it was necessary to capture two images for each sample because of the difficulty in establishing a single suitable threshold (in images of samples from different grades) when operating on the front-illuminated image alone. By segmenting the pepper berries from the background, the data extraction in the front illuminated image can be limited to the location of the pepper berries. Theoretically, this method imposes no constraints on the size of each sample that can be used although larger sample size will provide better grading accuracy when the percentage of black/dark gray berries is small. Figs. 2(a)-(b) show typical back- and front-illuminated images of the pepper berries.

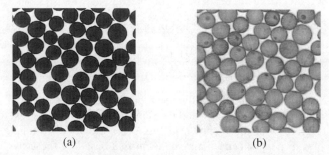

Fig. 2. Images of pepper berries after cropping: (a) back-illuminated, (b) front-illuminated.

The gray values of pixels at the location of pepper berries in Fig. 2(b) were extracted and saved as a data file for further processing using the Matrox Image

Processing libraries implemented in C/C++ codes. The back-illuminated image was used to identify the region of pepper berries in the front-illuminated image where the pixel gray values are extracted. This approach has been adopted instead of identifying the individual pepper berries by location because of problems associated with separating the touching berries and the partial views of the berries at the cropped boundary of the image.

The mean and standard deviation of the gray values of the pepper berries for each sample was calculated automatically in the same computer program. A total of 300 samples, 100 from each grade, were used in the study to extract the data to be used in training the neural network. An additional 60 samples, 20 from each grade, were recorded and used as the test data set. In the experiment, each sample was returned to the original population and mixed thoroughly before the next sample is taken (i.e. sample replacement). Fig. 3 shows a plot of pixel count against gray values of pepper berries for one particular sample set. The mean and standard deviation of gray values for grades A, B and C averaged over 100 samples in each grade set are shown in Table 1.

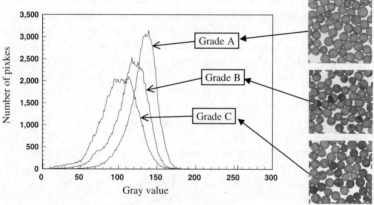

Fig. 3. Gray value distribution of white pepper.

Table 1. Mean and standard deviation of gray value for 100 samples.

Grade	A	B	C
Mean	156.2	148.1	139.2
Standard deviation	24.8	28.5	30.4

The data in Table 1 show that the mean gray value differ by about 7 to 8 gray values between the various grades and the standard deviation is largest for grade C. This result agrees with those shown in Fig. 3 for a particular sample. Thus, the various grades of pepper berries may be identified based on these two parameters alone without the need for RGB values from color images.

The mean pixel gray value and the standard deviation of the 60 test samples are shown in the scatter plot in Fig. 4. Slight overlap between the various grades is visible although they are scattered around a well-defined mean for each grade.

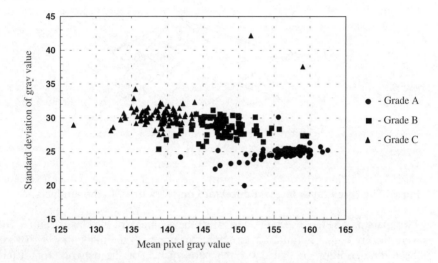

Fig. 4. Scatter plot of mean and standard deviation of gray value.

3. NEURAL NETWORK ANALYSIS

The grading of pepper berries was carried out using the multi-layer perceptron (MLP) network in the MATLAB Neural Network Toolbox [9]. The MLP network generally can be exemplified as a feed-forward type artificial neural network comprising a number of layers, namely the input layer, one or more hidden layers, and the output layer. Fig. 5 shows the three-layer MLP network used in the current work. The network has two nodes in the input layer corresponding to the two input parameters, three nodes in the hidden layer, and three nodes in the output layer corresponding to the three grades. The input parameters, i.e. mean and the standard deviation of gray value of the pixels, were normalized between 0 and 1. The output node of the network having the highest strength was set to one, whereas the others were set to 0. In this manner the output vector could be defined as [100] for grade A, [010] for grade B and [001] for grade C. The transfer functions used in the hidden and output layers were, respectively, log-sigmoid and linear, and the network was trained using the Levenberg-Marquardt algorithm [9]. This algorithm is reported to have the best performance compared to other training algorithms, such as Resilient Back-propagation, Scaled Conjugate Gradient and Bayesian Regularization [8].

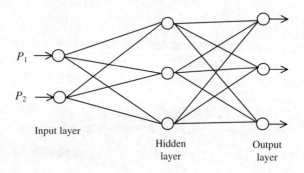

Fig. 5. The three-layer MLP feed-forward network used in the analysis.

Two main parameters were found to affect the accuracy of the classification. These were the error goal definition and number of nodes in the hidden layer. By using the error definition of 0 and 0.1 the performance of the network for different number of nodes in the hidden layer was investigated. The percentage accuracy was calculated and plotted as shown in Fig. 6. The maximum prediction accuracy was found to 96.7% when 2 nodes were used in the hidden layer. In this case, the classification accuracy for grade A was 100%, grade B was 90% and grade C was 100%.

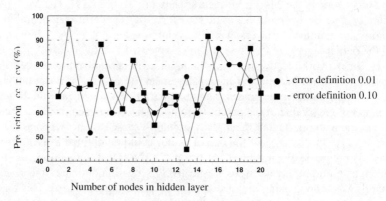

Fig. 6. Prediction accuracy of neural network analysis.

4. DISCRIMINANT ANALYSIS

Discriminant analysis (DA) is a multivariate technique that can be used to build rules to classify objects into appropriate categories. In this work, a discriminant rule was developed using the *SPSS* (Statistical Package for Social Sciences) software [10]. This was done by first running the input data (calibration data) in

SPSS to obtain the classification function coefficients of the Fisher's linear discriminant function. Then, the coefficients were substituted in equation (4.1) below to form the linear discriminant function $f(x_i)$ for each class or population,

$$f(x_i) = A + B_1 x_1 + B_2 x_2 \tag{4.1}$$

where i = population (1, 2 and 3), x_1 = parameter 1, x_2 = parameter 2, A = constant, and B_1, B_2 = classification function coefficients.

Sixty samples in the test set were substituted, one by one, into the three linear discriminant functions formulated for each population. The unknown sample (observation) was assigned to the class with the highest discriminant function value. Then, the percentage accuracy of classification was calculated. The classification function coefficients determined from the SPSS software are as shown in Table 2. Grade A was represented by 1.00, Grade B was represented by 2.00, Grade C was represented by 3.00 in the SPSS software.

Table 2. Fisher's linear discriminant functions.

Parameter	Grade		
	1.00	2.00	3.00
Parameter 1	11.977	11.037	10.144
Parameter 2	4.828	7.546	9.226
Constant	-996.678	-926.066	-847.297

Hence, the linear discriminant functions for the 3 classes are as follows:

$$f(x_1) = -996.678 + 11.977 x_1 + 4.828 x_2 \tag{4.2}$$
$$f(x_2) = -926.066 + 11.037 x_1 + 7.546 x_2 \tag{4.3}$$
$$f(x_3) = -847.297 + 10.144 x_1 + 9.226 x_2 \tag{4.4}$$

The parameter values from each sample were substituted into the discriminant function above and the value for each discriminant function (score) was obtained. The results of the discriminant analysis showed that 57 out of 60 samples were classified correctly, giving a performance of 95% accuracy.

5. CONCLUSION

The results of grading white pepper berries based on the amount of black/dark gray pepper berries by weight using grayscale images is presented. Two methods of classification were compared, namely the neural network and discriminant analysis methods. The neural network method produced classification accuracy of 96.7%, which is slightly higher than that predicated by discriminant analysis method (95%). The main setback of using the neural network method is the tedium involved in optimizing the network. This work demonstrates the possibility of grading the pepper berries using grayscale images, thus obviating the need for a color vision system for this particular application.

Acknowledgement
The authors wish to thank Universiti Sains Malaysia for supporting this work.

References
1. M. Nestler, H. Ganster, A. Pinz and G. Brasseur: Multispectral spatially registered unwrapping of 3D surfaces applied to industrial inspection of fruits, proceedings of 5th International Conference on Quality Control by Artificial Vision, 21-23 May 2001, Universite de Bourgogne, Le Creusot, France, pp 153-157.
2. R. Urena, F. Rodrýguez and M. Berenguel: A machine vision system for seeds germination quality evaluation using fuzzy logic, Computers and Electronics in Agriculture, 32, 2001, pp 1–20.
3. P.H. Heinemann, Z.A. Varghese, C.T. Morrow and H.J. Sommer III, R.M. Crassweller: Machine vision inspection of golden delicious apples, *Trans. ASAE*, v11 n6, 1995, pp 901-906.
4. C. Fernandez, J. Suardiaz, C. Jimenez and C. Vicente: Automated visual inspection application within the industry of preserved vegetables, proceedings of 5th International Conference on Quality Control by Artificial Vision, 21-23 May 2001, Universite de Bourgogne, Le Creusot, France, pp 134-138.
5. K. Nakano: Application of neural networks to the color grading of apples, Computers and Electronics in Agriculture, 18, 1997, pp 105-116.
6. M.Z. Abdullah, L.C. Guan and B.M.N. Mohd Azemi: Stepwise discriminant analysis for color grading of oil palm using machine vision system, Trans IChemE, Vol 79, Part C, 2001, pp 223-231.
7. Pepper Marketing Board, Malaysia: Specifications Of Sarawak Black & White Pepper, 2 January 2002, URL http://www.sarawakpepper.gov.my/sarawakpepper/
8. A. Abdesselam, R.C. Abdullah: Pepper berries grading using neural networks, Proceedings of TENCON 2000, 24-27 September 2000, Kuala Lumpur, Malaysia, pp 153-159.
9. H. Demuth and M. Beale: Neural Network Toolbox For Use With Matlab – User Guide Version 3.0, The Math Works Inc., 1998.
10. P.R. Kinnear and C.D. Gray: SPSS for Windows, Lawrence Erlbaum Associates Publishers, UK ,1994.

Autonomous Agricultural Robot

Mark W Phythian,
Faculty of Engineering & Surveying,
University of Southern Queensland, Toowoomba, Australia.

Abstract
The automation of agricultural machinery in Australia has tended towards refining the operation of existing equipment including tilling, spraying, and harvesting tasks. Little effort has been focused on developing an alternative to some of the labour intensive tasks, such as weed chipping, spot spraying and crop monitoring. It is proposed that an autonomous agricultural robot, utilising recent developments in machine vision and global positioning system (GPS) navigation, would provide a cost effective alternative to a hired hand for such repetitive tasks. This paper presents the design and current status of a robotic platform for row crop tending currently under development at the University of Southern Queensland, Australia. Details presented include the configuration of the prototype platform, control and drive systems, an overview of the navigation model and the low cost of GPS receivers and vision systems.

Keywords: *Autonomous, agriculture, GPS, navigation, vision, Kalman algorithm.*

1. DESIGN BRIEF

The Autonomous Robot Farmhand (ARF) concept, from which this project originated, was recently proposed by Professor John Billingsley of the University of Southern Queensland [1]. The challenge was to create a cost effective autonomous robot that could satisfactorily navigate a farm environment, locate a crop in a field and commence a specified task. The key is in the recent development of a low-cost GPS system with centimetre accuracy [2].

For heavier tasks of tilling and harvesting the tractor is the logical choice, but for tasks like weed chipping, spot spraying and crop monitoring it is becoming increasingly difficult to hire labour to complete such repetitive jobs. Such a robot would preferably be able to carry a selection of equipment for a range of tasks and to be able record where it has been. For monitoring aspects a field map could then be produced from any data collected.

At the very least, the robot should reliably navigate the crop rows performing its duties, but could be capable of navigating its way to and from a field. It should be large enough to do useful work yet small enough that it wont cause any significant damage to crop, equipment or personnel should anything go wrong. As might be expected a strong emphasis on safety is a key element of the project.

1.1 Requirements

The brief for the robot design included the following primary requirements:
- Autonomously navigate along a crop row to ±3cm of desired path.
- Identify the row end and turn back into a parallel row.
- Able to carry a selection of equipment for crop tending or monitoring.
- Incorporate a safety shutdown for a variety of conditions.
- Track GPS location and record selected data.

The brief also included the following desirable features:
- Set-up via a handheld interface.
- Navigate to/from the crop within a known field.
- Remote progress monitoring and control.
- Navigate to/from the field within a known farm configuration.
- A selection of ground speeds down to 0.25m/s.

1.2 Project Stages

Stage one of the project covers the design and commissioning of the robot platform. The scope of this stage includes the development of the chassis, drive system, control equipment, odometry and steer angle sensing, GPS navigation and vision input for row following. The tasks to be completed include: autonomous row navigation, row end identification and turning, location tracking and local operational safety. Field trials are planned to commence in July 2002. This paper reports on the current status of this first stage.

Stage two will involve expanding the navigation capabilities of the robot and incorporate remote monitoring and control. The tasks to be accomplished will include: coordinate system navigation, remote location monitoring and command control.

Stage three will involve the development of a spot spraying system utilising at least one additional vision source. The tasks to be completed will include: weed identification, spot spray control and data logging of weed distribution. The initial application crop is likely to be cotton.

2. ROBOT DESIGN

The configuration of the robot is based on a narrow central body with two outrigger wheels. The prototype is built around a standard set of motorcycle wheels - the front wheel used for steering and odometry and the rear wheel for traction.

2.1 Chassis design

The robot is just over two metres in length with a wheelbase of 1.4m. The side wheels are mounted on vertical supports one metre either side of the rear wheel. The right side wheel is fixed and the left side wheel is sprung to allow for variations in ground heights. The supports for the side wheels are 1.5m high and connect to the main frame at the top of the robot, each forming an arch over a row crop. Figure 1 is a front-on view of the robot showing the vertical workspaces.

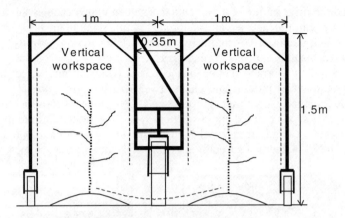

Figure 1. Front view of robot farmhand.

The vertical workspaces allow for crop monitoring, spraying and use of other equipment to two crop rows simultaneously. The horizontal workspace below the frame and between the main wheels allows for spray equipment or weeding mechanisms to service one furrow at a time. Additional workspace or materials storage space is available on the top of the main frame (1.4m x 0.35m x 0.65m). Figure 2 is a side-on view of the robot showing the horizontal workspaces. Figure 3(a) shows a photograph of the chassis prior to equipment installation.

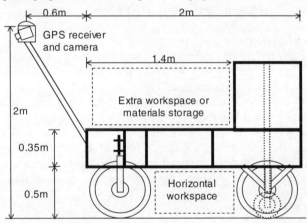

Figure 2. Side view of robot farmhand.

2.2 Drive System

Figure 3(b) shows a photograph of the drive system. The robot is powered by a 6kW air-cooled diesel engine, incorporating an electric start and a small generator. The output of the motor continuously drives a car alternator (lower left) for DC power and the drive pulley of a belt clutch. The belt clutch enables the transfer of power to the gear train and serves as a torque limiting point in the drive train.

The clutch output drives into a five-speed ride-on mower gearbox, providing five forward, neutral and one reverse gear. The output speed of the gearbox is reduced in a 20:1 worm reduction box before driving the rear wheel through a standard motorcycle sprocket and chain drive. This gear train provides the large speed reduction necessary to set the robot speed in the range 0.25m/s to 1m/s.

The motor speed is fixed and a solenoid is used to set the throttle to start and stop the motor. A windscreen wiper motor is used to tension the belt clutch using an idler pulley. Another wiper motor is used to make the gear selection.

Figure 3.　　　**(a) Empty chassis**　　　**(b) Drive system**

2.3　Control System

The control of the robot utilises two subsystems, Crop Task Control and Navigation Control. The computer hardware is duplicated in each subsystem and is comprised of a Personal Computer and a microcontroller.

Each PC is an ATX motherboard designed for use in robust Point-Of-Sale equipment. Each microcontroller is used as an interface device for the PC, gathering sensor input and maintaining low-level control of the robot's mechanisms. The microcontroller currently used is the MC68HC811E2, with support circuitry for analog measurement, digital sensor inputs and outputs for power switching. Figure 4 illustrates the control system configuration.

2.3.1　*Crop Task Controller*

As no crop task control is required in stage one of the project, the PC in this subsystem will initially be used to interface with the camera for the vision guidance. This PC runs a version of Real-Time Linux, called mini-RTL, that compresses to under 1.4Mb and uses "loadlin" to expand into an operational kernel. The microcontroller in this subsystem will later be installed to interface with future crop task equipment.

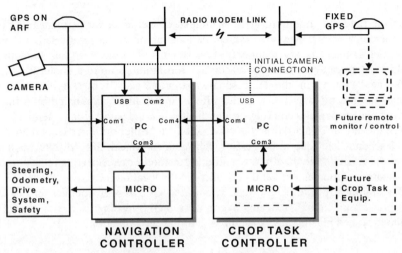

Figure 4. **Control System Configuration.**

2.3.2 Navigation Controller
The PC in this subsystem is responsible for the navigation of the robot and interfaces with the GPS receivers, its microcontroller and the Crop Task Controller. This PC currently runs MSDOS to support the VBDOS based GPS navigation code originally written for a tractor steering system.

This code will later be translated into C for the mini-RTL OS, at which time it is expected that the vision guidance code will also be transferred to this computer. The microcontroller in this subsystem provides a closed loop control on the steering and is responsible for the motion control and safe operation of the robot.

3. NAVIGATION

Many approaches have been developed for agricultural robots [3,4], but with the switching off of Selective Availability in April 2000 much interest have rightly focused on Global Positioning Systems. Even so, to achieve the positional accuracy of a few centimetres required for agricultural tasks, systems in commercial use today rely on precision receivers and inertial navigation that place the typical installation for a tractor at over US$35,000.

While specialised receivers have also been developed [5], the system proposed in [2] offers far greater flexibility with respect to handling GPS events and at a suitable cost.

3.1 The GPS System

The GPS system of [2] has been adopted and uses two low-cost (US$170) receivers (Garmin GPS 35-HVS), one mounted ahead of the front wheel of the robot, the other at a fixed point such as atop a barn roof or beside the field in use. The pseudorange data stream of the fixed receiver is transmitted to the robot via a radio modem. By processing the carrier phase information of the two GPS receivers, the robot can obtain a non-absolute location fix within a few centimetres that remains very stable while the system remains in lock with sufficient satellites.

In the field, the mobile receiver is initialised over a known permanent datum point such as a marker peg, after which the location of the receiver relative to the marker peg is tracked very accurately. Experimentally, it has been has shown [2] that this system can maintain ±2cm accuracy for durations up to an hour at a time before the "stretch" in the least squares solution caused by satellite motion becomes too great. Notably this stretch effect can be almost eliminated by using a surveyed marker peg and entering its coordinates into the navigation system.

GPS location fixes arrive at the Navigation Controller once per second but are up to 1 second 'late' due to processing delays. Also typical GPS 'events' such as the loss of a satellite lock, a new satellite acquisition or excessive 'stretch' in the least squares solution can effect the 'quality' of the location fix.

To minimise the effect of such events the GPS model operates in one of several modes including – carrying over a missed GPS reading; assessing loss of a satellite lock; and placing newly acquired satellites on 'probation'.

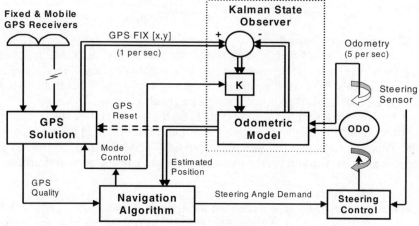

Figure 5.   **Combining GPS and Odometry Data.**

3.2   Combining GPS and Odometry Data

Figure 5 illustrates the technique used to combine the GPS and odometry data. Data from the steering sensor and odometry is used to maintain an odometric location model (x, y and heading) at 0.2sec intervals. The navigation algorithm uses this estimated position to close the control loop for steering angle demand. The micro-controller closes the loop between the demanded and measured steering angles.

The short-term accuracy of the odometric model complements the long-term accuracy of the GPS system. Where the odometric model tends to drift due to the integration of small datum errors, the regular GPS fixes reveal any discrepancy between the model and measurements. During periods of 'quality' GPS tracking, the datum for the odometric system can be adjusted. During GPS events that cause gaps in GPS data the odometry can carry on for short periods. When a GPS Reset is required the current location can be taken from the odometric model.

A Kalman algorithm [6] is used to combine the odometric model and GPS data to maintain an estimate of the 'true' location of the robot at 0.2sec intervals. Updates to the location model are made robust to GPS 'events' by linking the Kalman parameters to operational modes of the GPS system. It is this level of integration that makes this GPS technique superior to 'black-box' receiver systems which simply produce a location fix.

3.3 Vision Guidance

The vision guidance system detailed in [7] is adopted with the exception that the image source is updated to a USB compatible camera module and that the arrangement of rows differs slightly. The camera is mounted just under the GPS receiver, forward of the front wheel and 2m from the ground.

The algorithm processes data from several skewed 'viewports' positioned dynamically within the image, such that they each straddle a single crop row. The plants in the image are identified using the chrominance component and a dynamic threshold that is updated from frame to frame. (See Figure 6, courtesy of [7])

The displacement and angle of each row within each viewport is determined using an averaging technique akin to a moment of inertia calculation. Another measure, called 'quality', is also produced which is used to decide if the estimate from that viewport is to be trusted. A voting scheme based on the quality of detection of each row is used to provide robustness where the crop is patchy.

The primary error signal for steering is taken from the apparent displacement of the rows. By-products of these calculations are used to adjust the viewport positions and angles for the next frame. In the ARF application the vision guidance will be used as a demand input to the Navigation Controller during row following.

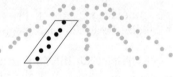 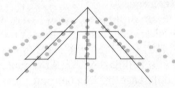

Figure 6. (a) Skewed viewports   (b) Row angles determined.

4. PROJECT STATUS

At the time of writing the Autonomous Robot Farmhand's chassis and drive system, control equipment and sensors are all but complete. The software for the navigation controller has been tested. The vision algorithm is yet to be adapted from the tractor steering application and no doubt presents some future challenges. Figure 7 shows a photograph of the completed Robot Farmhand assembly.

5. CONCLUSIONS

This project incorporates recent developments in GPS navigation with a proven vision guidance system in the design of an Autonomous Agricultural Robot. The mechanical design of the robot utilises off-the-shelf components where at all possible to keep the cost low and the maintainability high. The mechanical design is open enough to carry a range of equipment and the control system allows for future expansion, remote monitoring and control. The author is enthusiastic about the outcomes of this project and looks forward to presenting the results of field trials as soon as they become available.

Figure 7.   **The Autonomous Robot Farmhand.**

Acknowledgments

The author wishes to acknowledge the financial support of the University of Southern Queensland; the work of Prof. J. Billingsley (USQ) for the development of the GPS system and in association with M. Schoenfisch on the vision system. Thanks to Dr. R. Willgoss (UNSW) for work on the Kalman algorithm.

References

1. J. Billingsley, Low cost GPS for the Autonomous Robot Farmhand, Mechatronics and Machine Vision, Research Studies Press, England, September 2000, ISBN 0-86380-261-3, pp 119-1251.
2. J. Billingsley, An Affordable Farm Guidance System, 8th IEEE International Conference on Mechatronics and Machine Vision in Practice, City University of Hong Kong, Hong Kong, 2001, pp 75 - 78.
3. B. Rooks, Autonomous Vehicle for Plant Scale Husbandry, Service Robot, vol. 2, no. 3, 1996 MCB Univ Press Ltd, Bradford, England, pp 17-20, ISSN: 1356-3378.
4. J.N. Wilson, Guidance of agricultural vehicles - a historical perspective, Computers and Electronics in Agriculture, vol. 25, no. 1, Jan 2000, Elsevier Science Publishers B.V., Amsterdam, Netherlands, pp 3-9, ISSN: 0168-1699.
5. G. Elkaim, M. O'Connor, T. Bell, B. Parkinson, System identification and robust control of farm vehicles using CDGPS, Proceedings of ION GPS 97, Part 2 (of 2), Sep 16-19 1997, Kansas City, MO, USA, pp 1415-1424.
6. R. Willgoss, Steerage of a vehicle along a track using a blend of GPS and odometric modelling, Technical Report, USQ, Australia, May 2002.
7. J Billingsley, M. Schoenfisch, The successful development of a vision-guidance system for a tractor, Computers and electronics in agriculture (journal), Elsevier, Amsterdam, Netherlands, January 1997, pp 147-163.

Fuzzy Multivariable Control of a Meat Chiller

W.L. Xu, A. Cowie, and G. Bright
Institute of Technology and Engineering, College of Sciences
Massey University, Palmerston North, New Zealand

Abstract
This paper deals with the modelling of a meat chilling process in a freezer/chiller unit and design of a fuzzy logic controller (FLC) to control the cooling process. The entire control system was tuned and implemented in Matlab/Simulink. Simulations were carried out to test if a set of prescribed specifications have been met satisfactorily.

Keywords: *Meat chilling, multivariable control, fuzzy logic, heat transfer, mass transfer.*

1. PROBLEM DESCRIPTION

The chilling of meat is a critical process in freezing factories. Once a cattle has been killed, the carcass must be chilled until all the meat is below 10 degrees Celsius before the meat can be boned [1]. Being able to chill the meat as fast as possible is advantageous, as it allows production rates to be increased. However, the meat cannot be allowed to freeze at any stage as this reduces the quality of the meat. Meat freezes at –1 degree Celsius [2].

The rate that meat is cooled is the most important consideration, but not the only one. Also important is the amount of evaporative losses experienced by the sides of meat during the chilling process [3], because the meat cut from the carcass is sold on a weight basis and any water mass lost due to evaporation is mass that could have been sold as meat. This rate of evaporation is dictated not only by the difference in the air and the surface temperature but by the concentration of water on the surface of the meat and the amount of water in the air of the chiller (the humidity). By minimizing the difference between the water content of the air and the concentration of water at the surface of the meat, the loss of weight due to evaporation is minimised. There is another constraint that must be considered in this process and that is the microbiological spoilage of the meat surface. The bacteria that cause this spoilage thrives in damp conditions, therefore the surface of the meat must be kept at a low enough concentration so that the spoilage is minimized. It has been shown that microbiological spoilage is minimal as long as the 'Water Activity' (Aw) on the meat surface is kept below 0.9 [4,5].

2. MODELING OF MEAT CHILLING PROCESS

Heat energy is lost from the surface of the meat through two separate mechanisms, convection away from the surface and the energy used to evaporate water from the meat's surface. The rate of convection is dependent on the flow of air past the surface of the meat and the temperature difference between the meat surface and the air. The rate of heat loss due to evaporation is dependent on the rate of evaporation. The rate of evaporation is dependent on the velocity of the air passing the slabs surface, the temperature difference between the surface of the meat and the concentration of water in the air (humidity). The rate at which the inside of a slab of meat is chilled is dictated by the temperature gradient throughout the meat slab. If the temperature on the surface of the meat is much lower than the core temperature of the meat slab, a large (negative) temperature gradient is experienced. This results in a larger amount of heat energy being transmitted through the meat slab (towards the surface) by the process of conduction, than if a smaller temperature gradient was experienced. Figure 1 shows heat transferring and mass transferring mechanisms during meat chilling.

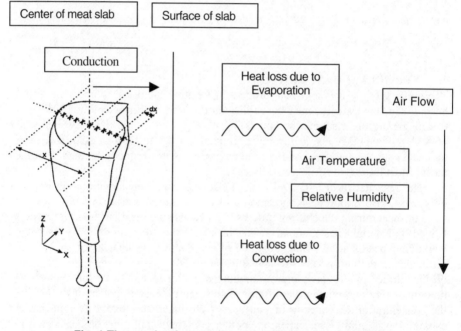

Fig. 1 The mechanisms that operate during meat chilling

The model to be created to test the chiller control system was based on a series of partial differential equations (PDE). Two sets of PDEs were established, one for mass transfer through water movement and the other for heat transfer [6,7]. These equations can be thought of as existing in two separate groups, those that calculate the temperatures throughout the slab of meat and those that calculate the concentration of water within the meat.

The PDEs work by separating the effect of the changes in temperature that are due to position in the slab from the effect of the progression of time. As shown in Figure 2, the slab is broken into many small portions or slices. These portions are defined by the position of 'nodes'. These slices are parallel to the surface of the meat. For any time step, the temperature at any point throughout the meat slab can be calculated from an equation involving the previous (one time step ago) temperature values of that same meat slice and the slices that are on either side of it. The values of temperature of the nodes on the surface and the center of the meat slab are of particular importance, and are called boundary conditions. The boundary nodes are the only nodes directly affected by the conditions in the chiller. The only way that the conditions in the chiller are 'felt' by the internal portions of the meat is by the effect being 'cascaded' on by the external portion of the meat slab.

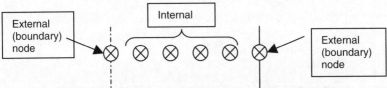

Fig. 2 The meat slab modelled using a series of nodes

3. CHARACTERISTICS OF MEAT CHILLING PROCESS

Given a relative humidity = 0.5 %, an air temperature = 0 degrees Celsius, an air velocity = 2m/s, the time-variations of the temperature of the center and the surface of the slab are calculated and plotted in Figure 3. It is found that the outer temperature of the meat drops under ten degrees quickly. The temperature of the slabs center drops down at a more consistent rate than the outside but does not drop to ten degrees in the time given. The surface temperature has not been brought down to the optimal value (-1 degree Celsius) to ensure the maximum amount of heat is removed from the meat center.

If a cut is taken through the surface at any one time we could observe the temperature gradient at any point. The temperature gradient drives the rate at which heat content and temperature is reduced. After the sides of meat have been in the chiller for half an hour, it can be seen that the temperature gradient across the slab is non linear. That means that the rate that heat is lost from the outside of the slab is greater than the rate that heat is lost from the inside of the slab. By the fifth hour that the slabs have been in the chiller, the temperature gradient is linear across the slab. This means that for any position within the slab, the rate that heat is being lost towards the outside of the slab is the same as the rate that heat is being conducted to that point from the inside of the slab.

Figure 4 shows the water activity on the surface of the slab over time. It is important to get the water activity value under 0.9 as soon as possible to reduce microbiological spoilage. From the plot it can be seen that the water activity value drops to be well under 0.9 in less than two hours. The water activity is well below 0.9. Being much lower than the critical value of 0.9 is counter-productive, as the lower surface water activity means that a higher rate of evaporation exists, which means water mass is being lost.

From initial testing of the performance of the chiller, many factors have become obvious: (i) To lower the temperature of the center of the meat slab to below ten degrees, while not allowing the outside to freeze, is the most challenging criteria to meet; (ii) The surface temperature responds quickly to changes in the chiller temperature; (iii) The heat loss is dependent mainly on the chiller temperature and the air velocity in the chiller; (iv) The water activity drops to well below 0.9 reasonably quickly; (v) The loss of mass due to evaporation is dependent mainly on the relative humidity of the air in the chiller.

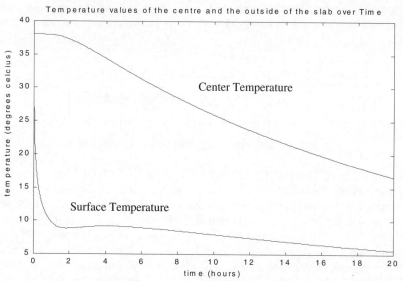

Fig. 3 Slab Center and Surface Temperature Variations over 24 Hours

4. DESIGN OF FUZZY LOGIC CONTROLLER

The objectives that should be achieved by the chiller control system are summarised below. (1) The core temperature must be brought down as quickly as possible; (2) The temperature of the meat should not go below –1 degree Celsius as at this temperature meat freezes; (3) Water Activity should be brought down to under 0.9 as quickly as possible to minimise microbiological spoilage; (4) The mass loss due to water evaporation should be minimised, as any weight loss is a loss of saleable product; and (5) The final temperature of meat throughout the side should be below 10 degrees Celsius as this temperature is cold enough for boning to take place. Other requirements for control of the cooling process in a boning room were mentioned in [8].

The fuzzy multivariable controller designed has three inputs: (1) Delta_Temp (i.e., the difference between the surface temperature and the ideal value of - 0.9 degrees), (2) Rate_Temp (the rate of change of temperature), and (3) Delta_Aw (surface water activity). Fuzzification of Delta_Temp through fuzzy sets was performed to cover a large range of inputs while providing very accurate control when the difference between the surface temperature and the set point became small. A triangular membership function was used for the positive small membership function to ensure smooth control action once the temperature had settled close to the ideal value. The membership functions for Rate_Temp were designed to provide fine control when the rate at which the temperature changes becomes very small. This is to allow the controller to adjust its output while the temperature rate changes, i.e. if the temperature is only slightly positive and the rate of change is largely negative, then control actions need to be taken. What gradient constituted a small or a large rate of change was found from experimentation with the model. Delta_Aw was fuzzified to provide accurate control once the water activity approaches its ideal value and triangular membership functions were used around the set point value to ensure a smooth control.

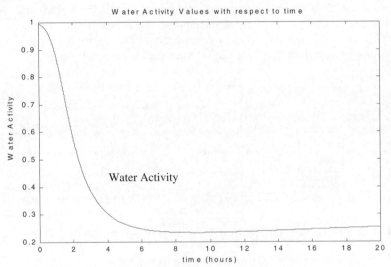

Fig. 4 Water Activity Variation Over Time

The three controls (i.e. the outputs) that the fuzzy controller could apply are: (1) Delta_AirTemp (the air temperature relative to zero), (2) Delta_AirVel (the air velocity relative to zero), and (3) Delta_RH (the relative humidity). The membership functions for Delta_AirTemp designed ensure a smooth transition from negative to positive changes in air temperature. Sensible values were chosen for the rate to make it possible for the air temperature to change. The membership functions for Delta_AirVel ensure a smooth transition from negative to positive changes in air velocity. Sensible values were chosen for the rate so that it was possible for the air velocity to change. For the Delta_RH, the membership functions ensure a smooth transition from negative to positive changes in relative humidity. Sensible values were chosen for the rate to allow for the humidity to change.

The rules for the fuzzy controller, that dictate the chilling regime related various fuzzy sets of the inputs and outputs, were designed around the following design objectives given above. The rules that related membership functions were designed around the following design objectives:
(i) Lowering the surface temperature close to −1 degree Celsius is of primary importance. This means that while Delta_AirTemp is positive, medium or higher, everything is devoted to lowering this temperature. This means that the air temp is as low as possible (-10 degrees Celsius), the air velocity is as high as possible (4 m/s) and the relative humidity is zero.
(ii) As the temperature gets close to the set point, the temperature and the air velocity is used to control it. This control is to take into account the rate that the temperature is changing to ensure the surface of the meat does not drop below −1 degree Celsius and to ensure that the temperature does not increase significantly.
(iii) Once the surface temperature of the meat is close to the set point and settled (the rate of change in surface temperature is small positive or small negative), the relative humidity is used to control the water activity to bring it to just below 0.9 (required to ensure against microbiological spoilage).

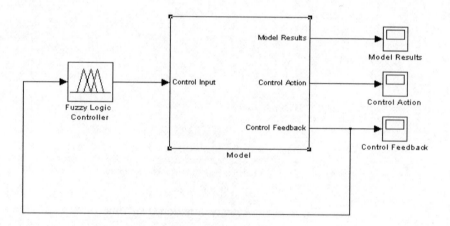

Fig. 5 Simulink Model of the Fuzzy Chilling Process

5. RESULTS AND ANALYSIS
The chilling process under fuzzy controller was implemented in Matlab/Simulink. The PDEs governing the chilling process were coded as m-functions and the fuzzy logic controller was designed and tuned in the Fuzzy Toolbox.

Time variation of the temperature at the center of the meat slab, the surface temperature and the surface water activity were plotted to check against the control objectives. It was found that the FLC acts to keep the surface temperature slightly above −1 degree, that once the temperature is settled at ~ -0.7 degrees the water activity at the surface is increased (by the increase in the Relative Humidity) without negatively effecting the surface temperature, as shown in Figures 6 and 7.

With respect to the control actions exerted by the FLC (Figure 8), the initial response of all three outputs is to maximize the rate at which the surface temperature drops to –1 degree Celsius. Once this is reached and settled, the relative humidity ramps up to increase the water activity on the meat surface and the air velocity decreases to minimize evaporative losses.

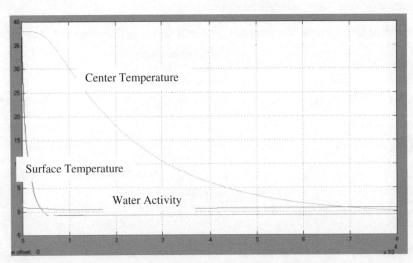

Fig. 6 Performance of the chilling process under fuzzy control

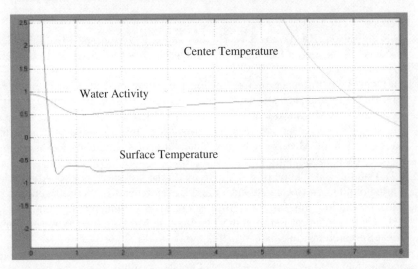

Fig. 7 Performance of the chilling process under fuzzy control (y-zoomed)

6. CONCLUSION

The model created, accurately replicates the response of a side of meat within a chiller. The FLC used to control this process is set up in a way that ensures near optimal control. Near optimal control, in this case, involves bringing the surface temperature down to a set point and then adjusting the parameters to reduce the total evaporative mass losses and maintain the water activity on the surface below the critical 0.9 threshold. This controller could be fine-tuned to ensure that the surface temperature is brought to a lower temperature than the ~ -0.7 degrees Celsius that have been currently reached.

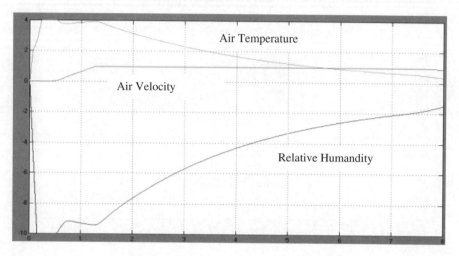

Fig. 8 The control actions taken by the FLC

References
1. N.R.P. Wilson, E.J. Dyett, R.B. Hughes and C.R.V. Joneset: Meat and meat products - factors affecting quality control, Applied Science Publishers, London, 1981.
2. E. Dransfield: Influence of freezing on the eating quality of meat, in Meat Freezing Why and How?, Meat Research Institute, Bristol, 1974, pp. 9.1-9.5.
3. B.G. Shaw, B.M. Mackey and T.A., Roberts: Microbiological aspect of meat chilling - an update, in Recent Advances and Developments in the Refrigeration of Meat by Chilling, International Institute of Refrigeration, Bristol, 1986, pp. 25-30.
4. J.D. Daudin: Calculation of water activity on the surface of hot boned muscles during chilling, in Recent Advances and Developments in the Refrigeration of Meat by Chilling, Int. Inst. of Refrigeration, Bristol, 1986, pp. 389-395.
5. C.L. Cutting: Current issues in meat chilling, in Meat Chilling Why and How, Meat Research Institute, Bristol, 1972, pp. 1.1-1.13.
6. O. Campanella and O.J. McCarthy: Engineering I - mass and energy balance, Lecture Notes, Massey University, New Zealand, 1996.
7. O. Campanella and O.J. McCarthy: Engineering principles - heat transfer, Lecture Notes, Massey University, New Zealand, 1996.
8. K. Murray: Temperature control in boning or cutting room, in Meat Chilling Why and How, Meat Research Institute, Bristol, 1972, pp. 19.1-19.3.

Control of the Sugar Cane Harvester Topper

Stuart G. McCarthy, John Billingsley, Harry Harris
National Centre for Engineering in Agriculture
University of Southern Queensland
Toowoomba, Queensland, Australia
Email: mccartst@usq.edu.au

Abstract
Operation of a mechanical sugar cane harvester is a skill and labour intensive task. The operator has many functions to manually control under difficult conditions. The focus of this paper will be one of these functions: the topper. The height at which the topper cuts the cane stalk has a considerable effect on the quality of the product, and ultimately the return to the grower. A sensor has been constructed that will measure and return a real-time signal of the height of cut on the cane stalk by the topper. It is proposed that that the introduction of some form of closed loop control system for the topper may reduce the responsibility of the harvester operator.

Keywords: Harvester, topper, sensor, refractometer, control.

1. INTRODUCTION

A majority of Australian growers prefer green cane harvesting because of benefits such as weed control and moisture conservation [1]; the alternative is to set fire to the cane and harvest it burnt.

With the widespread acceptance of this harvesting practice, higher demand has been placed on the sugar cane harvester's operation. It is the responsibility of the operator to control the functions of the harvester to maximise the quantity of sucrose to the sugar mill.

Modern mechanical harvesters are complex and ungainly machines that must perform numerous simultaneous functions. All functions are hydraulically powered. A majority of these functions have to be controlled under extremely difficult visual conditions, assuming they can be seen at all! Figure 1 illustrates the harvester operator's view of the topper out the front of the cabin.

Despite the obvious difficult operating conditions, cane harvesters of the near future will probably operate on similar principles to the machines in operation now, the only major differences being in the materials, the operator comfort and control achieved, and the technique used to maximise their performance [2].

The introduction of new technologies and functions to the harvester has complicated the operation of a harvester even more. To help alleviate the number of responsibilities expected of the operator, it is proposed that mechatronic systems should be introduced. Previous work [3] attempted to reduce the workload of the harvester operator by monitoring the amount of millable sugar cane was being ejected from the system by the primary onboard cleaning system of the harvester.

Fig. 1. View from the harvester cabin

1.1 The sugar cane plant

The area of the plant of greatest interest for the grower is the mature stalk because this is where the commercial product, sucrose, is stored internally in soft, fleshy juice-soaked fibres. Mature stalk juice can store up to 20% sucrose, or more commonly sugar.

The upper portion of the stalk is less mature, and thus contains sugar of a much lower concentration. The vertical distribution of sugar in a stalk is quite consistent until a particular section of stalk, where there is a significant transition from the maximum sugar concentration to a much lower value (up to a quarter of the maximum). This is considered to be the optimum topping height, and from this the grower must make a fiscal compromise between the costs of transporting harvested cane of low sugar content versus the return to the grower of this cane.

1.2 The topper

The function of the topper is to gather, sever and discard the leafy, immature and non-productive tops of the sugar cane stalk. Modern toppers are called 'shredder toppers' because of their ability to gather and shred the cane tops before discarding the waste back onto the field.

The operation of this function is particularly difficult because the operator cannot see the point of contact of the topper with the cane stalks. Even after the event, when the cane stalks are being fed into the harvester throat (Figure 1), there is too much waste material for any visual confirmation of the a correct height of cut to be established.

2. THE SENSOR

To establish the optimum topping height, a number of different techniques were considered. To identify the height, the cane stalk could be considered from either an external or internal perspective.

1. External indicators, such as machine vision, to identify cane stalk's maturity levels and acoustic sensor for an audible indicator of the cutting interface were considered. These techniques were not considered robust enough for the topper application because of inconsistencies of sugar cane stalks such as size of plants, varying crop density, the number of different crop varieties, and irregularities in growing patterns

2. The product of interest is the sugar inside the stalk, so the measurement of this was considered to be the most appropriate approach. Traditional techniques to measure sugar content are refractometry and polarimetry. More recently near-infrared (NIR) techniques are becoming more common in sugar mills as an accurate measure of sugar content.

2.1 Existing real-time sensor

In the past, work has been conducted on the development of a novel mechatronic sensor [4] to identify a chemical property of the cane stalk to indicate the correct cutting height. This sensor was a critical angle refractometer (Figure 2) that was constructed to be the active measuring device on the topper of the cane harvester.

This sensor proved to be a very accurate measure of sugar samples that had been artificially created, as well as from a sugar cane stalk. These tests consisted of the sugar samples being extracted and then placed on the sensitive surface of the refractometer.

Fig. 2. The novel critical angle refractometer with sensitive surface window visible

The major shortfall of the sensor was that raw sugar samples from a cane stalk could not be extracted and placed where required for a measurement without external (human) assistance.

To ensure that sufficient sugar sample could be returned from a cane stalk to the sensitive surface of the refractometer, a squashing mechanism was designed and constructed. This preliminary model still proved unsatisfactory, and so some modifications were made.

2.2 New and improved model

The sampling concept for the squashing mechanism was to test the juice extracted at particular heights by the mechanism, starting high in the leafy tops and gradually lowering the topper until the transition section was detected. The squashing mechanism was to be mounted from the topper arm, and this was the height of the crushing that was always going to be proportional to the lowest cutting plane.

Static load frame tests were performed along the length of cane stalks to determine the force required to deform the cross-section an amount that produced an adequate amount of juice to enable a measurement of sugar concentration to be determined. The amount of deformation required for this was up to approximately 30% deformation.

The load required for the 30% deformation varied by up to 400 N between the developing transition sections down to the mature cane stalk: from approximately 800 N to 400 N respectively.

This concept did not accommodate for the squashing mechanism to crush the very strong exterior of the mature cane stalk, however it allowed a conservative figure of 600 N to be chosen as the deformation force required for the "interesting" section of stalk.

A few alternate forms of mechanical device were available to be utilised for the squashing process. The principle design factor was that the device sampled the upper section of the topped cane stalks after the topper and before the throat of the harvester, and to ensure that this flow was not impeded. The forward ground speed of the mechanical harvester is up to 12 km/h, and so the contact time between the stalk and squashing device had to be kept to a minimum. The techniques investigated included impact devices, spring-loaded crushing wheels, and cam-driven crushing lever.

The latter of these was chosen for its ability to capture and crush stalks quickly and to create a juice flow for the refractometer.

Not all harvested cane is upright, and so the mechanism has a wide mouth that directs a selection of topped stalks into a narrower crushing area. The crushing arm is a lever that is on a cam. A heavy-duty DC motor drives the cam disc, and this disc was machined with many settings so that the throw of the lever arm could be adjusted to suit the diameter of the cane stalk variety being harvested. Captured stalks are crushed between the lever and the stationary back wall (Figure 3). The periodic throw generated by the cam allows stalks to be captured, crushed and released in a relatively quick motion.

A certain amount of force was required to deform the top section of stalks, and this torque was generated from the DC motor via a gear train to provide the necessary mechanical advantage.

Once the stalk has been crushed, the juice flowed down the back wall and onto the refractometer through a small window. The window has a sharp blade above it that penetrates the stalk when being crushed. This blade allows a very sufficient quantity of juice to flow.

Fig. 3. Improved squashing mechanism

Cane fields provide quite a hostile measuring environment, so to reduce the effect of impurities and other contaminants on the refractometer, a regular cleaning process has been introduced to periodically flush down the juice track and over the sensitive surface.

3. RESULTS AND DISCUSSION

Laboratory trials have been conducted on the squashing mechanism and refractometer using real stalks of sugar cane.

These trials consisted of dividing the length of the stalk in between nodes and then manually feeding the stalk section through the mechanism. Each stalk section was crushed and a flow of juice was identified, and this juice appeared on the sensor and a measurement of sugar concentration was taken. The sensor had been calibrated earlier by manually placing different concentrations of sucrose on the sensor.

During the testing process, the refractometer was regularly flushed with water. This also allowed the refractometer to be periodically recalibrated with a 0% sucrose measurement.

The quantities recorded when performing these tests included: node number, vertical height on stalk, sugar concentration, delay times from crush to measurement, and measurement time. From these quantities a vertical distribution of sugar in a cane stalk could be plotted, and this is indicted in Figure 4. The trend that is evident from this distribution is what was expected from literature, as well as previous manual testing performed with a stalk sugar and a laboratory refractometer.

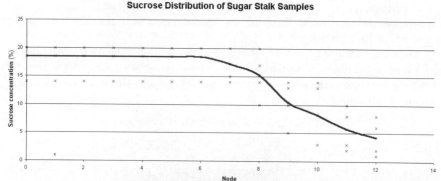

Fig. 4. Distribution of sugar concentration from refractometer with samples from stalk (left) up to top (right)

Field trials of the new sensor (squashing mechanism and refractometer) have yet to be satisfactorily completed.

4. CLOSED-LOOP CONTROL

Now that the sugar content can be measured in real-time from a cane stalk, the next step is to make the system mechatronic by introducing some form of closed-loop control.

There have been three alternatives that have been considered so far.
1. Information regarding the measured sugar content at the cutting height could be visually provided to the harvester operator. This could be in some form of a damped "too-high" or "too-low" display with an input to allow the reference point to be adjusted. A visual indicator of this sort would prove to be quite simple with only three outputs: high, good, and low.
2. The height of the topper may be automatically controlled. The measured height could be an input into a control microprocessor, and then some electro-hydraulic algorithms could be implemented. A possible control algorithm for this alternative could be non-linear so that the topper starts cutting high and is gradually lowered until the threshold sugar concentration is measured and then it is raised suddenly.
3. The integration of the indicator and automatic control to allow the height of the topper to be automatically controlled, wit the inclusion of an operator "over-ride" alternative.

5. CONCLUSIONS

The operation of the topper of a mechanical harvester is an integral task for the operator that must be performed under extremely difficult conditions.

A few mechatronic alternatives have been discussed here that could help alleviate the responsibilities of the operator. The most warranted of these, the critical-angle refractometer and crushing device, has proven to be quite successful in measuring and returning a signal for the sugar concentration at particular heights on a cane stalk.

The information that has been generated can be utilised to create a closed-loop control system for the topper. This has been demonstrated to be in either an automatic or indicative manner, or even an integration of the two.

Acknowledgments
The authors are grateful for the cooperation of the harvesting contractors in Murwillumbah, Northern NSW, Australia. The work was undertaken with the financial support of the Sugar Research and Development Corporation.

References
1. Schembri, M.G. and Garson, C.A. 1996, 'Gathering of green cane harvesters: A first study', *Proceedings of the Australian Society of Sugar Cane Technologists*, Mackay Queensland, pp. 145-151.

2. Kerr, B. and Blyth, K. 1992, They're all half crazy, Canegrowers, Brisbane, pp. 108-115.

3. McCarthy, S.G., Billingsley, J. and Harris, H. 2002, 'Development of an advanced cane loss monitoring system', *Proceedings of the Australian Society of Sugar Cane Technologists*, Cairns Queensland, pp. 180-183.

4. McCarthy, S.G., Billingsley, J. and Harris, H. 2001, 'Where the sweetness ends', *M^2VIP*, Hong Kong.

Mobile Robots and Navigation

Mobile robots, whether humanoid or not, have captured the imagination of most people for nearly a century. After many false starts, it now seems that they are beginning to leave the realm of science fiction and enter the real world. However, the current crop of robots does not look anything like the stereotypes of fantasy or fiction. Most resemble boxes on wheels or tracks. One problem that fiction writers always seem to ignore (maybe because it's not simple!), is how a mobile robot actually finds its way around.

Three of the papers in this section look at different methods of navigation. One describes an actual robot in a working environment, and the others a method for communicating with a mobile robot whilst it is underwater.

The first paper applies the common method of triangulation, albeit with a combination of laser and radio technologies. The system allows a positional accuracy of 1cm.

The second paper considers an old problem – how do you integrate data from many different types of sensors so that a robot can make sense of it all? Usually, the more sensors that are added to the system, for example, to improve accuracy or sensitivity, the more the complexity of the data analysis rises. The method proposed in this paper, however, suggests that instead of adding more sensors to the system, the temporal sequence of the data sets can be stored and utilised for the improvement of measurement.

The third paper describes a mobile service robot designed to carry items around a hospital, as well as interacting with staff and patients. Although there are similar systems currently in use they are functionally limited. The robot described in this paper uses hierarchical control with sub-sumption architecture, together with various intelligent control schemes.

The fourth paper is completely different in scope! Underwater autonomous vehicles (AUVs) are becoming more common, especially in sub-sea engineering applications such as the oil industry. However, all need umbilical systems for command and control. The communication systems described in this paper uses ultrasound as the communications medium.

The final paper is also concerned with underwater robots, but describes a reactive agent architecture that is capable of task delegation, data-directed execution, communication and planning. Delegating many of the control and decision making processes to a remote vehicle, instead of having all control functions at a base-station, means that only a low-speed communications system is required for monitoring. This can eventually lead to completely autonomous vehicles.

A New Beacon Navigation System

Eduardo Zalama Casanova*, Salvador Dominguez Quijada[§],
Jaime Gómez García-Bermejo*, José R. Perán González*
* ETSII, Dept. of Automatic Control, University of Valladolid
[§] Centre for the Automatisation, Robotics and Technology Information (CARTIF).

Abstract
In this paper a new beacon localisation system for mobile robots in indoor environments is presented. Laser and radiofrequency technologies are combined in order to determine the robot's position and orientation by means of a triangulation algorithm. The system allows the identification of the different beacons, which provides the robot's absolute position without requiring the maintenance of a position estimate. The system supplies up to 15 localisations per second to an accuracy of 1 cm in position and 0.1° in orientation.

Keywords:
Navigation, robot localisation, triangulation, beacon, mobile robot.

1. INTRODUCTION

One of the main problems in mobile robot navigation is the determination of the robot position. A common solution is to determine this position through dead reckoning [2,5,6]. Unfortunately, dead reckoning suffers from incremental errors, due to miscalibrations and wheel slippage. Therefore, other systems based on external sensors are required. Localisation systems that do not require a special preparation of the environment, based on computer vision, range finders or other sensors, are computationally expensive and not too robust. So, they are being used in research centres but are largely uncommon in industrial or service applications. Location through marks or beacons is usually preferred for these real applications [1,4,8]. The general principle is as follows: sets of marks or beacons are placed in known positions of the environment, and are detected by an optical system. The robot position is then obtained from a simple triangulation algorithm. Usually, laser telemeters [9,10] are employed for obtaining the distance to a set of reflectors placed on the walls. This solution leads to good precision in robot location but has two important drawbacks. On the one hand, the system is too sensitive to highly reflective surfaces in the environment (such as windows, mirrors, polished surfaces, etc). On the other hand, the system is not able to identify the different beacons. Therefore, it is necessary to maintain a position estimate of the robot, especially whenever the operation area is large, which means that detecting all the reflectors at the same time is not possible. This problem is overcome by the Conac system in the following way [7]. The system uses a rotary laser and a set of beacons that are activated by the laser beam. A stationary computer calculates the robot position and sends this position to the robot through a wireless modem. The main limitation of this solution is that a special wiring of all the beacons is required, in order to synchronise them. Furthermore the position has to be transmitted to the robot, which restricts the size of the operation area.

In the present paper, a low-cost system for the absolute positioning of mobile robots (and other moving objects) is presented. The system combines laser and radio frequency technologies to determine the position and orientation of the robot through a triangulation algorithm. The system is made up of the following elements: an infrared laser, a DC motor train reduction-encoder, a set of photosensitive beacons and a pulse-decoder box.

The laser, motor, and a pulse-decoder box are on-board the mobile robot. The system works as follows: the motor rotates the laser in a horizontal plane, so that the laser beam reaches a set of beacons placed along the walls. Each beacon is equipped with an infrared-sensitive photocell. Upon receiving a light pulse, a beacon issues its characteristic RF code. The decoder box receives this RF code and associates it with the laser angle obtained from the encoder. This information is transmitted to the on-board computer via an RS232 serial link. The computer can then calculate the position (and orientation) of the robot, by a triangulation procedure.

2. HARDWARE DESCRIPTION

In this section the hardware elements of the proposed localisation system are described.

2.1 Active photosensitive beacon

The beacon emits an identity code every time a sharp change of the incident light occurs, as a consequence of the incidence of the laser beam on a photocell. Figure 1 shows the developed beacon. The main elements of the circuit are:

- LM805CT 5v dcv regulator.
- Sensitivity circuit made up of a TSL250 photocell and a comparator that allows the adjustment of the photocell sensitivity.
- PIC16C84 microcontroller, which sends an identity code to the RF module when a light change occurs.
- AM-RT4-433 RF, 433Mhz transmission module.
- 8-bit switch dip to code the different beacons. 7 bits are employed for coding the beacon number, while one bit is employed for parity control.
- Two LEDs indicating power on and beacon transmitting, respectively.

Figure 1. An emitter beacon. The photocell is placed by the upper side of the beacon, and a visor protects the photocell from spurious light sources.

2.2 Pulse-decoder box

This system controls the motor speed, counts encoder pulses and receives data from the beacons. All this information is transmitted to the on-board host computer through an RS232C serial link.

The main elements of this module are:

- PIC16F876 microcontroller, which collects data transmitted by the beacons, checks parity and determines the angle of the rotating laser at the data reception. This microcontroller also controls motor speed by means of a Pulse With Modulation (PWM) technique. Furthermore, the laser is switched off during the beacon code reception in order to avoid interference from close-seen beacons.
- Power supply and amplification. The module is LM805CT dcv regulator equipped, and power to the motor and the laser is obtained from a power stage based on Mosfet IRF-540N and NPN BC547 transistors.
- AM-RRS3-433 433Mhz radiofrequency reception module, which receives information from the beacons.
- Serial communication module, based on the MAX232CWE chip which performs signal adaptation to the serial RS232C link to the on-board computer.

2.3 Laser rotating system

This system activates the beacons by means of a rotating, angle-encoded laser beam. The system is located at the end of a mast at a height of about 2.5 meters so that people in the environment do not mask the laser beam. The system is made up of the following elements:

- Infrared laser. A vertical 12°-stripe generation optic is employed to deal with small errors in the vertical location of the beacons as well as with irregularities of the floor flatness.
- DC motor-gear train reductor-encoder to rotate the laser in a horizontal plane (vertical rotation axis). The gear train has a 10:1 reduction, and the optical encoder delivers 500 pulses/lap; then 5000 encoder pulses are obtained in a complete laser rotation. This gives a resolution figure of 0.007 degrees/pulse.

3. COMMUNICATION PROTOCOLS

Two communication protocols are involved: at beacons to pulse-decoder box communication, and at pulse-decoder box to on-board computer communication.

3.1 Beacon to pulse-decoder box communication protocol

Communication is performed through RF because beacons are placed in the environment (walls) and the pulse-decoder box is on-board the robot. The following information is sent by a beacon every time its photocell is reached by the laser:

HEAD	BEACON CODE	PARITY	END OF TRANSMISSION
(1 long bit)	(7 short bits)	(1 short bit)	(two pulses)

Details are given in Figure 2. When the pulse-decoder box receives a head-long bit, the laser is switched off, while the angular position is registered from the encoder and the parity is checked. The laser is turned on again at the two end-of-transmission pulses reception.

3.2 Pulse-decoder box to on-board computer communication protocol

Communication between pulse-decoder box and the on-board computer is carried out through an RS232C serial link. A bi-directional communication is performed:
- Speed motor command (1 byte). This command is sent from the on-board computer to the pulse-decoder box to select the motor speed.
- Beacon command. This command is sent from the pulse-decoder box to the on-board computer, in the following way:

HEAD	BEACON CODE	ANGLE	TIME
(1 byte)	(1 byte)	(2 bytes)	(2 bytes)

The head field is constant valued FFh, the beacon code allows the computer to univocally identify the beacon, the angle is the laser angle corresponding to the given beacon and the time (microcontroller time) allows the instant in which the beacon was detected to be determined.

4. POSITION ESTIMATION

An initial calibration process with known-position beacons is required for robot localisation.

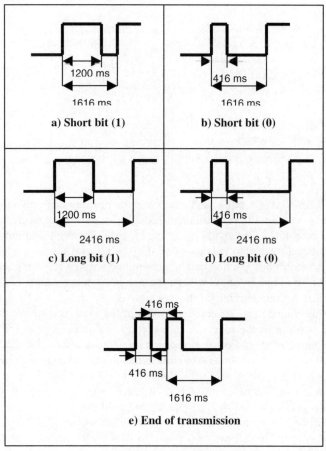

Figure 2. Beacon-pulse-decoder box: pulse codification for transmission.

Figure 3 (at the end of this paper) shows a scheme of the robot localisation process using three beacons. These beacons are represented by small circles, and the position *(x,y)* of the robot is represented by a small square, orientation *(ϕ)* being represented by an arrow. The parameters a, b, x_0, y_0, α can be easily obtained from the (known) beacon position. Furthermore θ, θ_1, θ_2 are the beacons relative angles, determined from the encoder readings. The remaining parameters are obtained from the following equations:

$$\sin \gamma_2 = \sqrt{\frac{a^2 \sin^2 \theta_2 \sin(2\pi - \theta_1 - \theta_2 - \beta)}{[b\sin\theta_1 + a\sin\theta_2 \cos(2\pi - \theta_1 - \theta_2 - \beta)]^2 + a^2 \sin^2 \theta_2 \sin(2\pi - \theta_1 - \theta_2 - \beta)}} \quad (4.1)$$

$$c = b\frac{\sin \gamma_2}{\sin \theta_2} \quad (4.2)$$

$$\beta_1 = \beta + \theta_2 - \pi - \gamma_2 \quad (4.3)$$

and robot position and orientation:
$$\begin{aligned} x &= x_0 + c\cos(\beta_1 - \alpha) \\ y &= y_0 + c\sin(\beta_1 - \alpha) \\ \phi &= \gamma_1 - \alpha - \theta \end{aligned} \quad (4.4)$$

At least three beacons are required to determine the position and orientation of the robot. So, beacons should be distributed through the environment in such a way that three (or more) beacons could be activated from every robot location. It should be noted that beacon location is known and that the activated beacon is univocally determined, so that robot location can be determined without requiring any further information (such as the instant at which a beacon becomes no longer visible).

The program that determines the position of the robot has two operation modes: execution mode and supervision mode. In the execution mode, the system provides the position of the robot (in cartesian co-ordinates (x,y), in millimetres), and the orientation of the robot (in decidegrees). The algorithm also evaluates the confidence of the results obtained from different subsets of beacons (when several subsets are possible), based on angles between beacons, and keep the results corresponding to the highest confidence. Another important feature is that the motor speed is self-regulated by the system: the speed decreases when the number of detected beacons is under a given threshold, or otherwise (gradually) increases to a nominal speed. This allows the system to deal with situations such as close beacons, large incident angles of the light, etc. Supervision mode is the other working mode of the system. In this mode, the triangulation process can be monitored by an operator, through statistics obtained from the collected data. Furthermore, in this mode, the system can accomplish a self-calibration process of the beacon location. This process consists in determining the position of an unknown-position beacon, from the location of the robot through time obtained from several known-position beacons. So, only a few known-position beacons are required, thus simplifying the beacon set-up.

The system performs up to 15 location measures per second, with a precision figure of 1cm and 0.1 degree. This is suitable for locating mobile robots travelling at moderate speeds (without requiring the use of Kalman filters [3]). The system is currently being employed for the location of a mobile robot in a Science Museum in Valladolid (Spain), which will be open to the public in a few months. The same system could also be employed for locating other moving artifacts, in outdoor and indoor environments, for pointing, guiding and motion control purposes.

5. CONCLUSIONS

A system for the localisation of a mobile robot has been presented. The main characteristics of the system are the following:

- It is a simple and low cost system. Common PIC microcontrollers are employed. The calculation of the position is carried out in an on-board computer, which also performs navigation tasks.
- Beacons are univocally identified, which allows the absolute position of the robot to be obtained without requiring the maintenance of a position estimate. Furthermore, location in environments with multiple rooms is possible.
- The system is not affected by highly reflective surfaces in the environment (while these surfaces strongly affect other reflective localisation systems.)
- The system performs up to 15 localisation measures every second, which is appropriate for locating vehicles travelling at moderate speeds (without requiring the use of the Kalman filter).
- Precision is about 1 cm and 0.1 degree, which is suitable for most mobile robot applications.
- The system can be easily mounted on a vertical mast, due to its small weight and size, so that people and objects in the environment do not constrain the visibility of the beacons.
- Beacon redundancy allows the computation of more confident values, and allows localisation when several beacons are occluded by walls, columns, etc.
- Localisation can be performed in large operation areas with a small number of beacons. For example, localisation can be efficiently performed in an area of $450m^2$ with only three beacons.

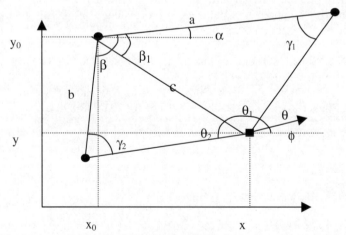

Figure 3. Robot localisation process using three beacons.

Acknowledgment

This work is supported by the Comisión de Ciencia y Tecnología project 1FD97-1580, and the Junta de Castilla y León "Programa de apoyo a proyectos de investigación 2002.

References

1. Betke, M. and Gurvits, L., 1994, "Mobile Robot Localization Using Landmarks." 1994 International Conference on Intelligent Robots and Systems (IROS'94). Munich, Germany, Sept. 12-16, pp.135-142.
2. Borenstein, J. and Feng, L., 1994, "UMBmark — A Method for Measuring, Comparing, and Correcting Dead-reckoning Errors in Mobile Robots." Technical Report, The University of Michigan UM-MEAM-94-22, Dec. 1994.
3. Brown, R. G. and Hwang, P. Y. C., 1992. "Introduction to Random Signals and Applied Kalman Filtering". Second Edition, John Wiley & Sons, Inc.
4. Cohen, C. and Koss, F., 1992, "A Comprehensive Study of Three Object Triangulation." Proceedings of the 1993 SPIE Conference on Mobile Robots, Boston, MA, Nov. 18-20, pp. 95-106.
5. Cox, I.J., 1991, "Blanche - An Experiment in Guidance and Navigation of an Autonomous. Mobile Robot." IEEE Transactions Robotics and Automation, 7(3), pp. 193-204.
6. Crowley, J.L. and Reignier, P., 1992, "Asynchronous Control of Rotation and Translation for a Robot Vehicle." Robotics and Autonomous Systems, Vol. 10, pp. 243-251.
7. MacLeode.N., Chiarella M., 1993, "Navigation and Control Breakthrough for Automated Mobility." *Proceedings, SPIE Vol. 2058, Mobile Robots VIII*, pp. 57-68.
8. McGillem, C. and Rappaport, T., 1988, "Infra-red Location System for Navigation of Autonomous Vehicles." Proceedings of IEEE International Conference on Robotics and Automation, Philadelphia, PA, April 24-29, pp. 1236-1238.
9. SIMAN - Siman Sensors & Intelligent Machines Ltd., MTI-Misgav, D.N. Misgav 20179, Israel, +972-4-906888.
10. TRC - Transitions Research Corp., "Beacon Navigation System." Product Literature, Shelter Rock Lane, Danbury, CT 06810, 203-798-8988.

Space and Time Sensor Fusion and Multi-Sensor Integration for Indoor Mobile Robot Navigation

Tae-Seok Jin[1], Jae-Pyung Ko[2] and Jang-Myung Lee[3]

[1,2]Dept. of El. Eng., Pusan National University, Jang-Jeon Dong, Keum-Jeung Ku, Pusan, Korea
[3]Dept. of El. Eng., Pusan National University, Jang-Jeon Dong, Keum-Jeung Ku, Pusan, Korea

Abstract
This paper proposes a sensor-fusion technique where the data sets for the previous moments are properly transformed and fused into the current data sets to enable accurate measurement, such as distance to an obstacle and location of the service robot itself. In the conventional fusion schemes, the measurement is dependent on the current data sets. As the results, more of sensors are required to measure a certain physical parameter or to improve the accuracy of the measurement. However, in this approach, instead of adding more sensors to the system, the temporal sequence of the data sets are stored and utilized for the improvement of the measurements. Its theoretical basis is illustrated by examples and the effectiveness is proved through the simulations. Finally, the new space and time sensor fusion (STSF) scheme is applied to the control of a mobile robot in an unstructured as well as structured environment.

Keywords: Space and time sensor fusion, multi-sensor integration, mobile robot.

1. INTRODUCTION

So far, a lot of research has been done on the spatial fusion technique. That is, multiple sensor data are utilized either for the purpose of providing complementary or redundant data to measure physical parameters. In other words, all of the current data from the sensors are integrated and fused to obtain a correct set of data.

In recent years interest has been growing in the synergistic use of multiple sensors to increase the capabilities of intelligent machines and systems. For these systems to use multiple sensors effectively, some method is needed for integrating the information provided by these sensors into the operations of the system.

In this new approach, the data obtained by the sensors are utilized until they do not have any efficiency for the measurement decision. The data set can be either (i) redundant to improve the accuracy or (ii) complementary for the measurement. For the later case, this space and time sensor fusion is essential for the measurement.

The space and time fusion is inevitable for the complementary case. Therefore, the effectiveness is very clear and the utilization method will be determined by the sensory data structure. However, for the redundant case, it is required to define how to fuse the previous data sets to the current data set. In this paper, we are going to utilize the minimum square solution for the fusion scheme without considering the error variance in the measurement, for simplicity.

2. SPACE AND TIME SENSOR FUSION

Multi-sensor fusion refers to any stage in the integration process where there is an actual combination (or fusion) of different sources of sensory information into one representational format.

2.1. A General Pattern of Sensor Fusion

Figure 1 represents a general pattern of multi-sensor integration and fusion in a system. In this figure, n sensors are integrated to provide information to the system.

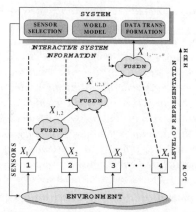

Fig. 1. General pattern of multi-sensor integration and fusion in system.

The output x_1 and x_2 from the first two sensors are fused at the lower left-hand node into a new representation $x_{1,2}$. The output x_3 from the third sensor could then be fused with $x_{1,2}$ at the next node, resulting in the representation $x_{1,2,3}$, which might then be fused at nodes higher in the structure.

2.2. Sensor Fusion Transformation

Let us define the k-th moment data set provided by the i-th sensor as $\mathbf{z}_i(\mathbf{k})$, and the k-th measurement vector as $\mathbf{x}(\mathbf{k})$. Then the conventional sensor fusion technique provides the measurement as

$$\hat{x}(k) = \sum_{i=1}^{n} W_i x_i(k) \qquad (1)$$

where $x_i(k) = H_i z_i(k) \in R^m$, H_i represents the transformation from the sensory data to the measurement vector, and $W_i \in R^{m \times m}$ represents the weighting value for *i*-th sensor.

Note that in the measurement of $z_i(k)$, the low-level fusion might be applied with multiple sets of data with known statistics [2]. The determination of H_i is purely dependent on the sensory information and the decision of W_i can be done through the sensor fusion process. Later, this measured data are provided to the linear model of the control/measurement system as current state vector, $x(k)$. In this approach, we propose a multi-sensor data fusion using sensory data, $Tz_i(j)$, as

$$\hat{x}(k) = \sum_{i=1}^{n} W_i \{ \sum_{j=1}^{k} P_j Tz_i(j) \} \qquad (2)$$

where $\sum_{j=1}^{k} P_j = 1$.

Note that when each of sensor information can provide the measurement vector, that is, the redundant case, $Tz_i(j)$ can be expanded as

$$Tz_i(j) = T_j + H_i z_i(j) \qquad (3)$$

where T_j represents the homogeneous transformation from the location of the *j*-th to the *k*-th measurements.

However, when the multi-sensors are utilized in the complementary mode, the transformation relationship cannot be defined uniquely; instead it will be defined depending on the data constructing algorithm from the measurements. For example, a single image frame captured by a camera on a mobile robot cannot provide the distance to an object until the corresponding object image is provided again from a different location. This algorithm will be described in detail in the section 3.1.

Figure 2 illustrates the concept of this multi-sensor temporal data fusion. Estimation of each parameter may provide the measurement vector at each sampling moment. The verification of significance and adjustment of weight steps are pre-processing stages for the sensor fusion. After these steps, the previous data set will be fused with the current data set, which provides a reliable and accurate data set as the result of multi-sensor temporal fusion. Significance implies how much the previous data set is related to the current data. An arbitrary value of significance may cause the problem to be complex. Therefore, here we may consider whether it corresponds to the same data or not, that is 1 or 0. When the significance is 0, the weight can be adjusted simply to be 0. However, when the significance equals 1, the adjustment of weight should be properly performed to provide reliable and accurate data. In the following sub-section, we introduce a simple methodology for the weight adjustment.

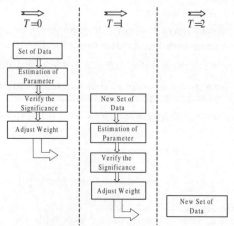

Fig. 2. Concept of Space and Time Sensor Fusion.

2.3. Auto-correlation for Estimation Techniques

Each previous data set is transformed to the k-th (current) sampling location, and represented by the measurement vector, $T_{z_i}(j)$. Now how can we fuse the k data sets into a reliable and accurate data set? In Eq. (2), W_i can be determined by the geometrical relationship among sensors, in other words, by the spatial sensor fusion.

While the estimating sensor is tracking a feature, it generates a stream of measurements. When there is relative motion between the feature and the sensor, the processes then cease to be stationary. As an illustration of gathering the model information from the sensor, we shall only consider the stationary case, in order words, there is no motion between the sensor and the feature being tracked. Our interest in the proceeding analysis lies only in determining whether the process noise is white or not.

Random processes are defined in terms of their ensemble averages and these can be estimated. Our model shall be in terms of such averages. In practice, we require to estimate these averages from finite sequences. We consider a process y_k as realized (estimated) by the finite sequence $y(k)$, for $0 \leq k \leq N-1$. That $y(k)$ is an estimate of the random process y_k is made plausible by a consideration of ergodic processes. From $y(k)$, we can, therefore, estimate the averages for the process, the mean is estimated by $\hat{\mu}_x = \frac{1}{N}\sum_{k=0}^{N-1} x(k)$ and the variance by $\hat{\sigma}_x^2 = \frac{1}{N}\sum_{k=0}^{N-1}(x(k)-\mu_y)^2$,

A biased estimate of the autocorrelation is given by
$$\tilde{\phi}_{xx}(m) = \frac{1}{N}\sum_{k=0}^{N-|m|-1} x(k)\cdot x(k+m) \qquad (4)$$

with a variance of $\tilde{\sigma}_{\tilde{x}} = \frac{1}{\sqrt{N}} \tilde{\sigma}_x^2$, (5)

where $|m| < N$. Similarly, an unbiased autocorrelation is estimated by

$$\hat{\phi}_{xx}(m) = \frac{1}{N-|m|} \sum_{k=0}^{N-|m|-1} x(k) \cdot x(k+m) \qquad (6)$$

and the variance of the unbiased autocorrelation is given by

$$\hat{\sigma}_{\hat{\phi}} = \frac{\sqrt{N}}{N-|m|} \hat{\sigma}_x^2 \qquad (7)$$

for the biased autocorrelation and similarly for the unbiased one. Therefore, determination of P_j is the final step for the temporal sensor fusion. Note that this expands the dimension of sensor fusion from one to two.

As one of solid candidate, we propose here to use the auto-correlation as an index for the weight adjustment and have the form,

$$\Psi_j = \sum_{k=-\infty}^{+\infty} x_i(k) x_i(j-k). \qquad (8)$$

Depending on the correlation, P_j will be determined as

$$P_j = \frac{\Psi_j}{\sum_{j=1}^{k} \Psi_j} \qquad (9)$$

3. APPLICATIONS TO MOBILE ROBOTS

3.1 Complementary Usage for 3D Vision

If the image for an object is well matched to one model in the database, the position of the object can be obtained directly. In a well-structured environment, it may be the case. However, when the mobile robot is navigating in an unstructured environment, it needs to recognize the position/orientation of an object located in the middle of its path, which is not known to the robot *a priori*.

As a typical geometrical model for a camera, a pinhole model is widely used in vision application fields, as shown in Figure 3. At the k-th sampling moment, a scene point O(X,Y,Z) is captured by a camera on the mobile robot. The vectors from the scene point to the k-th and *(k-1)*th camera perspective center are represented by V_k and V_{k-1}, respectively.

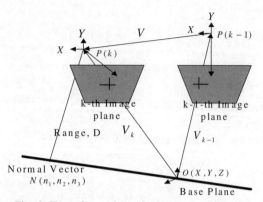

Fig. 3. Transformation of camera coordinates.

The motion of the mobile robot from *(k-1)*th moment to *k*-th moment is represented by **V**. Now we can write the vector relationship as

$$V_{k-1} = V_k - V . \tag{10}$$

This can be represented as a matrix form,

$$\alpha \begin{bmatrix} x_{k-1} \\ y_{k-1} \\ -f \end{bmatrix} = \beta \begin{bmatrix} r_{11} & r_{12} & r_{13} \\ r_{21} & r_{22} & r_{23} \\ r_{31} & r_{32} & r_{33} \end{bmatrix} \begin{bmatrix} x_k \\ y_k \\ -f \end{bmatrix} - \begin{bmatrix} v_1 \\ v_2 \\ v_3 \end{bmatrix} \tag{11}$$

where (x_k, y_k, -f) and (x_{k-1}, y_{k-1}, -f) represent the projection of the scene point onto the camera image planes; V(v_1, v_2, v_3) represents the translational motion of the mobile robot; r_{ij} is an element of the rotation matrix, R represents the relative rotation between the two camera frames; α and β are constants.

Now consider the reference base plane passing through the scene point P with a direction vector N(n_1, n_2, n_3); then the range value, *D*, can be represented as

$$D = V_k \cdot N . \tag{12}$$

This can be represented again as

$$D = \beta(n_1 x_k + n_2 y_k - n_3 f) . \tag{13}$$

Now, Eq. (11) is reformulated as

$$\alpha \begin{bmatrix} x_{k-1} \\ y_{k-1} \\ -f \end{bmatrix} = \beta \begin{bmatrix} r_{11} & r_{12} & r_{13} \\ r_{21} & r_{22} & r_{23} \\ r_{31} & r_{32} & r_{33} \end{bmatrix} \begin{bmatrix} x_k \\ y_k \\ -f \end{bmatrix} - \frac{\beta}{D} \begin{bmatrix} v_1 \\ v_2 \\ v_3 \end{bmatrix} \begin{bmatrix} n_1 & n_2 & n_3 \end{bmatrix} \begin{bmatrix} x_k \\ y_k \\ -f \end{bmatrix} \quad (14)$$

$$(\alpha/\beta) \begin{bmatrix} x_{k-1} \\ y_{k-1} \\ -f \end{bmatrix} = \begin{bmatrix} a_{11} & a_{12} & a_{13} \\ a_{21} & a_{22} & a_{23} \\ a_{31} & a_{32} & a_{33} \end{bmatrix} \begin{bmatrix} x_k \\ y_k \\ -f \end{bmatrix} \quad (15)$$

where $a_{ij} = r_{ij} - (v_i \cdot n_j / D)$.

Expanding the matrices and dividing rows one and two by row three gives

$$D(R_3 x_{k-1} + R_1 f) = C_3 x_{k-1} + C_1 f \quad (16)$$
$$D(R_3 y_{k-1} + R_2 f) = C_3 y_{k-1} + C_2 f \quad (17)$$

where $R_i = r_{i1} x_k + r_{i2} y_k - r_{i3} f$ and $C_i = v_i (n_1 x_k + n_2 y_k - n_3 f)$.
In matrix form, these equations can be expressed as

$$AD = B \quad (18)$$

where $A^T = [a \ b]$, $B^T = [c \ d]$, $a = R_3 x_{k-1} + R_1 f$, $b = R_3 y_{k-1} + R_2 f$, $c = C_3 x_{k-1} + C_1 f$, and $d = C_3 y_{k-1} + C_2 f$.

Use of the pseudo-inverse matrix enables computation of the range value D, which is associated with image point (x_k, y_k), and is written as,

$$D = (A^T A)^{-1} A^T B \text{ or} \quad (19)$$
$$D = \frac{(ac + bd)}{a^2 + b^2}. \quad (20)$$

So far, we have shown that using the consecutive two image frames, the distance information of the scene point can be obtained as using the stereo images at a certain moment.

3.2   Space and Time Fusion Filter

The space and time fusion consists of combining information acquired at different instants and then deciding the data. It implies that the system must be able to predict objects state at each instant (see Figure 4).

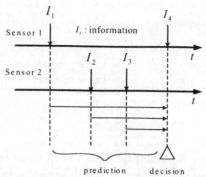

Fig. 4. An example of Space and Time fusion.

Given two estimators $\hat{\Theta}_1$ and $\hat{\Theta}_2$ of Θ, the task is to fuse then together to form one single "optimal" estimate $\hat{\theta}$. Here, estimators are stochastic variables and are denoted by capital letters.

Assume that $\hat{\Theta}_1 - \Theta$ and $\hat{\Theta}_2 - \Theta$ are independent Gaussian distributed with zero mean and covariances P_1 and P_2, respectively. Now let $X = \Theta - \hat{\Theta}_1$ and $Y = \hat{\Theta}_2 - \hat{\Theta}_1$. Then $\Sigma_{xy} = P_1$ and $\Sigma_{yy} = P_1 + P_2$. Hence

$$\hat{x} = P_1(P_1 + P_2)^{-1}(\hat{x}_2 - \hat{x}_1) \tag{21}$$

$$\hat{\theta} = \hat{\theta}_1 + P_1(P_1 + P_2)^{-1}(\hat{\theta}_2 - \hat{\theta}_1) \text{ or} \tag{22}$$

$$\hat{\theta} = [P_1^{-1} + P_2^{-1}]^{-1}[P_1^{-1}\hat{\theta}_1 + P_2^{-1}\hat{\theta}_2], \tag{23}$$

with covariance

$$P_1 - P_1(P_1 + P_2)^{-1}P_1 = [P_1^{-1} + P_2^{-1}]^{-1} . \tag{24}$$

The fusion formula just means that estimates should be weighted together, with weights inversely proportional to their qualities/variances. It is easy to modify the fusion filter to handle correlated estimators.

4. ROBOT TYPE IN EXPERIMENTS SETUP

The mobile robot used in the experiments is an *IRL-2001* developed in the *IRL*, PNUwhich is designed for an intelligent service robot.

This robot is shown in Figure 5 along with some of its sensory components. Its main controller is made on a system clock 600 MHz, Pentium III Processor. The 16-ultrasonic sensors and a robust odometry system are installed on the mobile robot. Ultrasonic sensors and infrared sensors in eight sides(25°) sense obstacles of close range, and the main controller processes this information.

Fig. 5. *IRL*-2001 robot.

For visual information, a CCD camera is mounted on top, in order to sense obstacles or landmarks of the side and rear of the mobile robot. DC servomotors are used for steering and driving.

5. EXPERIMENTAL RESULTS
5.1 Robot Localization useing Landmark Pattern Recognition
The *IRL*-2001 robot is commanded to follow the environment, as shown from (a) to (f) of Figure 7. We performed the experiment for two cases.

To begin with, the 2-D landmark used by *IRL-2001* is shown in Figure 6. The primary pattern is a 10cm black square block on a white background and a 5cm square block. The major reasons for choosing the square blocks are:

- The projection of a square block in the image plane can always be approximated by an ellipse, which is easy to recognize using the elliptical Hough transformation technique.
- A circular pattern does not get mixed up easily with the majority of patterns frequently seen in an indoor environment, such as circular disk, polygonal objects, etc.
- A square pattern is more robust to noise and occlusion than circular, polygonal patterns during template matching process, even though all these patterns can be detected by using Hough transformation technique.

Fig. 6. The landmark pattern and size used by *IRL-2001*.

The image corners are then automatically extracted by the camera parameters, and displayed on Figure 6 and the squares around the corner points show the limits of the corner finder window. The corners are extracted to an accuracy of about 0.1 pixel.

The extrinsic parameters and relative positions of the landmark with respect to the camera, are then shown as a 3D plot in Figure 8. In Figure 9, every camera position and orientation are represented by pyramids, therefore we can see the location and the orientation of a mobile robot in the indoor environment.

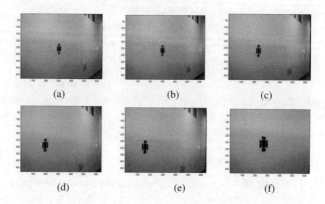

Fig. 7. A landmark locations detected by camera.

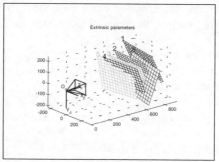

Fig. 8. Relative positions of the landmark w.r.t. the camera.

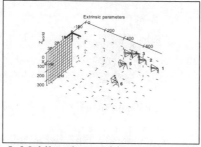

Fig. 9. Mobile robot position and orientation.

To measure the relative distance of the landmark from the mobile robot, we first measure the distance of the image from the fixed position in the *IRL* laboratory corridor. The predefined values of the landmark defined in this section are given as follows: the origin of coordinates is equal to the origin of mobile robot, a Y-axis is fit to the front of mobile robot and an X-axis is perpendicular with Y-axis.

Table 1 lists the data measured in the corridor. The left direction marks negative. From table 1, we find the maximum and the minimum error on distance is 0.32 m and 0.13m, respectively.

It shows that the distance error becomes less and less by frames, which composes the environment map. In this way, we can use it to measure the relative distance of the mobile robot.

Table 1. The result of relative distance (Dim.:m).

Frame Number	World Coordinate Distance	Image Coordinate Distance	Error
1	7.81	8.13	0.32
2	7.02	7.30	0.28
3	6.28	6.53	0.25
4	5.06	4.89	0.17
5	5.52	5.39	0.13
6	6.32	6.46	0.14

5.2   Mobile Robot Navigation

Conventional fusion and STSF (a space and time sensor fusion) have first been tested with a simulation, to show the usefulness of STSF in two environments respectively. Starting at 0.3m, 5m, 0 degree, a virtual robot was driven around a virtual square corridor first. The walls in the artificial environment are denoted by the real map, the *IRL* corridor of the PNU.

In each round, the robot stops a total of 12 times to rescan the environment. The size of the given map is 12m x 8m, the total distance traveled is 12 + 8=20 meters, and the total number of scanning points is 38. The comparison of the simulation position and direction at all stops is shown in Figure 10 and Figure 11.

Figure 10 shows the plots of the pointing vector based upon only current readings using conventional sensor fusion, i.e. spatial fusion. This robot was made to move randomly within the confines of the above setup and at the region *Cc*. There is little difference between conventional fusion and the new STSF, but in the region *Ac*, the robot moves not keeping the distance between robot and wall constant, and has some difficulty in trapping the local minimum problems.

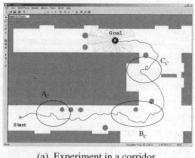

(a). Experiment in a corridor.

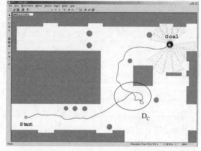

(b). Experiment in a corridor with wide space.

Fig. 10. Simulation for pointing vector based upon current readings.

Figure 11 shows how the multi-sensor STSF scheme is applied for the measurement. The results are compared to show the superiority of the proposed scheme. The robot was allowed to move keeping the distance between robot and obstacles constant at the region A_T and B_T.

The region B_T shows the improvement in steering at corners. The simulation experiments show that a mobile robot, utilizing our scheme, can avoid obstacles and reach a given goal position in the workspace of geometrical complexity. Experiments results, using the new STSF, show the robot can avoid obstacles (boxes and trash can) and follow the wall. Figure 10 through Figure 11 demonstrate one of many successful experiments. The algorithm is very effective in escaping local minima encountered in laboratory environments.

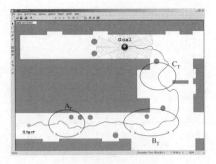

(a). Experiment in a corridor.

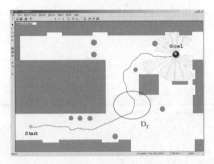

(b). Experiment in a corridor with wide space.

Fig. 11. Simulation using a STSF scheme.

The mobile robot navigates along a corridor with 3m width and with some obstacles as shown in Figure 10 and Figure 11. It demonstrates that the mobile robot avoids the obstacles intelligently and follows the corridor to the goal.

Also notice that especially at the region A_T, the errors of the robot position converge to zero for the same reason, referring to the simulation result and experimental result in Figure. 10-(a) and 10-(b) respectively, Figure 11-(a) and 11-(b) represent the reference of the robot's direction produced by the proposed STSF. Finally, the robot is tested to follow the whole trajectory from start position to the final position as shown in Figure 10 and Figure 11.

The simulation and experimental results of the robot status under such control strategy are given in Figure 12 and 13.

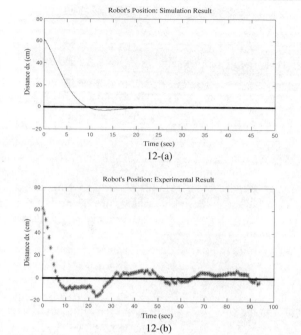

Fig. 12. Robot's position simulation (a) and experiment (b) results.

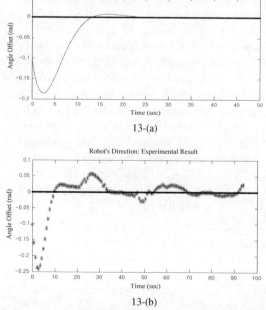

Fig. 13. Robot's direction simulation (a) and experiment (b) results.

6. CONCLUSIONS

In this paper, a new sensor fusion concept, STSF(space and time sensor fusion), was introduced. The effectiveness of STSF was demonstrated through examples, simulations and experiments. To generate complete navigation trajectories without *a priori* information on the environment, not only the data from the sensors located at different places but also the previous sensor data are inevitably utilized. Although we have tried using the sonar system for map building and navigation in an indoor environment, the result from the above experiments clearly shows that by utilizing both systems and applying active sensing to a different situation, a high level of competent collision avoidance behavior by STSF can be achieved.

This scheme may require more memory space and computing power in the navigation system, which becomes achievable with the rapid price drop of ICs. Sonar system and visual systems are utilized for collision avoidance, based upon STSF, such that a mobile robot was successfully navigated in an unstructured as well as in a structured environment. Based on these results, further experiments will aim at applying the proposed tracking technique to the multi-sensor fusion scheme for the control of a mobile robot in an unstructured environment. The STSF will be applied on a landmark-based real-time robot guidance, including visual servo control of the *IRL-2001* mobile robot for autonomous navigation.

7. REFERENCES

[1] R. C. Ruo and K. L. Su, "A Review of High-level Multisensor Fusion: Approaches and applications," *Proc. Of IEEE Int'l. Conf. On Multisensor Fusion and Integration for Intelligent Systems*, pp. 25-31, Taipei, Taiwan, 1999.

[2] J. M. Lee, B. H. Kim, M. H. Lee, M. C. Lee, J. W. Choi, and S. H. Han, "Fine Active Calibration of Camera Position/Orientation through Pattern Recognition," *Proc. of IEEE Int'l. Symp. on Industrial Electronics,* pp. 100-105, Slovenia, 1999.

[3] M.e Kam, X. Zhu, and P. Kalata, "Sensor Fusion for Mobile Robot Navigation," *Proc. of the IEEE*, Vol. 85, No. 1, pp. 108-119, Jan. 1997.

[4] P. Weckesser and R. Dillman, "Navigating a Mobile Service-Robot in a Natural Environment Using Sensor-Fusion Techniques," *Proc. of IROS*, pp. 1423-1428, 1997.

[5] J. Llinas and E. Waltz, *Multisensor Data Fusion*. Boston, MA: Artech House, 1990.

[6] D. Hall, *Mathematical Techniques in Multisensor Data Fusion*. Boston, MA: Artech House, 1992.

[7] L. A. Klein, *Sensor and Data Fusion Concepts and Applications*, SPIE Opt, Engineering Press, Tutorial Texts, vol. 14, 1993.

[8] H. R. Beom and H. S. Cho, "A sensor-Based Navigation for a Mobile Robot Using Fuzzy Logic and Reinforcement Learning," *IEEE Trans. on system, man, and cybernetics*, Vol. 25, No. 3, pp. 464-477, March 1995.

[9] A. Ohya, A. Kosaka and A. Kak, "Vision-Based Navigation by a Mobile Robot with Obstacle Avoidance Using Single-Camera Vision and Ultrasonic

Sensing," *IEEE Transactions on Robotics and Automation,* Vol. 14, No. 6, pp. 969-978, December 1998.

[10] S. Atiya and G. D. Hager, "Real-time vision-based robot local-ization," *IEEE Trans. Rob. Autom.* 9(6), pp. 785–800, 1993.

[11] R. M. Haralick and L. G. Shapiro, *Computer and Robot Vision,* Addison-Wesley, Reading, MA, 1992, Vol. 1, pp. 582–584.

[12] J. Illingworth and J. Kittler, "A survey of the Hough transform," *Comput. Vision Graphics Image Processing* 44, pp. 87–116, 1988.

Multi-purpose Autonomous Robust Carrier for Hospitals (MARCH): Design and Implementation

P. Sooraksa[a], B. L. Luk[b], S. K. Tso[b], and G. Chen[c],
[a]Faculty of Engineering, King Mongkut's Institute of Technology,
Ladkrabang, Bangkok, Thailand.
[b]Centre for Intelligent Design, Automation and Manufacturing
City University of Hong Kong, P. R. China.
[c]Department of Electronic Engineering
City University of Hong Kong, P. R. China.

Abstract
This paper describes the basic principles, design, and implementation of the second phase of a multi-purpose autonomous robust carrier for hospitals (MARCH) — March II. It is an autonomous robot, working as a mechatronic assistant for medical devices. The aim of the research is to develop more reliable, maintainable and intelligent functions for MARCH II. The robot is expected to be able to perform the following tasks by the end of the current phase of research: line tracking, wall following, collision avoiding, remote operating, communicating (with the central unit), navigating, and even entertaining (the elderly and children, with games, music and video-on-demand (VOD)). Hierarchical control with subsumption architecture and various intelligent control schemes is employed to enable progressive system development and achievement of the MARCH II objectives. Experiments of mechatronic and visual subsystems have been carried out to validate the system.

Keywords: *Mobile robot, intelligent control, mechatronic system, medical equipment.*

1. INTRODUCTION

One recent trend in robotics is to equip service robots with intelligent functions such as image and sound recognition, fault detection, self-diagnosis, and comprehensive capabilities of interacting with human. Service robots will assist human in various ways, such as keeping chores at home, cleaning floors in streets, carrying luggage at airports, cooking at restaurants [1]. In particular, service robots are urgently needed for many kinds of routines and tedious work in households, clinics, and hospitals [2, 3]. For example, Care-O-bot is a mobile service robot developed by Fraunhofer IPA, Germany, which is designed as a futuristic home care system [10]. This robot assists nursing activities for the elderly and dependent people, to provide food and drinks, supplies and disposals, etc.

Another example of an hospital robot is HelpMate, which is a trackless robotic courier system developed by the Pyxis Corporation [4]. The robot is designed to perform material transport tasks under the hospital environment to work twenty-four hours a day, 365 days a year. HelpMate transports pharmaceuticals, lab specimens, equipment, supplies, meals, medical records, and radiology films, back and forth in between the supporting departments and nursing floors.

In this paper, the second phase of a multi-purpose autonomous robust carrier for hospitals (MARCH) — MARCH II, functionally similar to HelpMate, is described. Differing from the HelpMate, MARCH II has more advanced functions such as a visual system for room and landmark recognition, entertainment on demand such as games and music for elderly and children, and a lift interface mode (LIC) enabling the robot to communicate with the lift interface controller via RF transmitter/receiver.

The paper presents the basic principles, design, and implementation of this robot system. MARCH II is an autonomous robot, working as a mechatronic assistant in hospitals. The aim of the research is to develop more reliable, maintainable and intelligent functions for MARCH II. The robot is expected to be able to perform the following tasks by the end of the current phase of research: line tracking, wall following, collision avoiding, remote operating, communicating (with the central unit), navigating, and even entertaining (the elderly and children, with games, music and VOD).

The paper is organized as follows: Section 2 briefly reviews the initial phase of the development of a prototype called MARCH I, and discusses what we have learned in order to improve the system in the second phase for MARCH II. The basic design, planning, hardware organization, and control architecture of MARCH II are presented in Section 3. Section 4 reports the current state of the development and future work of the project. Section 5 provides the concluding remarks.

2. MARCH I: THE LEARNING CURVE

MARCH I was the original machine developed by the Centre for Intelligent Design, Automation and Manufacturing (CIDAM), City University of Hong Kong [9], as an hospital robot. For more than 2 years, the project has been continuously developed, aiming to provide a multi-purpose autonomous robot for delivery of materials inside the hospital. This would benefit the staff in the sense that they do not need to perform the time-consuming delivery tasks thereby having more time to increase their activities and productivities for better services to the patients. The robot weighs 80 kg and can carry material up to 300kg. It has a size of 0.8m (L) x 0.8m (W) x 1.5m (H) [9]. Figure 1 is a drawing of the mechanical frame of the robot and a photo of its appearance is shown in Figure 2.

The robot can navigate along a corridor by using a tricycle mechanism, and can carry a trolley loaded with materials, such as drugs, linens, food or medical equipment, in its rear. It can travel at a normal speed of 0.5m/s from one place to another by combining the odometry, landmark recognition, line following and wall following navigation control schemes. A vision system (CCD camera) is also implemented on the robot, thus the operator in the control center can monitor and control the robot remotely, through a 2.4 GHz wireless LAN [9]. However, MARCH I was in need of a improved speed response and more development on various communication and control aspects. There were electronic parts and components that needed to be upgraded, so as to keep pace with current engineering and technological advances. Hence, MARCH II was designed to enhance the existing capabilities, and to fully utilize an intelligent control system, as discussed further in the next section.

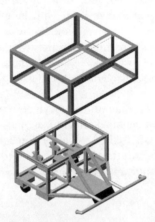

Fig. 1. The mechanical and structural frame of MARCH I: The top layer holds circuits and electronic control components; the bottom layer is used for power supply, photo-sensors, and actuators.

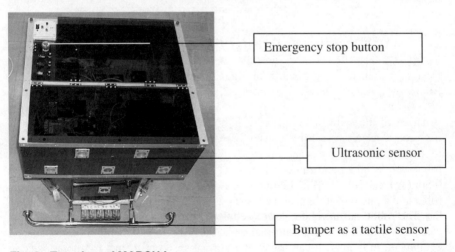

Fig. 2. Top view of MARCH I.

3. MARCH II: DESIGN UPGRADING

In redesigning the existing MARCH I, current research for MARCH II was aimed at a robust layered control paradigm, as a fundamental principle. Only the mechanical parts and components of MARCH I are retained, including the frame and compartment. In the new version, the control architecture was redesigned, electronic circuits re-built and the visual system updated.

The overall system consists of four major levels categorized by the levels of intelligent competency: an executive level with human-in-the-loop, a high level implemented by computer control, a middle level executed by 87C196CA controllers via a CAN-bus, and a low level managed by 89C51 modules. Each level runs independently, so that a hierarchical complex system does not have any effects on the speed of independent motions and behavior of each functional module.

Figure 3 shows a block diagram of the redesigned system. Note that the middle and the top levels of the block diagram are the same as the configuration in the existing structure of MARCH I — only the lower levels are redesigned. Additional modules in the lowest level may be added, as desired, which will be facilitated by the CAN technology [11].

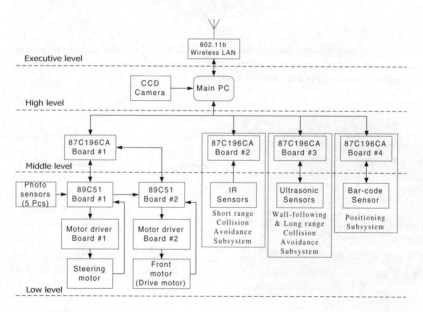

Fig. 3. Block diagram of the overall system hardware architecture of MARCH II.

The classical management strategy will also be adopted, implemented, and executed. This scheme consists of planning, organizing, communicating, commanding and controlling functions, categorized in a layered fashion. In planning and organizing for routes and job assignments, the main PC and the manual remote access can perform all the tasks. In the communication module, such as uploading, downloading, and monitoring, the tasks will be initiated via two-way communication between high and low levels. Command and control are implemented by using subsumption algorithms, in which the higher control levels (i.e. the more intelligent levels) can temporarily suppress the signals commanded from the lower levels [6].

The following subsections discuss the design and implementation principles about MARCH II.

3.1 Low level

Regarding the control function at the low level, a fuzzy control scheme with simple two-straight-line-type membership functions is used for both line tracking and wall following, which can significantly reduce error dynamics in the drive and steering subsystems. Five photo-electric sensors are installed, providing reference angles for tracking of the drive and steering subsystems. Given position errors from the sensors, the outputs of the subsystems, such as position and speed of MARCH II, can be regulated by a simple fuzzy PI or fuzzy PI+D controller [7,12]. The tracking tasks can be temporarily interrupted by the collision avoidance subsystem, when reflexive signals are received from ultrasonic sensors (for long range) or IR sensors (for short range) or from limit switches at the bumper, which provide triple layers of safety warranty during the operation. The collision avoidance subsystem is attainable by using a rule-base controller. Bar code readers are used for identifying the robot's location (such as a patient's bed) inside wards, whereas the vision system is used for room identification within the hospital. The bar code readers can also be used for reading off the patient's medical data or test tube identifiers.

3.2 Middle level

At the middle level, two hardware configurations have been used: one is a stand-alone CAN controller attached to the external address and data bus of the main processor (the PC); the other is a microcontroller integrated with a CAN controller, which provides the CAN controller with access over the lower-level microcontroller's internal address. Since poorly-designed decisions for data traffic may cause loss or bursts of messages or information in the CAN network, as more control modules are added to this level, one must try to minimize such occurrences. Given inputs as a data flow and outputs as an optimized queue, the Just-In-Time (JIT) principle along with an on-line optimization method may be used in design.

3.3 High level

High-level control is addressed to perform more intelligent tasks by the main processor (the PC). Given the control parameters such as control gains for lower-level subsystems, off-line optimization for tuning the parameters can be carried out by using say genetic algorithms [13] or other search methods. For the visual system, given image objects as inputs, image recognition may be achieved by using a fuzzy recognition algorithm. The "dead reckoning" technique can also be incorporated with this visual subsystem in the navigating mode of the system.

3.4 Executive level

In case of subsystem failures, an alarm is flashed on the screen of the central computer, indicating the location where the failure has occurred. This can be monitored to inform the human operator via wireless LAN at the executive level, so as to enable on-line supervising, control, and reprogramming. The incident can be monitored remotely via the CCD camera. This level enables the functions of remote access, data acquisition, and interaction among the vehicle, the patient, and the operator.

Real-time response is a crucial issue to be considered. Since data may be contaminated with noise during the uploading process, an extended Kalman filtering algorithm may be used to resolve the problem.

4. TEST RESULTS OF MECHATRONIC AND VISUAL SUBSYSTEMS

In this section, experiments on both mechatronic and visual subsystems in the operating environment are described. Due to space limitation, only two test results are reported.

4.1 Mechatronic subsystem

Because the ability to follow a referent point is important for movement accuracy of the robot, selected testing was carried out for a low level tracking referent command in the automatic mode. In this test, the robot was assigned to rotate 60 degree to the left. Bang-bang control implemented in the old MARCH I yields significant overshoots, steady-state errors, and chattering effects. Yet, for MARCH II, since the effective Proportional-Integral-Derivative (PID) and fuzzy PI control [7] schemes are implemented in the platform via software burned into a 8051 microcontroller for DC motor drives, it performs much better as described below.

Performance comparison between the two approaches is summarized here. In this experiment, control parameters are: $Kp = Ki = Kd = 1$ for the PID scheme, and $Kp = 5$, $Ki = Kui = 1$, and $L = 10$ for the fuzzy PI algorithm. Notice that the control gains must be integers due to the limitation of a 8051 microcontroller. One typical test result is shown in Figure 4, where the output signals are measured using an Agilent 54622D oscilloscope. In the line-tracking mode, a set point for a 60-degree turn can be assigned by sensing a strip on the floor, when one of the five photo-switches located at the front of the robot is out of the track-line. Similar set point can be assigned by changing the photo-switches to be ultra-sonic sensors. As can be seen from Figure 4, the result of PID control has an overshoot and steady-state error (left), whereas that of fuzzy PI control is much better (right).

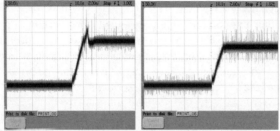

Fig. 4. Performance comparison between conventional PID control (left) and fuzzy PI control (right) for motor drives.

4.2 Visual subsystem

For the visual subsystem, the development at the higher level is proceeding in parallel with the implementation of the mechatronic subsystems at the low level. Color recognition for establishing room-coding assignment at the high-level subsystem has also been implemented. Color recognition for sets of color codes in assigning room numbers, as in an hospital, is being addressed for robot's positioning, which can be used as landmarks. Introducing an RGB color system incorporating hue and brightness, one can assign up to millions of rooms, at least in theory. By using a color filter and the method of Support Vector Machine (SVM) [8, 14], high robustness color recognition that can cope with the disturbance or noises from illumination variation is achieved.

In this test, the robot was assigned to run with a speed equivalent to human walking and to read a color code located on the wall near the door of a patient's room. The light condition was regular under the indoor environment. To demonstrate the effectiveness of the implementation, the aforementioned visual algorithm was written in Visual Basic. Only 4 colors were used for testing, which have 24 combinations. Numbers 0, 1, 2, and 3 are used to represent blue, yellow, white, and red, respectively. The robot could identify the color code accurately, as indicated in Figure 5 by the numbers 3210 (left) and 0123 (right), for colors corresponding to two given serial numbers of colors. It works well even when the robot moves at normal typical human walking speeds.

The recognition time is within 0.3s, at the speed of human walking. More colors can be used for more codes to be assigned as landmarks or room numbers. The same method can be employed to recognize faces of patients and hospital staffs, as demonstrated in [14].

Fig. 5. Test of color code recognition for landmark identification.

5. CONCLUDING COMMENTS

MARCH II is an autonomous robot designed for working in hospitals as a mechatronic assistant for delivery tasks. This paper has addressed the design principles based on subsumption architecture and hierarchical intelligent control configuration for the integrated robot. Current state-of-the-art development and implementation of the system have been described.

References
1. R. D. Schraft and G. Schmierer: Service robots, A K Peters Ltd, Germany, 2000.
2. http://robotics.jpl.nasa.gov/accomplishments/surgery/surgery.html
3. http://physicaltherapy.about.com/cs/medicalrobots/
4. http://www.pyxis.com/products/newhelpmate.asp
5. L. C. Wai, L. C. Keung, L. K. Wai, Y. M. Wai, and W. W. Ling: MARCH Project, Dept. of Mechanical Engineering and Engineering Management, City University of Hong Kong, 2001.
6. R. A. Brooks: A robust layered control system for a mobile robot, IEEE Trans. on Robotics and Automation, vol. 2, no. 1, pp. 14-23, 1986.
7. W. Tang, G. Chen and R. Lu: A modified fuzzy PI controller for a flexible-joint robot arm with uncertainties, Fuzzy Sets and Systems, vol. 118, pp. 109-119, 2000.
8. M. Pontil and A. Verri: Support vector machines for 3-d object recognition, IEEE Transactions on Pattern Analysis and Machine Intelligence, vol. 20, no. 6, pp. 637-646, June 1998.
9. B. L. Luk, S. K. Tso and K. L. Lee: Development of two service robots for hospital applications, Proc. of CIDAM Workshop on Service Robotics and Automation, pp. 18-24, June 2000.
10. http://www.ifr.org/pictureGallery/servRobAppl.htm
11. M. Farsi and M. Barbosa: CANopen Implementation: applications to industrial networks, Research Studies Press Ltd, Baldock, Hertfordshire, England, 2000.
12. D. Misir, H. A. Malki and G. Chen: Design and analysis of a fuzzy proportional-integral-derivative controller, Fuzzy Set and System, vol.79, pp. 297-314, 1996.
13. K. S. Tang, K. F. Man, G. Chen and S. Kwong: A GA-optimized fuzzy PD+I controller for nonlinear systems, IECON'01: The 27th Annual Conf. of the IEEE Industrial Electronics Society, pp. 718-723, 2001.
14. P. Palasuthikul, P. Sooraksa, and A. Lasakul: Real time detection on color image by support vector machine, Proc. of 2002 FIRA Robot World Congress, pp. 427-429, 2002.

Communication with an Underwater ROV Using Ultrasonic Transmission

Eric Law[1], Robin Bradbeer[1], L F Yeung[1], Li Bin[2], and Gu Zhongguo[2]

[1] Department of Electronic Engineering, City University of Hong Kong, Hong Kong.
[2] Institute of Acoustic Engineering, Northwestern Polytechnical University, Xi'an, 710072, China.
Email: tmlaw@ee.cityu.edu.hk

Abstract

Communication with underwater remote operated vehicles (ROV) is usually done by using umbilical cables. These sometimes cause problems with the control of the vehicle, as well as its use in areas where fouling of the cable can take place. This paper describes the prototype of an ultrasonic communications system that can transmit colour, still and video, pictures from such an ROV.

The system uses multicarrier modulation for underwater acoustic communications, and has been successfully tested at a data rate up to 10kbps over 1km. It is designed for operating in multipath fading environments. The transmitter and receiver use digital signal processors which control the modulation and transmission, and synchronization and demodulation, respectively.

The system algorithm generates 48 frequencies for transmitting 48 parallel bits of data in each packet. A long transmitted signal sequence is combined with synchronisation, zero gap and information packets. The long multi-frequency signal packets have been implemented to minimise the effect of multipath fading. To acquire the starting point of the transmitting sequence, the Linear Frequency Modulation (LFM) signal is used for synchronisation. In order to reduce noise, the adaptive threshold packets are used to set up a suitable signal.

Experimental results from sea-trials have shown that the system can cope with multipath fading environments and is characterised by its simplicity and robustness. Various system architectures, such as the training pulse, channel identification, etc. are described.

Keywords: *underwater communications, modem, image transmission.*

1. INTRODUCTION

Over the years, much research has been carried out to obtain a reliable high data rate underwater acoustic communication system. However, the underwater acoustic channel is an unforgiving wireless communication medium. The strong amplitude and phase fluctuations cause multipath fading. Due to limited capacity of bandwidth, maximum data rate on the available bandwidth should be used. Therefore, channel equalization techniques have been used. Previous work at City University of Hong Kong [1 - 4] has shown that it is possible to send data reliably through liquid-filled pipes, where the multipath problem is considerable. However, the open water environment is a different situation.

As mentioned in [4], there are three methods of underwater acoustic communication systems. The first is a 'no diversity' technique; the second is 'only explicit diversity reception' such as time, frequency diversity, etc.; the third is 'at least implicit diversity' processing. This method spectrally spreads the signal over a single transmission band so that bandwidth is much larger than the coherence bandwidth of channel.

The first method of underwater acoustic communication systems uses a no diversity technique so that it can implement the system easily. However, it has low data rate and reliability over a short range because of the multipath fading problem. The second method uses explicit diversity reception. The advantage of this system is a higher data rate and reliability with longer transmission range. But its complexity and power consumption are also increased. The third method uses 'at least implicit' diversity reception. As described in [5], it can provide the highest reliability, speed and power efficiency, but it requires the most complex system to implement.

In the following table, different methods are shown based on current experimental results (Table 1).

Table 1. Existing results of underwater acoustic communication

Developed by	Water depth (m)	Carrier frequency (Hz)	Distance (km)	Modulation	Data rate (bps)
[6]	100-200	25k	3	QPSK	10k
[7]	6-18	25k	0.7	MFSK	5k
[8]	6-20	20k	0.75	128-FSK	10k
[9]	~ 18	10k	5 (maximum)	QPSK / BPSK	4k
[10]	~ 40 feet	3.5k	~ 6.5 knots	1870-coded QPSK	1250 symbols per second

In this paper, we describe the multicarrier modulation system with a data rate of 10kbps over 1km which uses 48 frequencies in 43kHz to 53kHz. Section 2 describes the structure of the data sequence. Synchronisation and multicarrier modulation are described in section 3 and 4 respectively. Section 5 mentions channel identification. System configuration and experimental results are discussed in section 6 and 7 respectively.

2. DATA SEQUENCE

The data sequence is in packet form. The sequence contains synchronisation packets, gap packet, adaptive threshold packets and information data packets. These packets are used to allow synchronisation and noise reduction. Each packet is formed by 48 frequencies within 43kHz to 53kHz, which represent 48 bits with a duration of 5.12ms.

The sequence begins with the linear frequency modulation (LFM) signal packet which is used to synchronise the receiver to the start of the data. The details of the LFM signal will be discussed in the next section. Then a packet of the gap signal follows the LFM packet so that synchronisation can be performed in this period. Eight adaptive threshold packets (ATP) are transmitted for receiver-training purposes. The training packets act as a reference of the transmitted signal block so that channel estimation can be calculated from the reference packets. Also, the long training sequence can be sufficient for the system convergence. 800 information data packets (IDP) follow the adaptive packets. At the end of the data sequence is also a gap signal packet which can minimise the effect of multipath fading between two signal blocks. The time duration for one signal block is (1+1+8+800+1)*5.12ms = 4.15s. Fig. 1 shows the data sequence.

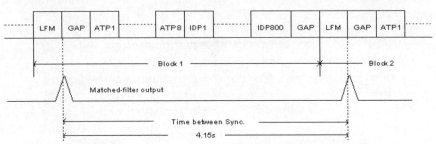

Fig. 1. Data structure with LFM signal for synchronisation

3. SYNCHRONISATION

For many current systems, a linear frequency modulation (LFM) signal is used for frame synchronisation. The most significant property of the linear frequency modulation signal is its symmetry in time and frequency. In general, the expression for a linear frequency modulation signal, also referred to as a 'chirp', is mentioned in [11] as:

$$s(t) = \cos(2\pi f_0 t + \pi k t^2) \qquad (3.1)$$

The instantaneous frequency can be obtained by differentiation

$$f(t) = \frac{1}{2\pi} \cdot \frac{d}{dt}(2\pi f_0 t + \pi k t^2) = f_0 + kt \qquad (3.2)$$

where f_0 is initial frequency, $|k| = \dfrac{B}{T}$, B is the bandwidth and T is the signal duration. Hence the LFM chirp described in [11] is characterised by its starting frequency (f_0), stopping frequency (f_1), and time duration (T) as:

$$|k| = \frac{|f_1 - f_0|}{T} = \frac{B}{T} \qquad (3.3)$$

The resolution of the time depends on the BT product.

At the receiver, a matched filter is used to indicate the arrival of the LFM chirp. The output of this correlation allows selection of the channel with the most energy for synchronisation. The impulse-like auto-correlation function of the LFM signal in Fig. 2 allows synchronisation to be achieved by linear cross-correlation between the received signal and a known LFM signal. Therefore, the starting position of the data sequence can be found by output of the matched filter.

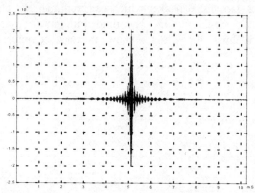

Fig. 2. Ideal output of the matched filter of the synchronisation signal

4. MULTICARRIER MODULATION

Multicarrier modulation divides a channel into a set of parallel independent subchannels [14]. The signal-to-noise ratio (SNR) of each subchannel is measured and a suitable number of bits is then assigned to each channel. There are two reasons for choosing multicarrier modulation (MCM) in the system. According to [12], the MCM signal can be processed in a receiver without the enhancement of noise or interference that is caused by linear equalisation of a single-carrier signal. Another is that the long symbol time used in MCM produces a much greater immunity to impulse noise and fast fades. According to [12] and [13], the input data is Mf_s b/s. They are grouped into blocks of M bits at block (symbol) rate of f_s. Therefore

$$f_{c,n} = n\Delta f \quad \text{for } n = n_1 \text{ to } n_2 \qquad (4.1)$$

$$M = \sum_{n=n_1}^{n_2} m_n \qquad (4.2)$$

where $N_c = n_2 - n_1 + 1$, $f_{c,n}$ = carrier frequency, Δf = frequency separation and N_c = number of carriers.

In our system, the modulation and demodulation techniques used are the inverse fast Fourier transform (IFFT) and the fast Fourier transform (FFT). The IFFT and FFT are well-known efficient algorithms and significantly reduce the complexity of implementing the modulation and demodulation functions. The benefit in system implementation using the IFFT and FFT are mentioned in [14] and [15]. However, the resulting signalling filters have relatively large overlapping sidelobes (-13dB), and this causes a deviation from the ideal multicarrier scheme of independent carriers.

In our system, the binary input data are parsed to each subchannel with fixed number of bits for system initialisation. The number of bits for each subchannel is determined by measuring the SNR of each subchannel during startup. As our system is half-duplex there is no feedback signal back to the transmitter. Therefore, a fixed number of bits is used instead of varying bit numbers. Details will be discussed in section 6.

Mathematically, the discrete complex multicarrier modulation signal [16] can be represented for n^{th} sample by:

$$s(nT) = \frac{1}{N} \sum_{k=0}^{N-1} A_k e^{j(2n\pi f_k T + \phi_k)} \qquad (4.3)$$

The sampling frequency is $1/T$ and the period of one data symbol is $2NT$ with inter-modulation frequency domain. The frequency, f_k, is given by:

$$f_k = f_0 + k(\Delta f) \qquad (4.4)$$

where f_0 is the lowest frequency of the signal spectrum and $\Delta f = 1/NT$ is the frequency separation between carriers.

5. CHANNEL IDENTIFICATION

As mentioned in [14] [17], a periodic training sequence x_k with period M is needed once for every transmitted block of bits which is equal to or slightly larger than the length of the channel pulse response. The receiver measures the corresponding channel output averaging over L cycles, and then divides the FFT of the channel output by the FFT of the known training sequence. The channel estimate in the frequency domain is

$$\hat{H}_n = \frac{1}{L}\sum_{i=1}^{L}\frac{Y_{i,n}}{X_n} \tag{5.1}$$

$$y_n = |H_n|x_n + n_n \tag{5.2}$$

where $Y_{i,n}$ is the n th element of the FFT of the channel output on i th cycle and X_n is the n th element of the FFT of the input training sequence. x_n and y_n are input training sequence and channel output in the time-domain, respectively. n_n is noise in n th element. $H_n = h_0 + h_1 e^{-j(2\pi/N)n} + h_2 e^{-j(2\pi/N)n} + \cdots + h_v e^{-j(2\pi/N)vn}$.

6. SYSTEM CONFIGURATION

Fig. 3 shows a block diagram of the system. In the transmitter, a serial-to-parallel (s/p) buffer, adaptive threshold packets, MFSK modulator, LFM signal packet and parallel-to-serial (p/s) buffer are generated by a digital signal processor (DSP) TMS320C542. Also synchronisation, s/p and p/s buffers, MFSK demodulator and threshold learning are performed by a DSP unit in the receiver.

In the transmitter, the input data is first transmitted from a personal computer to the underwater acoustic modem via an RS232 link. The data is then converted from serial to parallel form by a serial-to-parallel buffer. The binary input data are parsed to each subchannel with one bit. Therefore, there are 48 parallel bits represented by 48 frequency components, so the parallel data bits can be modulated by multicarrier modulation.

To overcome the multipath fading, eight adaptive threshold packets are added at the front of input data sequence before transmission. Each packet contains 48 bits. The odd packets are [101010…101010] and even packets are [010101…010101]. This is used as a reference signal for the receiver to estimate the channel, and then the IFFT algorithm is used for modulation.

As mentioned above, there are 48 frequency components within 43kHz to 53kHz. They are equally distributed in the frequency range, so that the frequency separation between carriers is $\Delta f = \dfrac{53k - 43k}{48} \approx 200 Hz$. The choice of number of carriers is determined by the implementation in the DSP. The TMS320C542 is a 16 bits (one word) fixed point DSP, so that multiples of 16 bits are used due to easy implementation. In the system, 48 bits (3 words) are used for each data packet [8]. Also, the choice of the number of tones is a trade-off between the system sensitivity to multipath and practical implementation constraints.

After IFFT modulation, the data is converted from the frequency to time domain. At that time, the LFM signal is implemented at the front time slot of the data sequence. This is used for synchronisation in the receiver.

In the receiver, some of the noise is reduced using a bandpass filter of 40kHz to 60kHz. The matched filter captures the LFM signal from the received signal, which indicates the start of the data sequence. The output of this correlation provides an estimation of the channel that is used to select the pulse with the most energy for synchronisation. Then the data sequence is demodulated by the FFT algorithm. The data packets are converted from the time-domain to frequency-domain so that decoding can be take place in the error correction algorithm.

Once synchronisation has been achieved, threshold learning can be used to calculate the channel characteristic. Because the acoustic channel is a time varying channel, a fixed threshold cannot work well in the system. Therefore, the threshold of the detector is changed by transmitted reference packets at each data sequence. These change every 4.15s of the data sequence. Adaptive threshold in frequency-domain is used and the training signal is sent out repeatedly after each time the LFM signal was send. At the receiver, the frequency-domain's threshold is calculated from the known training signal and the received training signal. According to the received training signal's frequency-domain characteristic, the adaptive threshold is calculated. This can be used to estimate the channel characteristic [8]. Depending on the acoustic channel, an equaliser/echo canceller may be inserted into system at this point; otherwise the output of the FFT is passed to an error-correction algorithm.

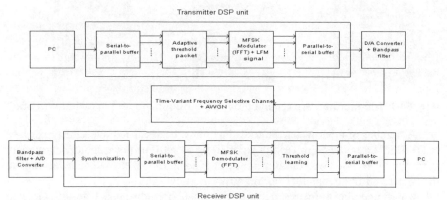

Fig. 3. System Configuration

7. EXPERIMENTAL RESULTS

The system has been tested in City University's swimming pool and in a coastal area near Hong Kong. The pool has dimensions of 50m x 25m, and the transducers are at a depth of 0.6m. Fig. 4a shows the synchronisation signal and Fig. 4b is the signal after the matched-filter. In Fig. 4b, there is serious multipath fading problem, and the reflected signals from the side walls and the bottom of the pool can be seen. Fig. 5a-f shows the results of sending a raw bit-mapped signal of 24 bits 80*60 pixels. The left hand picture (Fig. 5a) is the sent signal. The others are from the received signal.

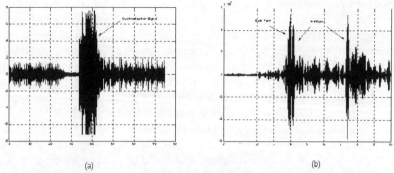

Fig. 4. (a) Received the synchronization signal (b) Matched filter output of the synchronization signal

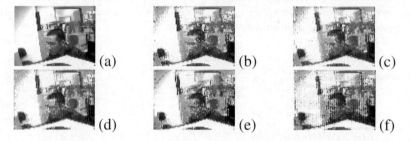

Fig. 5a-f. The pictures are received by the acoustic modem in swimming pool

A number of open water sea trials have also been carried out. One took place at Da Mei Do, in the New Territories of Hong Kong. The geographic environment is shown in Fig. 6. Two transducers are at a depth of 4 metres. The receiver is at the wharf and the transmitter on the boat. The depth of the water is 5.6 metres at the wharf and the 8 metres deep at the boat. The distance between the two transducers is 820m, measured via a geographic positioning system. Fig. 7a-e shows the received pictures using the same sent signal as in Fig. 5. The raw bit error rate is less than 5%.

Fig. 6. The map of Da Mei Do

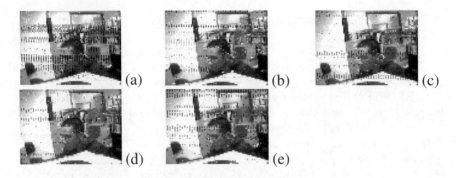

Fig. 7a-e. Received pictures from sea trial

8. CONCLUSION

In this paper, a multicarrier modulation underwater acoustic system has been developed and demonstrated the ability of real time underwater acoustic communication. The system performed at a data rate of 10kbps over 1km.

Current work in developing the system involves using error correction techniques, adaptive filtering to achieve a higher data rate and reliability. The error rate can also be improved by implementing channel equalisation and channel coding.

9. REFERENCES

[1] D. Z. Liao; S. O. Harrold and L.F. Yeung, "An underwater Acoustic Data Link for Autonomous Underwater Vehicles", IEEE Int. Conf. in Circuit and Systems, pp. 28-33, Singapore, July 1995.
[2] S. O. Harrold, D. Z. Liao and L. F. Yeung, "Ultrasonic Data Communication Along Large Diameter Water-filled Pipes", IEEE Int. Conf. M^2VIP, Hong Kong, 1996.
[3] Li Yinghui, S. O. Harrold and L. F. Yeung, "Experimental Study On Ultrasonic Signal Transmission With The Water-Filed Pipes", IEEE Int. Conf. M^2VIP, Australia, Sept 1997.
[4] Li Bin, S. O. Harrold, R. S. Bradbeer and L. F. Yeung, "An Underwater Acoustic Digital Communication Link", in Mechatronics and Machine Vision, (J. Billingsley (Ed)), Research Studies Press, UK, pp. 275-282, 2000.
[5] H. V. Poor and G. W. Wornell, *Wireless Communications: Signal Processing Perspectives*, New Jersey, Prentice-Hall Inc., pp. 353-356, 1998.
[6] M. Stojanovic, L. Freitag and M. Johnson, "Channel-Estimation-Based Adaptive Equalization of Underwater Acoustic Signals", OCEANS '99 MTS/IEEE, Riding the Crest into the 21st Century, vol. 2, pp. 985-990, 1999.
[7] L. E. Freitag and J. A. Catipovic, "A Signal Processing System for Underwater Acoustic ROV Communication", Proceedings of the 6th International Symposium on Unmanned Untethered Submersible Technology, pp. 34-41, 1989.
[8] J. A. Catipovic and L. E. Freitag, "High Data Rate Acoustic Telemetry for Moving ROVS in a Fading Multipath Shallow Water Environment", Proceedings

of the Symposium on Autonomous Underwater Vehicle Technology, pp. 296-303, 1990.

[9] H. K. Yeo, B. S. Sharif, A. E. Adams and O. R. Hinton, "Multiuser Detection for Time-Variant Multipath Environment", Proceedings of the 2000 International Symposium on Underwater Technology, pp. 399-404, 2000.

[10] H. A. Leinhos, "Block-Adaptive Decision Feedback Equalization with Integral Error Correction for Underwater Acoustic Communications", OCEANS 2000 MTS/IEEE Conference and Exhibition, vol. 2, pp. 817-822, 2000.

[11] A. W. Rihazek, *Principles of High-Resolution Radar*, Peninsula Publishing, pp. 226-231, 1985.

[12] J. A. C. Bingham, "Multicarrier Modulation for Data Transmission: An Idea Whose Time Has Come", IEEE Communications Magazine, pp. 5-14, 1990.

[13] J. G. Proakis, *Digital Communication*, 3rd Edition, McGraw-Hill,Inc, New York, pp. 689-690, 1995.

[14] I. Lee, J. S. Chow and J. M. Cioffi, "Performance Evaluation of a Fast Computation Algorithm for the DMT in High-Speed Subscriber Loop", IEEE Journal on Selected Areas in Communications, pp. 1564-1570, 1995.

[15] A. D. Rizos, J. G. Proakis and T. Q. Nguyen, "Comparison of DFT and Cosine Modulated Filter Banks in Multicarrier Modulation", IEEE Global Telecommunications Conference, pp. 687-691, 1994.

[16] W. K. Lam and R. F. Ormondroyd, "A Coherent COFDM Modulation System for A Time-Varying Frequency-Selective Underwater Acoustic Channel", 7th International Conference on Electronic Engineering in Oceanography, pp. 198-203, 23-25 June 1997.

[17] J. S. Chow, J C. Tu & J. M. Cioffi, "A Discrete Multitone Transceiver System for HDSL Applications", IEEE Journal on Selected Areas in Communications, pp. 895-908, 1991.

Reactive Agent Architecture for Underwater Robotic Vehicles

J.H. Ho, G. Seet, M.W.S. Lau, E. Low,
Robotics Research Center,
School of Mechanical & Production Engineering,
Nanyang Technological University,
Nanyang Avenue, Singapore 639798,
Email : mglseet@ntu.edu.sg

Abstract

This paper describes a reactive agent architecture for an Underwater Robotic Vehicle (URV) which could be incorporated into a control system in supervisory mode. An agent can be described as a software object that is capable of task delegation, data-directed execution, communication and planning. Agents can be designed to help pilots in structured tasks such as pipeline tracking and moving to an absolute position. The reactive system operates in a sense-decide-act cycles, where agents receive sensing data from the sonar system and produce planning results for the pilot module. Agents in the architecture are divided into two layers, namely the mission level and the basic level. The mission level consists of agents designed specifically for a particular task. The basic level incorporates the sense-decide-act events. Simulation was carried out to verify algorithms in the reactive system and promising results were obtained.

Keywords: *Underwater robotic vehicles, pilot assistant, supervisory control.*

1. INTRODUCTION

The Underwater Robotics Vehicles (URV) industry has been expanding rapidly since it was first introduced in the 1950s. This is mainly due to the increasing research and development from the offshore oil and gas industry. The annual value of the sub-sea industry is expected to exceed US$20 billion in the year 2003 [1]. This provides strong motivation for continued research on improving work capabilities of URVs.

In general, URVs can be divided into two classes, namely Remotely Operated Vehicles (ROVs) and Autonomous Underwater Vehicles (AUVs). As far as the control scheme is concerned, most of the ROVs are tele-operated whereas the AUVs are fully autonomous from a control perspective. We are interested in building a control architecture that allows tele-operation as well as exhibits a certain degree of autonomy. We refer to this as a hybrid AUV-ROV supervisory control system.

A hybrid system will exhibit both AUV and ROV properties. ROVs are capable of performing more tasks than AUVs [2] due to the higher payloads available on this class of vehicles. Simple but long duration missions, such as pipeline tracking, may lead to pilot fatigue. These tasks are normally performed under structured conditions and thus are good candidates for automation.

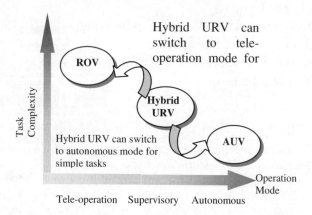

Figure 1 Comparison of Task Complexity for Various URVs

In a hybrid URV system, figure 1, some autonomous behaviours are implemented while maintaining the work capabilities of the ROV. It is not a fully autonomous system but there are some intelligent agents designed to assist pilots in structured tasks. This will relieve the pilot's workload tremendously. In this case, pilots adopt a supervisory role instead of directly operating the vehicle. Pilots will choose one or more agents to assist them in autonomous mode. If the task becomes complex or when an emergency is encountered, pilots will take over and control the vehicle in direct tele-operation mode. It is believed that with the supervisory control mode in the proposed hybrid URV system, pilots will be able to work for longer hours with lower levels of fatigue.

2. URV MULTI-MODE TOPSIDE CONTROL SYSTEM

Tan et al. (2000) proposed a multi-mode topside control system for the RRC underwater vehicle [3]. It comprises of four separate modules. These are identified in figure 2:

- QNX message-handling module
- Operator interface panel
- Vehicle and thruster dynamic simulator module
- Virtual environment module

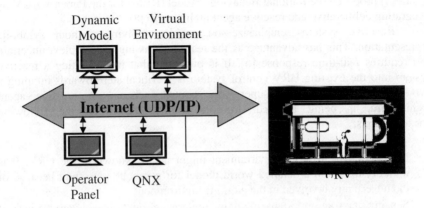

Figure 2 URV Multi-mode Topside Control System

Each module resides on a separate computer and has a specific function in the control system. The message-handling module runs on the QNX4 real-time operating system. It serves as a temporary data storage buffer, providing the other modules with data when requested. Data transfers between various processes are achieved using the User Datagram Protocol (UDP/IP). The pilot interacts with the system via the operator panel. This is achieved, either, through a joystick or virtual buttons and switches on the touch-screen panels. The dynamic simulator module consists of a mathematical model of the vehicle and thrusters. Finally the virtual environment module consists of the virtual underwater world that serves as the basis for the implementation of the off-line simulation and training capabilities of the URV system.

The URV control system adopts a modular architecture. This allows the upgrading of the existing system with new functionality, with relative ease. Every module in the control system is working on a separate computer and the communication among modules is through the QNX message router. This configuration has the advantage of allowing new module to be added. Adding functionality is achieved by simply allocating another computer and setting up the communication and data flow at the QNX module. This reduces the need for extensive code change.

The control system is fully functional in tele-operation mode, and in the hybrid URV/AUV system mode with limited autonomous. A number of behaviours have been implemented assisting URV pilots in structured tasks.

3. REACTIVE AGENT SYSTEM

Advances in reactive agent technology (Brooks 1990 [4], Agre & Chapman 1987 [6]) have motivated researchers to investigate the possibility of implementing reactive system for the URV. Bonasso (1991) proposed a reactive agent architecture, which could be used in an underwater environment [7]. Brutzman et al. (1998) proposed the Rational Behaviour Model (RBM) for the Phoenix AUV by integrating deliberative and reactive agent architecture [8].

Reactive systems emphasize on problem solving without symbolic representation. This has advantage as the real world is highly complex, uncertain and requires real-time response [6]. It is proposed that incorporating a reactive system into the existing URV control system is a critical step towards building a hybrid URV system with supervisory control scheme. A reactive agent architecture, is being adopted, for our vehicle based on the following considerations:

- In general, the working environment might not be known to the URV. It is unlikely to obtain a detailed world model for symbolic reasoning because of the uncertainty involved in the working environment.
- Sensing plays an important role in the underwater operations due to the lack of information about the working environment. The URV has to react in real-time based on the limited sonar data. In this case, computing time is crucial. Global planning might not be useful as there are thousands of predicate logic to be resolved. Furthermore, the rapidly changing environment makes global planning a less efficient approach since re-planning is needed frequently. A better strategy is to let the URV react with the sonar data by planning, ahead, the next step or next few steps.
- Modular architecture is preferred, as it will provide room for functionality expansion in future.
- The agent system should exhibit good human interaction behaviour. The idea of a supervisory control mode is to build a man-in-the-loop system whereby the pilot is the supervisor of the system. Therefore, it is very important to have a user-friendly system such that the pilot can issue commands easily. Meanwhile the system will prompt for higher level command from the pilot when agents are unable to make decisions in a particular situation.

4. PROPOSED AGENT ARCHITECTURE

A general framework can be constructed based on considerations discussed previously. The architecture can be divided into 3 modules, namely sensing module, agent module and piloting module. This forms a sense-decide-act cycle, figure 3.

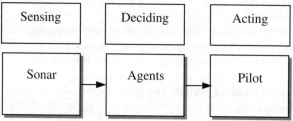

Figure 3 Sense-Decide-Act Cycle

The sensing module collects sonar data to determine the distances between the URV and targets. The agent module is the heart of the system where all decision-making processes will be done. This module can be further divided into many smaller modules, with each being responsible for a specific agent behaviour, such as moving to a specific position, pipeline tracking and obstacle avoidance. The acting process is taken care by the piloting module. It is basically a pilot program which sends motion commands to the URV control system.

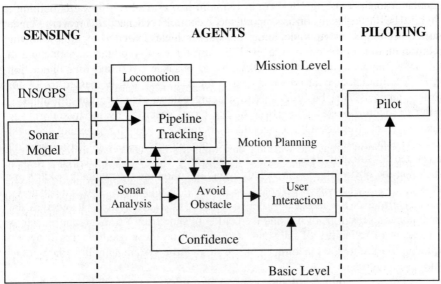

Figure 4   Proposed URV Agent Architecture

The complete agent architecture is shown in figure 4. At the current stage of development, the architecture has been implemented and tested under simulation mode. The sensing module contains a sonar model that simulates sonar response in the virtual underwater environment. In addition a INS/GPS module provides absolute position and orientation information, of the URV, to the agent module for analysis. The layered approach was adopted in the agent module. This architecture does not commit, fully, to the original sub-sumption competence proposed by Brooks (1986) [5].

The whole module is further divided into two levels, namely basic level and the mission level. Levels in our architecture cannot be regarded as level of competence since each level is not a completely workable level. Our goal is to build a control system with supervisory mode. Layered competence is not a critical issue here since our system is not a fully autonomous system. In addition, the structured tasks are relatively simple and predictable and thus allow algorithms to be developed easily.

The mission level consists of agents designed specifically for certain task. This level receives the simulated sonar data from the sensing module. There are two autonomous behaviours implemented at this level, namely locomotion behaviour and pipeline tracking behaviour. The locomotion behaviour takes care of driving the URV to a specified location. This behaviour is useful when the URV is tasked to move to the work site from the surface vessel or vice versa. The pipeline tracking behaviour helps the pilot to track the pipeline autonomously based on the simulated sonar data.

The basic level consists of all the basic analysis needed in every sense-decide-act loop. For example, sonar analysis is always needed regardless of the mission chosen in the mission level. There are three sub-modules implemented in this level, namely sonar analysis (SA), avoid obstacle (AO) and user interaction (UI). All simulated sonar data received at mission level will be passed to SA for further analysis. The basic idea is to execute pattern classification algorithm in order to identify obstacles, pipeline or any target specified by the module in the mission level. Results will be sent back to mission level for planning the next location that the URV should go. The AO module receives results from mission level and SA module. If an obstacle is detected, AO will try to re-plan the path such that the obstacle can be avoided instead of following the planning result from mission level. Otherwise it will pass the planning results to the UI module. The UI module provides the interaction between user and agent system. It receives the final planning result from AO and the confident state from SA. Confidence state is a measure of reliability of the sonar data. Based on these results, UI will decide whether to execute the planning results, or to halt the system and allow the pilot to take over control.

Finally the piloting module consists of a pilot program which sends motion commands to the URV multi-mode control system. It receives planning results from the agent module that indicate where the URV should go.

5. SIMULATION RESULTS

At the current stage of development, the architecture was implemented and tested on a simulated URV pipeline tracking sequence. Figure 5 shows the simulation of pipeline tracking. The URV starts to track the pipeline close to the oil rig structure and follows path A. When the URV loses track of the pipeline at the junction, it prompts the pilot for intervention. The pilot responds with a high level command directing the URV to the desired heading. The URV subsequently continues to track the pipeline and follows path B. The system halts again when the URV encounters the buried section of pipeline. The pilot intervenes again and makes a decision for the next action.

In order to test the robustness of the tracking algorithm, a highly uneven terrain was constructed at region B with a curved pipeline. Results showed that path B has larger deviation than path A. Nevertheless the URV was still able to track the pipeline until it reached the buried region.

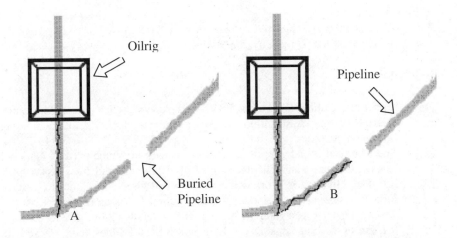

Figure 5 Simulation Results: Showing Autonomous Pipeline Tracking

6. CONCLUSION

The reactive agent architecture for supervisory URV control, described in this paper, provides the necessary automation needed for the URV multi-mode control system. Simulation results indicate that pilots could rely on the agent system to help them in structured tasks such as pipeline tracking and in moving to a desired location. In such a scenario, pilots will spend minimum effort in monitoring the process. They are only required to make decisions at critical moment such as when the pipeline was buried.

The pipeline tracking module has been successfully integrated into the "Multi-Mode Control System" [3]. As would be expected the pilot is relieved from extended periods of direct teleoperation. Pilot intervention is only required when the system has difficulty in identifying the centreline of the pipeline. Tracking reliability is highly dependent on the quality of the sonar scanner.

The modularity of the architecture provides for ease of functionality expansion. Agents in the mission level are designed specifically for certain tasks. A new agent can be implemented in the architecture by adding additional sub-modules in the mission level.

Additional tasks for automation are being reviewed and suitable agents being developed. These include vehicle docking, station keeping and intelligent NDT evaluation of underwater structures.

Acknowledgement

This project is in part supported by British Gas Asia Pacific Ptv. Ltd. and the Nanyang Technological University, Singapore.

References

[1] "What the Future Holds for RoVs and AUVs", Underwater Magazine, Journal of the Association of Diving Contractors International, Inc. May/June 2000.

[2] The National Academy of Sciences, "Undersea Vehicles and National Needs", Chapter 1 and Chapter 2, pp. 7-46, National Academy Press, 1996.

[3] K.C.Tan, "Integrated Work-cell Simulator and Graphical Control Interface for a Tele-operated Underwater Robotic Vehicle", Masters Thesis, Robotics Research Center, Nanyang Technological University, 2000.

[4] R.A.Brooks, "Elephants Don't Play Chess", in P.Maes (ed.), *Designing Autonomous Agents*, pp. 3-15, The MIT Press, 1990.

[5] R.A.Brooks, "A Robust Layered Control System for a Mobile Robot", in *IEEE Journal on Robotics and Automation*, vol. RA-2(1), pp. 14-23, 1986.

[6] P.E.Agre, D.Chapman, "Pengi: An Implementation of a Theory of Activity", in *Proceedings of the Sixth National Conference on Artificial Intelligence (AAAI-87)*, pp. 268-272, 1987.

[7] R.P.Bonasso, "Underwater Experiments Using A Reactive System for Autonomous Vehicles", in *Proceedings of the 1991 National Conference on Artificial Intelligence (AAAI-91)*, pp. 794-800, July 1991.

[8] D.Brutzman, T.Healey, D.Marco, B.McGhee, "The Phoenix Autonomous Underwater Vehicle", in D.Kortenkamp, R.P.Bonasso, R.Murphy (ed.), *Artificial Intelligence and Mobile Robots*, pp. 323-360, AAAI Press/The MIT Press, 1998.

Robot-human Interaction

The final section of this book may be considered by some to contain the more interesting applications of mechatronics and machine vision, the interaction between robots and human beings. The ease with which human beings interact with robots will define how society accepts robots in everyday life. These interactions range here from facial recognition, both on the part of the robot and the human, to the use of mechatronics to aid medical examination, diagnosis and treatment.

The first paper addresses an artistic different application, the design and teaching of a robot to perform the ancient art of Chinese calligraphy. The aim is not merely to teach calligraphy skills to the robot, where five styles of Chinese character image patterns written by famous calligraphers are used, but also to preserve the calligraphy culture and the more general teaching of its skills.

The second paper offers a new approach to the problem of counting the number of people in a crowd by the use of a video image. Employing a camera pointing vertically downwards, the system views heads from the top. A template is obtained from a model projection and a head search and count is made.

The next paper describes a method for recognising human facial expressions. This uses an image transformation based upon a quadtree partition scheme. (Perhaps my computer will one day notice that I am scowling at it!)

The converse problem is tackled in the next paper. A robot has been designed to mimic human facial expressions and is to be used as an interactive museum guide. The robotic head has nine degrees of freedom and allows a great variety of facial expressions to be performed.

Interfaces can also be designed to recognise hand gestures. One method, described in the next paper, uses a glove with sensors embedded in it. This can detect various gestures and hand-motion. The resulting data can be fed to the robot which can interpret the gesture. The data-glove has been used to command robot toys.

Finally in this section, high frequency ultrasound is used for the non-invasive treatment of breast tissue. The mechatronic system described enables the sound beam to be focused and directed at the appropriate point being treated.

Development of a Chinese Character Calligraphy Robot

Fenghui Yao[+], Guifeng Shao[++], Ryoichi Takauc[+], Akikazu Tamaki[+]
[+]Graduate School, University of East Asia
[++]Department of Commerce, Seinan Gakuin University

Abstract
This paper describes a Chinese character calligraphy robot that can be categorized as an art robot. The whole system consists of a calligraphy dictionary, robot arm, robot hand, writing brush and system controller. The calligraphy dictionary includes five styles of Chinese character image patterns written by famous calligraphers in Chinese history. When the character to be written and the style are given, the system starts to search the calligraphy dictionary, and outputs all image patterns registered in the dictionary for the assigned character. Then, the contour detection and thinning are performed, based on the character image pattern designated from the output image patterns. These two features, together with writing order information, are sent to the robot to write the characters as a human calligrapher would. The aim of this work is, firstly, to teach calligraphy skills to the robot, and secondly, in order to preserve the culture of character calligraphy, have the robot teach any beginner calligraphy skills

Keywords: Calligraphy robot, character image pattern, contour detection, thinning, calligraphy skills, writing brush control.

1. INTRODUCTION

Recently, in addition to industrial robots, a variety of robots such as humanoid robots (e.g. ASIMO[4] [13], SDR-4X[14]), pet robots (e.g., AIBO[15]), entertainment robots [5], soccer robots [16], guard robots (e.g., Guard Robo C4 [17]), rehabilitation robots [2] [10], human care robots [1], [6], [8], [9], [12] and so on, have been developed. From now on, research in the robotics field will become even more vigorous and more diverse. In the very near future, robots will fill every corner of our human community. In every aspect of our daily life, robots will co-exist and co-operate with people, behaving as intelligent and friendly partners. This paper describes a Chinese character calligraphy robot that can be categorized as an art robot. Chinese character calligraphy has more than four thousand years' history. Nowadays, it is also becoming popular in east and south-east Asian countries. Generally, Chinese character calligraphy can be roughly classified into five styles, which are *block, angular, ancient, semi-cursive* and *cursive style*, as shown in Fig. 1. Different styles need call for different skills to write and to arrange every stroke. Usually, people have to keep training and practising for several to tens of years to reach a good level of Chinese character calligraphy as there are many skills involved.

A good calligrapher combines these skills adroitly. The representative skills are: (1) brush pressure, (2) brush speed, and (3) brush rotation angle. By pressing the writing brush heavily, the calligrapher can produce a thick stroke. By moving the brush quickly, the calligrapher can write a blurred stroke. By rotating the brush, the calligrapher can form a smooth part of a stroke. The purpose of this research is to teach the skills of Chinese character calligraphy to the robot, and in turn, have the robot teach the beginner calligraphy skills, in order to preserve this ancient culture. As the first step in this research, this paper focuses on how to teach the calligraphy skills to the robot. The contents include system configuration, the construction of the calligraphy dictionary, the feature extraction from image pattern, and robot control strategies to realise the above three skills.

Fig. 1. Five different writings of the Chinese character for "bird"; (a) block, (b) angular, (c) ancient, (d) semi-cursive, (e) cursive.

2. SYSTEM CONFIGURATION

The whole system consists of a calligraphy dictionary, robot arm, robot hand, writing brush and system controller. When a character to be written and its style are given, the system starts to search the dictionary, and outputs the character patterns registered in the dictionary. For the designated character and writing style, there exist plural ways of writing. They are also registered in the dictionary. The contour is extracted for a designated character pattern from the output of the dictionary search, and then the thinned character is generated. These two features, together with the writing order information, are sent to the robot. Based on these three pieces of information, the robot is led to write the character as a calligrapher would. The prototype calligraphy robot is shown in Fig. 2.

3. DICTIONARY COMPOSITION

The calligraphy dictionary consists of three sub-dictionaries, a Chinese character pattern dictionary (CPD), an index of the CPD (ICPD) and a writing order dictionary (WOD). The compositions of these three sub-dictionaries are described below.

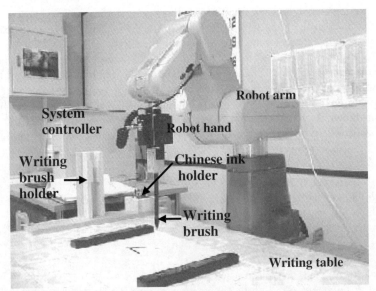

Fig. 2. Prototype of Chinese character calligraphy robot.

3.1 Chinese Character Pattern Dictionary (CPD)

The CPD is constructed on the basis of characters written by famous calligraphers in Chinese history since the Tang Dynasty [7]. A scanner is used as the input device. The characters are extracted from the input image, and are registered in the CPD, after being normalised into 64×64 dots, as shown in Fig. 3 (a). The recording structure of the CPD consists of four fields: *code*, *style*, *ratio*, and *character pattern image*, as given in Fig. 3 (c), which are described in the following:

(i) *Code*: Chinese character code, two bytes.
(ii) *Style*: one byte, 0, 1, ..., 4 for ancient, angular, block, semi-cursive style and cursive, respectively, and 5 for all styles;
(iii) *Ratio*: one byte. To keep the original character shape, the ratio H/W is necessary. To be able to record this ratio with one byte, the integer H/W×100 is used.
(iv) *Character pattern image*: Because the normalised character pattern is a binary image, one byte can represent 8 dots. The total size of one character pattern image is 8×64 bytes. Therefore, the size of one record in the CPD is fixed; its length is 516 bytes.

3.2 Index of CPD (ICPD)

The ICPD provides the information for CPD searches. Its records consist of three fields: *code*, *number* and *pointers to CPD*, as shown in Fig. 3 (d). Its fields are as follows:

(i) *Code*: Chinese character code, two bytes.

(ii) *Number*: two bytes, number of character patterns corresponding to the character in the *code* field, registered in the CPD. That is, different versions of the same character are all registered in the CPD.

(iii) *Pointers to CPD*: pointers to the CPD of the character pattern corresponding to the character in the *code* field. Each pointer takes 4 bytes, so the length of a record is *Number*×4+4 bytes.

In this way, any new writing of a character that is already registered in the CPD or the writings for a new character can be added to the end of the CPD. Note that it is necessary to update the ICPD if the CPD is updated by an operation such as adding new writings for a new character.

3.3 Writing Order Dictionary (WOD)

The WOD is constructed based on the points of the folded-line, as shown in Fig. 3 (b), which are assigned in coordinate system *XOY*. The fields of the WOD are as follows (refer to Fig. 3 (e)):

(i) *Code*: Chinese character code, two bytes.
(ii) *Style*: same as the *style* field in the CPD.
(iii) *Ratio*: same as the *ratio* field in the CPD.
(iv) *Pointer to CPD*: four bytes, position in the CPD for the character in the *code* field.
(v) *Length*: two bytes, length of the folded-line, that is, the number of points in the coordinate system XOY (refer to Fig. 3(b));
(vi) *Points*: *x*- and *y*-coordinates of points in XOY. Each point takes 4 bytes.

Therefore, the size of a record is 8+ *Length*×4 bytes.

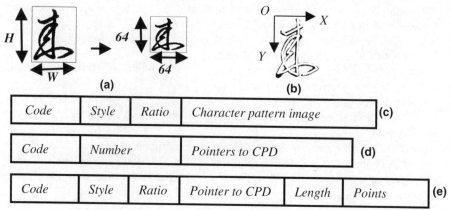

Fig. 3. (a) Normalisation of the extracted character for "come"; (b) writing order represented by points in the coordinate system XOY; (c) record structure of CPD; (d) record structure of ICPD; (e) record structure of WOD.

4. FEATURE EXTRACTION AND ROBOT CONTROL

4.1 Feature Extraction

The image pattern of the designated character is retrieved from the CPD. The original character pattern can be obtained by employing the interpolation according to the H/W *ratio* in the CPD. Fig. 4 (a) shows the interpolated image pattern for "mountain". From this image, the contour is detected as shown in Fig.4 (b) according a contour tracing algorithm [3]. The thinned pattern is generated as given in Fig. 4 (c), by using a thinning algorithm [3]. The writing order is obtained from the WOD, as shown in Fig. 4 (d). This character consists of three strokes, *i.e.*, (x_0,y_0) to (x_2,y_2), (x_3,y_3) to (x_6,y_6) and (x_7,y_7) to (x_9,y_9). The end of each stroke is marked by (-1,-1). The contour, thinned character pattern and writing order are used for robot control.

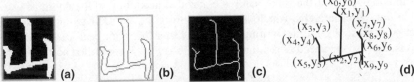

Fig.4. (a) Original character pattern from CPD which means "mountain"; (b) contour; (c) thinned pattern; (d) writing order.

4.2 Robot Control

Chinese character calligraphy requires many techniques, such as the pressure of the writing brush, the moving speed and the rotation of the hand, method of writing the beginning of the stroke, method of finishing the ending of the stroke, and so on. These techniques are very difficult for people to master, and for robots, too. The following describes the pressure control and the rotation control of the writing brush, and the moving speed of the robot hand. The open-loop control strategy is employed here. This is much like what the human beings do when they learn calligraphy, referring to samples written by famous calligraphers - that is by imitation. Imitation is also important for the calligraphy robot and is realised as follows. Upon being shown a calligraphy sample, the robot firstly separates it into strokes according to the information registered in the WOD. Next the changes of the stroke width and stroke inclination are extracted from the sample. Then, the distribution of the Chinese ink in each stroke is calculated. Finally, the interior angles at the points on the folded-line of each stroke are calculated. The following subsections discuss these features in detail.

4.2.1 Pressure Control of Writing Brush

Generally, the thickness of a stroke of a Chinese character is not uniform. It changes to make the calligraphy beautiful. Fig. 5 (a) is a sample of a stroke whose width changes from thin to thick. The thinned stroke is expressed by a line with a one-dot width, that is, a set of points $S_P = \{P_0, P_1, ..., S_{L-1}\}$, where L is the length of the stroke. The width of the stroke along the thinned pattern, as shown in Fig. 5 (a) by the short blue lines that are perpendicular to the thinned pattern in the support region in question, is denoted as a set $S_W = \{w_0, w_1, ..., w_{L-1}\}$. The pressure of the writing brush at P_i is controlled according to

$$h_i = \begin{cases} Z_{def} & \text{(if } w_i < w_{thres}) \\ w_i / w_{max} \times Z_{max} & \text{(otherwise)} \end{cases} \quad (4.1)$$

where h_i is the depth to which the writing brush is to be pressed at P_i, Z_{max} is the deepest distance for the writing brush to be pressed, Z_{def} the default depth, $w_{max} = \max\{w_0, w_1, ..., w_{L-1}\}$, and w_{thres} is the threshold value. At present, w_{thres} is set at 5 dots.

4.2.2 Moving Speed of Robot Hand

The moving speed of the robot hand is controlled according to the coordinates of the writing order in XOY, which are obtained from the WOD. The longer the line from point P'_i to P'_{i+1} of the writing order, the faster the robot hand can be moved. That is, the moving speed of robot hand from P'_i to P'_{i+1} is controlled according to

$$v_i = \begin{cases} v_{max} & \text{(if } \|P'_{i+1} - P'_i\| \times v_{def} > v_{thres}) \\ \|P'_{i+1} - P'_i\| \times v_{def} & \text{(otherwise)} \end{cases} \quad (4.2)$$

where v_i is the moving speed of the robot hand from P'_i to P'_{i+1}, v_{max} the maximal moving speed of the robot hand, v_{def} the default moving speed, v_{thres} the threshold value, and $\|\cdot\|$ represents Euclidean distance. The writing order P'_i is shown by the x- and y-coordinates in XOY (see Fig. 5 (b)).

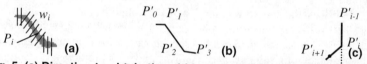

Fig. 5. (a) Direction to obtain the width of the stroke along the thinned pattern; (b) folded-lines from WOD; (c) rotation angle.

4.2.3 Rotation Angle of Writing Brush

The rotation angle of the writing brush is also determined according to the coordinates of the writing order in XOY. Fig. 5 (c) shows three consecutive points of the writing order of a stroke. The writing brush is made to move from P'_{i-1} to P'_i, and is rotated α degrees at P'_i and then is made to move toward P'_{i+1}.

The rotation angle α is given by

$$\alpha = 180° - \cos^{-1} \frac{\|P'_i - P'_{i-1}\|^2 + \|P'_{i+1} - P'_i\|^2 - \|P'_{i+1} - P'_{i-1}\|^2}{2 \times \|P'_i - P'_{i-1}\| \times \|P'_{i+1} - P'_i\|} \quad (4.3)$$

where ‖ ‖ represents Euclidean distance.

At every point of the writing order, the writing brush is controlled according to the combination of the pressure, moving speed and rotation angle shown above.

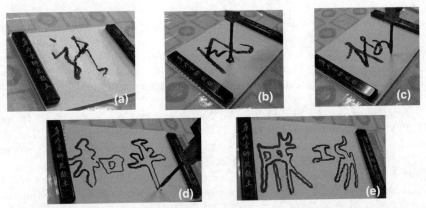

Fig. 6. Robot calligraphy. (a) "dragon" in cursive style; (b) "phoenix" in cursive style; (c) "plum" in cursive; (d) "peace" in block style; (e) "success" in ancient style.

5 EXPERIMENTAL RESULTS

The whole system is coded by MS-Visual C++ ver. 6.0 working on a Windows 98 platform. The prototype system was demonstrated at the Japan Expo, in Yamaguchi (Yamaguchi KIRARA Expo, Aug. 24-26, 2001).

Fig. 6 shows some characters written by the calligraphy robot. Fig. 6 (a) shows the character for "dragon" in cursive style; Fig. 6 (b) "phoenix" in cursive style; (c) "plum" in cursive; (d) two characters that mean "peace" in block style; (e) two characters that mean "success" in ancient style.

6 CONCLUSIONS

This paper describes a Chinese character calligraphy robot. At present, its CPD includes 29,712 characters in five styles, *i.e.*, block, angular, ancient, semi-cursive and cursive style.

The direct application of this robot is the design and printing of signboards. However, the main purpose of this research is to preserve and develop the ancient culture of character calligraphy. Nowadays, with the spread of the computer word processor, such as Word, Ichitaro, and so on, more and more people get used to them and do not like to write characters even by pen or pencil, not to mention the writing brush. Therefore, the number of competent calligraphers is becoming smaller and smaller, day by day. If this situation continues for several decades, the character calligraphy culture may face extinction. If the robot can master all the skills of a professional calligrapher, it can do creative jobs such as making new works of art. Further, the robot can instruct people in the study of calligraphy, thus help to preserve and develop this culture. This is the final goal we are working towards.

References
1. Y. Adachi, S. Nakanishi, Y. Kuno, N. Shimada and Y. Shirai: Intelligent wheelchair using visual information from human face, J. of the Robotics Society of Japan, vol. 17, no. 3, pp. 423-431, 1999.
2. G. Bourhis and P. Pino: Mobile robotic and mobility assistance for people with motor impairments: Rational justification for the VAHM project, IEEE Trans. Rehab. Eng., vol.4, no. 1, pp. 7-12, 1996.
3. J. Hasegawa, H. Koshimizu, A. Nakayama and S. Yokoi, ed.: Image Processing on Personal Computer, Gijyutsu-Hyoron Co. Ltd., Tokyo, 1986.
4. M. Hirose, T. Takenaka, H. Gomi and N. Ozawa: Humanoid robot, J. of the Robotics Society of Japan, vol. 15, no. 7, pp. 23-25.
5. H. Ishiguro, T. Ono, M. Imai, T. Maeda, T. Kanda and R. Nakatsu: Robovie: A robot generates episode chains in our daily life, Proc. of 32nd Int. Symp. on Robotics, Korea, April, 2001, vol. 2, pp. 1356-1361.
6. R. L. Madarasz, L. C. Heiny, R. F. Cromp, and N. M. Mazur: The design of an autonomous vehicle for the disabled, *IEEE J. Robotics and Automat.* vol. RA-2, no. 3, 1986.
7. N. Matsuda, ed.: Gotaijikan, Kashiwashobo Publishing Co. Ltd., 1996.
8. M. Mazo, F.J. Rodriguez, J. Lazaro, J. Urena, J.C. Garcia, E. Santiso, P. Revenga and J.J.Garcia.: Wheelchair for physically disabled people with voice, ultrasonic and infrared sensor control, *Autonomous Robot*, vol. 2, pp. 203-224, 1995.
9. E. Prassler, J. Scholz, and P. Piorini: Navigating a robotic wheelchair in a railway station during rush hour, *Int. J. Robotics Res.*, vol. 18, no. 7, pp. 711-727, Jul. 1999.
10. T. Sakai, R. Hirata, S. Okada, N. Hiraki, Y. Okajima, N. Tanaka and S. Uchida.: Rehabilitation robot for stroke patients (TEM, therapeutic exercise machine), Proc. of 32nd Int. Symp. on Robotics, Korea, 19-21 April 2001, vol. 3, pp. 1587-1591.
11. A. Takanishi: Human communication oriented humanoid robot, J. of the Robotics Society of Japan, vol. 15, no. 7, pp. 11-14.
12. F.H. Yao, G.F. Shao, H. Yamada and K. Kato: The Visual Feedback System for an Outdoor Intelligent Powered Wheelchair, Proc. of 32nd Int. Symp. on Robotics, Korea, vol. 1, pp. 592-597, April 2001.
13. http://www.honda.co.jp/ASIMO
14. http://www.sony.co.jp/SonyInfo
15. http://www.aibo.com/
16. http://www.robocup.org/
17. http://www.sok.co.jp/r_d/index.html

Automated People Counting using Template Matching and Head Search

Grantham Kwok-Hung Pang, Chi-Kin Ng,
Department of Electrical and Electronic Engineering,
The University of Hong Kong.

Abstract
People counting using image processing has been carried out for years. Conventional methods can count people accurately when only a few isolated people pass through a counting region in a non-crowded situation. In this paper, the emphasis is on people counting in a crowded environment and a method using head search and model matching is described. A camera is mounted vertically downwards viewing the people heads from the top. People head search can be used to locate some passengers. In addition, templates obtained from the perspective projection of the human model are used to locate and isolate individual person. Our approach aims at dealing with a congested situation where occlusion is a major problem. This paper describes a real-time, high-accuracy, automated people counting system that has been developed. Experimental results are illustrated and the effectiveness of the developed method for real-time application is verified.

Keywords: Automated people counting, real-time image processing, template matching.

1. INTRODUCTION

Everyday, a large number of people move around in all directions in buildings, on roads, railway platforms and stations. The information on passenger flow is very important to public transport operators and it can help them to control the flow and manage the traffic effectively.

One conventional people counting method is based on turnstiles, but the mechanical contact of turnstiles is inconvenient and uncomfortable to the passengers. It is also impossible to install the gates on every platform and escalator to count the number of people. Counting method using optical beam [1] fails to count correctly when several passengers cross the beam at the same time.

People counting based on image processing on either the spatial images [2-7] or spatial-temporal images [8-9] have been proposed. For spatial images, the counting methods included mathematical morphology [2], shape model filtering on the round shape of people head [3], block matching with clustering [4], supervised split and merge [5] and optical flow [6,7]. Mathematical morphology with averaging-thresholding methods [8] and template matching in stereo images [9] are used to count the number of people in spatial-temporal images. Their counting accuracy is satisfactory when passenger flow rate is low. When people walk in a crowded group, the accuracy can be greatly decreased.

In this paper, an automated people counting system that can run in real-time on a PC-based platform is described. The algorithm is based on a novel two-stage people isolation method using head search and projected model templates. The system can provide high counting accuracy even in a crowded and occluded situation.

2. SYSTEM OVERVIEW

The people counting system is targeted to operate in an indoor environment with limited variation of the illumination level. A color digital camera is placed vertically above the passengers viewing downwards to the head. The camera is mounted near the ceiling, which is at around 3 to 5 meters from the floor. The detection region viewed from the camera would be larger than 2m along a walking direction. Normal plain floor background is assumed. The aberration and spherical distortion of the camera is assumed negligible. The frame rate has been set to 5 fps, but it can vary according to the hardware configuration. People would walk in all directions with a nominal speed of around 1.5m/s. In the developed algorithm, a person should be seen for five or more successive frames in an image sequence. After image acquisition, the system would consist of the following stages: image segmentation, cluster isolation, cluster tracking and counting.

3. IMAGE SEGMENTATION

Segmentation is used to discriminate the people from the background. The input for segmentation are images of M pixels in column and N pixels in row. Each pixel is a 24-bit true-color RGB pixel. In order to reduce the computation in the latter stage, the segmented image is processed in sub-blocks consisting of BxB pixels.

Before segmentation, threshold values are found as follows. Two background images are acquired from the scene. They are converted into the HSV color model, and the maximum range of variation of the hue (H) value is obtained. Hue is used for the representation of the color feature of the background for segmentation because it is invariant to images containing high saturation, even in the presence of shading, shadows and highlights [10].

Hence, in order to segment an image, the image is converted into the HSV color space. When the hue value of a pixel within each sub-block of the image lies outside the threshold value range of $[H_{min}, H_{max}]$, the corresponding pixel is considered as part of a person. Otherwise, the pixel is considered as background or shadow. When more than 50% of the pixels within a sub-block is occupied, the sub-block will be considered as part of a person. Otherwise, the sub-block is considered as background or shadow.

4. CLUSTER ISOLATION

From a segmented image, the number of people and their positions should be isolated to perform counting. In a crowded situation, the isolation of an individual person is a difficult problem. The occlusion due to perspective projection will make the isolation even more difficult. A two-stage method based on head search and template matching has been developed. Combining the two methods, cluster isolation is divided into head search, template model matching and cluster removal as shown in Fig. 1.

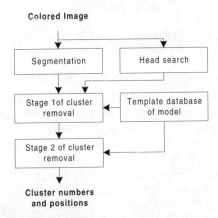

Fig. 1. Block diagram of cluster isolation

4.1 Head search

Head search is used to find the potential locations of people heads. As the camera is viewing vertically downwards, the heads are usually seen circular in shape. The hair color can help in the search (e.g. dark or black color in most part of Asia; silvery grey, brown or golden color in North America). However, it is only a heuristic approach that is used in stage one of cluster removal. If this heuristic method fails, we still have stage 2 of cluster removal to identify people. In our test scene, the people hair is mostly dark in color. The head search is thus based on finding the circular regions that are dark.

The color image is first divided into sub-blocks, where the sub-block mean is calculated by summing up each RGB color value within each sub-block. The sub-blocked mean image will be convolved with a circular template mask. The resulting data represents the sum of the color values over the circular template mask. When the sums are below a threshold value (T_{head}), the regions are considered as heads and the regional centres will be treated as the head centres.

4.2 Template database

The usual template matching technique with standard masks of the template can isolate people, but it fails to isolate the varying size of people due to the perspective projection and occlusion. Therefore, in this paper, template masks from the human model is proposed. Simple geometric shapes are used to model people, and their perspective projections on the view plane are obtained as templates. These templates can be used to find and isolate each cluster from the segmented image where each cluster represents a person.

In order to obtain the perspective projection, a person is modeled by a set of simple geometric shapes. In this paper, a person is modeled by an ellipsoid as head, a cylindroid as the body and a flat ellipse on the top of the cylindroid as the shoulder (Fig. 2).

Fig. 2. Two side views and top views of a model of the passenger (left) and the three-dimensional view of the model (right)

To find the perspective projection of the model into the view plane, the camera and the scene are modeled as shown in Fig. 3.

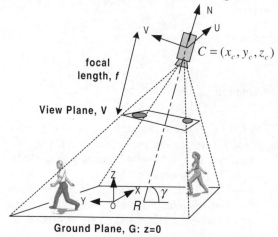

Fig. 3. Modeling of camera and the scene for finding perspective projection of object points

The image viewed from the camera is divided into (P-1) columns and (Q-1) rows to form a grid. There are, in total, P×Q grid points. Their corresponding positions on the 3D space are calculated. The world co-ordinate system in XYZ is transformed into camera co-ordinate system in UVN. It can be done by translating XYZ world space into the view point C and then rotating to match the UVN view space.

$$\begin{pmatrix} u \\ v \\ n \\ 1 \end{pmatrix} = \mathbf{M} \begin{pmatrix} x \\ y \\ z \\ 1 \end{pmatrix} = \begin{pmatrix} 1 & 0 & 0 & 0 \\ 0 & \cos\gamma & \sin\gamma & 0 \\ 0 & -\sin\gamma & \cos\gamma & 0 \\ 0 & 0 & 0 & 1 \end{pmatrix} \begin{pmatrix} 1 & 0 & 0 & -x_c \\ 0 & 1 & 0 & -y_c \\ 0 & 0 & 1 & -z_c \\ 0 & 0 & 0 & 1 \end{pmatrix} \begin{pmatrix} x \\ y \\ z \\ 1 \end{pmatrix}$$

View point C is a distance of f in front of the plane of projection along the N-axis. Suppose the view rectangle at the view plane is bounded from $-V_u$ to V_u in the U-axis and from $-V_v$ to V_v in the V-axis. Grid points can be represented as (p, q) where $p \in [1..P]$ and $q \in [1..Q]$. Each grid point corresponding to a point (u_p, v_q) on the view plane can be calculated as follows:

$$(u_p, v_q) = ((\frac{p-1}{P-1} - \frac{1}{2}) * V_u, (\frac{q-1}{Q-1} - \frac{1}{2}) * V_v)$$

At each position on the P×Q grid points, the model projects from the 3D space into the view plane in a perspective way. Head position and body position in the view plane are found as the projected position of the ellipsoid centre of the head model, and the bottom center of the cylindroid of the body model respectively. The templates with their corresponding head and body positions are stored in a template database.

Fig. 4. Cluster isolation using human model templates

4.3 Cluster removal

Cluster removal is performed in two stages. In the first stage, the template with the head position nearest to the positions from the head search process are used to remove the occupied regions from the segmented image. In the second stage, the remaining clusters are searched, located and removed by using the templates in the template database. An example of people and their corresponding human model templates are shown in Fig. 4.

In the first stage, the template with the nearest distance between the head position of the template and the potential head position in the head search process is used. The nearest distance is calculated by the following formula:

$$D = \min_{i \in [1..P], j \in [1..Q]} \sqrt{(xh_{i,j} - xh)^2 + (yh_{i,j} - yh)^2}$$

where (xh, yh) is the head position in segmented image and $(xh_{i,j}, yh_{i,j})$ is the head position of the template which is the nearest to the head position.

From the head position of the template, the corresponding body position of the template can be retrieved from the template database. The sub-blocks of the segmented image covered by the template are removed. The steps would be continued until all the templates corresponding to the potential positions of the heads are used.

In the second stage, the sub-block values of the template at (p, q) position is multiplied with the corresponding sub-block values on the remained binary image and the sum is obtained (Fig. 5a). Totally, there are P×Q sums when all the templates are used (Fig. 5b). These sums indicate the probability of presence of a person within the template mask. In order to reduce the effect of occlusion due to the body viewed from the camera, the template has higher weighting on the head and shoulder rather than the body, so the template values are 3 for head, 2 for shoulder, 1 for body and 0 for all others. If the sum is larger than a certain coverage percentage (CP), it should be considered as the presence of a person.

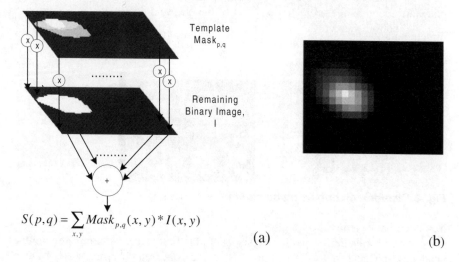

$$S(p,q) = \sum_{x,y} Mask_{p,q}(x, y) * I(x, y)$$

(a) (b)

Fig. 5(a) The sum calculation between template at (p, q) and remained binary image; (b) Sums of all S(p, q)

5. CLUSTER TRACKING AND COUNTING

The tracking algorithm is based on a matching process of the tokens between two successive frames. It includes estimation and matching. A token is also determined to be a coming-in token, coming-out token, tracking token or false detection. People counting is done by counting the number of tokens coming out of the tracking region. A tracking of the tokens between two successive time instants is carried out. Estimation of the token positions is by linear prediction from previous walking displacement. Matching is performed between the tokens within the tracking and alerting regions. People are counted as coming out of the tracking region when they pass through the top or down count line. The determination of a new comer is by a coming-in token in the tracking region. False detection is alarmed when a token in the previous time instant no longer matches any token in the current time instant within the tracking region. The test results show that the algorithm is capable of tracking the trajectory of the people and thus count the number of people.

6. EVALUATION AND RESULTS

The output of the people counting system is the counting results, which give the number of people passed through a counting region within a specified period of time, and the direction of travel. The developed system has been evaluated in many different scenarios, and in both crowded and non-crowded cases, with satisfactory results. The result in one scenario is given here. It is also found that the method can be implemented in real-time on a PC-733 MHz platform.

An image sequence obtained at five frames per second is processed by the system. The image resolution is 320x240. The snapshots of the scenario are shown in Fig. 6. The type of background is a corridor and the people can travel bi-directionally. The camera is mounted 5m from the floor and the coverage area is 4.14m(W)x3.09m(H).

(a)   (b)   (c)

Fig. 6. Snapshots of one test scenario

Some test results are shown below:

Condition	Number of frames	Expected no. of counting	No. of counting from the system	Absolute error	Absolute error with direction of travel	Overall counting error
Crowded Connected Uni-directional walking	1-894	Top=0 Down=234 Total=234	Top=6 Down=213 Total=219 False detection =47	Top=+6 Down=-21 Total=-15	11.53%	6.41%
Crowed Connected Bi-directional walking	1-459	Top=54 Down=59 Total=113	Top=51 Down=61 Total=112 False detection =17	Top=-3 Down=2 Total=-1	4.42%	0.88%
				Average	7.96%	3.65%

Table 1. Test results in a scenario with original background

Most of the errors are mainly due to the poor segmentation of the people. Also, when many people come together, their shadows sometimes have significant changes to the background. At the moment, the tracking algorithm uses simple geometric information to estimate and match tokens between two frames. A missing token along the path of walk would generate false detection and cause under-counting.

In a more extensive system evaluation, the average counting error is 4.23%. This result is comparable to or better than other similar approaches in a spatial domain. Especially in crowed situation with people walking bi-directionally, the above is a high-accuracy result. For comparison, the accuracies achieved by other researchers in [2], [3], [4] are around 94%, 87% and 90% respectively.

The developed algorithm has also been analysed for real-time performance. The worst-case computational time for all the system modules is given below:
1. Image acquisition t_{acq} : 2.3×10^{-6}s
2. Segmentation t_{seg} : 0.05s
3. Head search t_{head} : 0.00637s
4. Cluster removal by human model templates (one person) $t_{cluster}$: 0.00263s
5. Tracking and counting (for 40 tokens) $t_{trackcount}$: 85.7×10^{-6} s

Assuming that the maximum number of people in each image is 40 (N_{max}). The time required in the worst case for processing one image in the sequence is

$$= t_{acq} + t_{seg} + t_{head} + N_{max} * t_{cluster} + t_{trackcount}$$
$$= (2.3 \times 10^{-6} + 0.05 + 0.00637 + 0.00263*40 + 85.7 \times 10^{-6})s = 0.1617s$$

Hence, the system is capable of processing over six frames per second. The performance was evaluated using an Intel Pentium III 733MHz PC with 256MB RAM, running Visual C++ 6.0 under the Microsoft Windows 98 platform.

7. CONCLUSIONS

An automated people counting system has been developed. The algorithm of the system is mainly based on a novel two-stage people isolation to find the positions of individual person across an image sequence. The first stage of isolation would perform head search to locate some people using the features of circular black regions. The second stage of isolation uses model templates to locate the remaining people. These model templates are obtained from the perspective projection of a three-dimensional human model into a two-dimensional image plane. The use of the projected model templates considers the shape variation of the people located at different positions of a scenario.

The developed method has been tested with images captured at real scenarios at 5m camera-mounting height relative to a plain floor. People can walk in a bi-directional way with different degree of crowdedness. The overall system counting accuracy is around 95%. The worst-case computation time of the algorithm is also evaluated for the real-time feasibility of the system running on PC.

References

1. L. Khoudour, L. Duvieubourg and J. P. Deparis, "Real-time passenger counting by active linear cameras", Proceedings of SPIE on real time imaging, v. 2661, 1996, pp. 106-117.

2. R. Glachet, S. Bouzar, F. Lenoir and J. Blosseville, "Counting Pedestrian in the Subway Corridors using Image Processing", Proceedings of SPIE, Applications of Digital Image Processing XVIII, v. 2564, 1994, pp. 261-270.

3. X. Zhang and G. Sexton, "Automatic human head location for pedestrian counting", The Sixth Int. Conference on Image Processing and its Applications, vol. 2, 1997, pp. 535 –540.

4. M. Rossi and A. Bozzoli, "Tracking and counting moving people", Proceedings of IEEE International Conference on Image Processing 1994 (ICIP-94), vol. 3, 1994, pp. 212 –216.

5. A. Rourke and M. G. H. bell, "An image-processing system for pedestrian data collection", IEEE Conf. on Road Traffic Monitoring and Control, April 1994, pp. 123-126.

6. F. Bartolini, V. Cappellini and A. Mecocci, "Counting People Getting In and Out of a Bus by Real-time Image-sequence Processing", Image and Vision Computing, Vol. 12, No. 1, 1994, pp. 36-41.

7. P. Nesi and A. Del Bimbo, "A vision system for estimating people flow", Jorge L. C. Sanz, "Image Technology", Berlin, Springer-Verlag, 1996, pp. 170-201.

8. J. P Deparis, L. Khoudour, B. Meunier and L. Duvieubourg, "A device for counting passengers making use of two active linear cameras: comparison of algorithms", Proceedings of IEEE Int. Conf. on Systems, Man and Cyb., vol. 3, 1996, pp. 1629 –1634.

9. K. Terada, D. Yoshida, S. Oe and J. Yamaguchi, "A Method of Counting the Passing People by Using the Stereo Images", IEEE Conf. on Image Processing, 1999, pp. 338- 342.

10. F. Perez and C. Koch, "Toward Color Image in Analog VLSI: Algorithm and Hardware", International Journal of Computer Vision, Vol. 12, 1994, pp17-42.

Automatic Facial Expression Recognition for Human-Robot Interaction

A.Z. Kouzani,
School of Engineering and Technology, Deakin University.

Abstract
This paper presents a method to recognise facial expressions in images obtained by a camera. The system can be utilised in a humanoid or a home robot enabling the robot to receive commands from a person, or to perceive the person's mood or intention. The proposed method implements a global-local decomposition of the image under examination; it then recognises the expression. Experimental studies are reported, and a comparison with a number of existing techniques is presented.

Keywords: *Human-robot interaction, vision-based, expressions, global-local, quadtree.*

1. INTRODUCTION

Human-robot interaction is a growing area of research [3,7]. The goal is to develop robots with the ability to communicate and learn through social interaction with humans. A number of methods can be used to enable robots to interact with people. One approach is to use a speech recognition system. In this case, the robot is equipped with an auditory sensory system, as well as hardware and software required to recognise speech. A person could provide verbal commands to the robot via the sensor. There are a number of factors that limit utilisation of a speech recognition system. These include: sensor sensitivity and behaviour, complexity of audio signals, the need to train vocabulary for individual users, etc.

A different approach is to use a vision-based facial expression recognition system. The system includes a camera and a processing unit. The robot would receive commands from a person, or perceive the person's mood or intention, by processing facial images of the person.

Facial expression is one of the most effective ways for human beings to communicate their intentions and emotions. Facial expressions are produced by facial muscles. The muscles pull the skin, temporarily distorting the shapes of the eyes, brows, and lips, and the appearances of folds, furrows, and bulges in different patches of skin. Numerous methods exist for measuring facial movements resulting from the action of muscles [4]. The majority of the systems for automatic recognition of facial expressions attempts to characterise expressions into a few categories, such as anger, happiness, or disgust. Research in psychology has indicated that at least six emotions are universally associated with distinct facial expressions [4,11]. The six principal emotions are: happiness, sadness, surprise, fear, anger, and disgust. The evidence for universal facial expressions does not imply that these measurement categories are sufficient to describe all facial expressions [2]. Real face images consist of hundreds of distinct expressions, for which a gross categorisation is insufficient.

It is reported in the literature [1,5,6,8-9,11] that the methods which recognise facial expressions based on the local regions of the image have achieved higher success rates than those of the methods which recognise those expressions based on the entire image. A method is proposed in this paper for classification of an input image into one of the possible facial expression classes. This method implements a global-local decomposition of the input face image.

This paper is organised as follows. In Section 2, the exiting methods are reviewed. In Section 3, the proposed quadtree principal components analysis method is presented. In Section 4, experimental results are given. These results are then discussed in Section 5. Finally, concluding remarks are given in Section 6.

2. REVIEW OF EXSITING METHODS

A number of methods have been developed for automatic recognition of facial expressions. Some of the distinct existing methods are given in the following.

2.1 Feature-Based

Feature-based and template matching methods have been used for classification of facial actions. Bartlett et al. [1] explored classification of upper facial actions using feature measurements, such as the increase of facial wrinkles in specific facial regions and the degree of eye opening. A difficulty with the feature-based approach is that the selected features may fail to capture sufficient discriminative information.

2.2 Neural Networks

Neural networks can be trained to categorise facial expressions. Rosenblum et al. [10] developed a radial-basis-function-network-based human-emotion-detection system. They trained the network so that the network was able to learn the correlations between facial feature motion patterns and specific emotions. Neural network approaches can be advantageous for face processing because the physical properties relevant to classification need not be specified in advance. The network learns the relevant features from the statistics of the image set.

2.3 Appearance-Based

Another method which has been successfully applied to recognition of facial expressions is the Principal Components Analysis (PCA) [6]. Penev and Atick [9] developed a topographic representation based on the PCA. The representation is based on a set of kernels that are optimally matched to the second-order statistics of the input ensemble. The kernels are obtained by performing zero-phase whitening of the principal components, followed by a rotation to topographic correspondence with pixel location. This technique is called Local Feature Analysis (LFA). Bartlett et al. [1] has employed this technique for the classification of facial expressions.

2.4 Local Spatial Filters

Representations based on local spatial filters may be superior to spatially global representations for image classification. Padgett and Cottrell [8] improved the performance of their facial expression recognition system by performing the PCA on 32×32 sub-image patches of the face images, and using these 32×32 principal component images as convolution kernels. This finding is supported by Gray et al. [5] who obtained a better performance for visual lip-reading using the principal components of sub-image patches instead of the full images. Gabor filters, obtained by convolving a 2D sine wave with a Gaussian envelope, are local filters that resemble the responses of visual cortical cells. Representations based on the outputs of such filters at multiple spatial scales, orientations, and spatial locations, have shown to be useful for recognising facial expressions.

2.5 Discussions

The PCA has been widely used for recognition tasks [6]. The PCA typically obtains basis images which are non-local from a training set. However, recent studies suggest that the image structure in local regions of an image may be important to the classification tasks. As a consequence, it has been proposed that decompositions which use a local basis image are preferable [5].

Gray et al. [5] have compared the Local Principal Components Analysis (LPCA) versus the PCA and the Independent Components Analysis (ICA)-based image decompositions on automatic visual lip-reading task. They have reported that image decompositions with local basis images outperform decompositions with global basis images. Similar results have also been obtained by Padgett and Cottrell [8] on a facial expression classification task using a neural-network architecture. This supports the idea that the local basis may be a better approach for this classification task.

There are two issues associated with the LPCA-based image decomposition that need to be considered. These issues are: (i) selection of the size of local image regions, and (ii) selection of the location of image regions. A review of the existing literature shows that no attempt has been made for finding an appropriate size for local image regions. Different fixed-region sizes have been chosen by different methods. Whereas Padgett and Cotrell have used 32×32 sub-image patches, Gray et al. and Bartlett have used 12×12 and 15×15 sub-image patches, respectively.

Regarding the selection of the image region locations, two different methods have been discussed in the literature: (i) random location and (ii) fixed location. In the random location method, image patches are selected from random locations within the image. In the fixed location method, however, image patches are chosen from fixed pre-selected locations.

The method proposed in this paper implements a global-local decomposition of the input face image, in which the above-stated issues associated with both the PCA and the LPCA need not be considered.

3. PROPOSED METHOD

The proposed method is an image transformation that takes its name from the quadtree partition scheme on which it is based. A quadtree partition is a representation of an image as a tree in which each node, representing a square portion of the image, contains four sub-nodes that correspond to the four quadrants of the square. The root of the tree is the initial image.

Algorithm: Quadtree Principal Components Analysis (QPCA)
The QPCA method consists of the following operations:
1. An initial set of N n-dimensional face images, $\{I_1, I_2, ..., I_N\}$ (Training Set 0), is acquired. This forms the first level of the quadtree partition.
2. The principal components of the distribution of face images in Training Set 0 are calculated [6]. The basis vectors (images) are named $\{b_1^0, b_2^0, ..., b_N^0\}$. Any n-dimensional input face image I can be projected onto the basis images through the following operation: $w_k = b_k^0 (I - \mu)$ $(k = 1, ..., N)$, where μ is the average face. The weights w_k describe the contribution of each basis image to the input face image representation and can be used for reconstruction of the input face image. The reconstruction operation is implemented by $\hat{I} = \mu + \frac{1}{N} \sum w_i b_i^0$, where $\hat{I}$ represents the approximation of I by a global combination of the basis images. The error image can be obtained using $e = I - \hat{I}$.
3. For each face image I_k $(k = 1, ..., N)$ in Training Set 0, the following operations are performed:
    a. The image I_k is omitted from Training Set 0 and the PCA is performed on the rest of the face images to obtain $(N-1)$ basis images.
    b. The weights are calculated based on I_k and the basis images.
    c. $\hat{I}$ is reconstructed.
    d. The error image e_k is calculated.
    e. The error image is divided into four non-overlapping equally-sized sub-images. These sub-images are used to construct Training Set 1 (the second level of the quadtree partition).
4. The PCA is performed on the face images of Training Set 1 to obtain the associated basis images $\{b_1^1, b_2^1, ..., b_N^1\}$.
5. For each image in the latest training set the following operations are performed:
    a. One image is omitted from the training set and the PCA is performed on the rest of the images to obtain basis images.
    b. The weights are calculated by projecting the omitted image onto the basis images.
    c. The projected image is reconstructed.
    d. The error image is calculated.
    e. The error image is divided into four non-overlapping equally-sized sub-images and a new training set is constructed.

6. The PCA is performed on the images of the new training set to obtain the associated basis images.
7. A jump to Step 5 is always performed unless the sizes of images in the new training set are less than four pixels.

4. EXPERIMENTAL RESULTS

The QPCA together with five existing counterparts are implemented to compare their relative performances for expression recognition. The implemented methods are the PCA, the LPCA Random, the LPCA Fixed, the ICA, and the Fisher Linear Discriminant (FLD) [11]. In the LPCA Random and the LPCA Fixed, the size of sub-image patches are set to 15×15 pixels. Experiments are carried out on the following two test sets of face images.

Test Set 1 contains 765 front-view face images of 15 individuals synthesised from a selected set of images from the *Yale Face Database*. Figure 1 shows sample face images from this database. Ninety images of 15 subjects, containing the 6 facial expressions, are selected from this database. For each combination of two images of a subject, 3 extra images with different facial expressions are synthesised using the image morphing technique [6]. Therefore, the total number of images per subject will increase from 6 to 51, representing 51 different facial expressions. Figure 2 shows samples of the facial expressions synthesised in this way.

Test Set 2 contains 320 face images of 20 individuals selected from the *Second CMU Face Database*. Each image contains one of four different facial expressions: neutral, angry, happy, and sad. There are 16 images per subject, four front-view, four left-view, four right-view, and four up-view images.

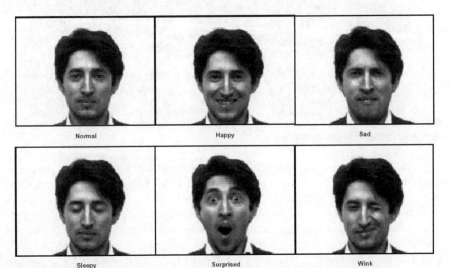

Fig. 1. Sample face images of a subject with six different facial expressions from the *Yale Face Database*.

Fig. 2. Samples of the intermediate facial expressions synthesised from the left and right images in each row.

The face area is manually extracted and automatically aligned using a pixel-based correspondence technique [6]. In Test Set 1, the faces are grouped based on the expressions they represent. Six groups are built and a reference image is selected for each group. In Test Set 2, the face images are grouped based on their poses and expressions. Sixteen groups are formed and a reference image is selected for each group.

The methods under examination are trained and tested using a leave-one-out cross-validation procedure [6] which makes maximal use of the available data for training. In this procedure, all of the images of that subject are reserved for testing.

This procedure is repeated for each of the 15 subjects in Test Set 1 and 20 subjects in Test Set 2. A simple nearest-neighbour classifier is used for classification. The classification results are displayed in Table 1.

Table 1. Classification of facial expressions for Test Sets 1-2.

Method	Test Set	Correct Classification	Classification Rate
Ideal System	1	765	100%
	2	320	100%
PCA	1	586	76.6%
	2	180	56.2%
ICA	1	569	74.3%
	2	185	57.8%
FLD	1	603	78.8%
	2	192	60.0%
LPCA Random	1	603	78.8%
	2	198	61.8%
LPCA Fixed	1	639	83.3%
	2	209	65.3%
QPCA	1	297	91.1%
	2	236	73.8%

5. DISCUSSIONS

According to Table 1, the best performance is obtained using the QPCA which achieves 91.1% and 73.8% correct classification for Test Set 1 and Test Set 2, respectively. The two LPCA methods outperform the PCA and the ICA. The FLD improves the classification performance obtained by the PCA and the ICA, but fails to compete with the QPCA. The results discussed here are based on the experiments performed on two small test sets containing a total of 410 face images. The face images do not contain all the possible facial actions. A larger collection of test sets would allow a more reliable evaluation of the different existing methods.

A problem associated with the data collection is that it is extremely hard to get the subjects to make facial actions, even the so-called universal expressions. Expressions of happiness and surprise are easy to elicit, whereas expressions of anger, disgust, and fear are difficult for most of the subjects. This data acquisition problem seriously limits the ability to properly characterise human emotional expressions. A direction for future research would be to use the data acquisition method in a controlled environment. Visual stimuli can be presented to the subjects, and their expressions recorded and analysed.

6. CONCLUSIONS

This paper explores a vision-based human-robot mutual communication system. A novel quadtree principal components analysis method is proposed to perform facial expression recognition. The method implements a global-local decomposition of the input face image. This solves the problems associated with the existing PCA and LPCA methods. Experimental studies suggest that the QPCA outperforms the other methods.

References

1. M.S. Bartlett, *Face Image Analysis by Unsupervised Learning and Redundancy Reduction*. PhD Thesis, Science and Psychology, University of California, San Diego, 1998.

2. C. Breazeal, A. Edsinger, P. Fitzpatrick, and B. Scassellati, "Active vision for sociable robots", *IEEE Transactions on Systems, Man and Cybernetics, Part A*, pp. 443-453, vol. 31, issue 5, Sept. 2001.

3. L. D. Canamero and J. Fredslund, "How does it feel? Emotional interaction with a humanoid LEGO robot", *Socially Intelligent Agents. Papers AAAI Fall Symp.*, Menlo Park, CA, pp. 23-28, 2000.

4. P. Ekman, "Facial expressions", *Handbook of Cognition and Emotion*, Wiley, New York, pp. 301-320, 1999.

5. M. Gray, J. Movellan, and T. Sejnowski, "A comparison of local versus global image decomposition for visual speechreading", *Proc. 4th Joint Symposium on Neural Computation*, pp. 92-98, 1997.

6. A.Z. Kouzani, *Invariant face recognition*. PhD Thesis, Flinders University, Adelaide, Australia, 1999.

7. H.G. Okuno, et al., "Human-robot interaction through real-time auditory and visual multiple-talker tracking", *Proceedings of IEEE/RSJ International Conference on Intelligent Robots and Systems*, pp. 1402 -1409, vol. 3, 2001.

8. C. Padgett and G. Cottrell, "Representing face images for emotion classification", Eds. M. Mozer, M. Jordan and T. Petsche, *Advances in Neural Information Processing Systems*, MIT Press, 1997.

9. P. Penev and J. Atick, "Local feature analysis: a general statistical theory for object representation", *Network: Computational in Neural Systems*, pp. 447-500, vol. 7, no. 3, 1996.

10. M. Rosenblum, Y. Yacoob and L. Davis, "Human expression recognition from motion using a radial basis function network architecture", *IEEE Transactions on Neural Networks*, pp. 1121-1138, vol. 7, no. 5, 1996.

11. Y.I. Tian, T. Kanade, J.F. Cohn, "Recognizing action units for facial expression analysis", *IEEE Transactions on Pattern Analysis and Machine Intelligence*, pp. 97 -115, vol. 23, issue 2, Feb. 2001.

Development of an Expressive Social Robot

Salvador Dominguez´Quijada[§], Eduardo Zalama Casanova[*], Jaime Gómez García-Bermejo[*], José R. Perán González[*]

[*] ETSII, Dep. of Automatic Control, University of Valladolid
Paseo del Cauce, s/n, 47011 Valladolid, Spain
E-mail: {eduzal, jaigom, peran}@eis.uva.es

[§] C.A.R.T.I.F. Parque Tecnológico de Boecillo. Parcela 205.
Boecillo, Valladolid, Spain,
E-mail: *saldom@cartif.es;*

Abstract

In this paper the development of a robotic head for a mobile robot with expressive capacity for its use as guide robot in a museum is presented. The robotic head can perform different facial expressions: happy, sad, angry, frightened, astonished, etc. using a caricatured face and speech generation.

Keywords: *Moving object localisation, vehicle localisation, triangulation.*

1. OVERVIEW

The design of agents with personality and emotions has been one of the most prominent challenges during the last years [2,4,11]. Different authors are working in the development of robots with capacity to displays gestures and to interact with people, in what are known as *social robots* [1,3,5,6,7]. In fact, performing movements and expressions with the body plays a crucial role in social interaction and communication (language of the body, facial expressions, oral communication, etc). The Minerva robot, developed by S. Thrun [3], is probably most similar robot to the one proposed in this paper [9]. Minerva is an interactive tourguide robot that displays four basic expressions: neutral, happy, sad and angry, using a caricatured face and simple speech. The robot's face has 4 degree-of-freedom (DoF): one DoF to control each eyebrow and two DoF to control the mouth. However, the different emotions Minerva can show are not an integral part of the robot's architecture.

This paper presents the development of a robotic head with nine degrees of freedom that allows multiple face expressions to be performed: happy, sad, angry, sleepy, frightened, astonished, etc. The movement of the neck is accomplished through three servos and permits multiple movement combinations: rotation, frontal nod and lateral nod. The different expressions of the face are procured through the remainder servos, which drive mouth, lips, eyes and eyelids. Two microcameras have been introduced into the eyeballs, so that what the robot is looking at can be displayed. A voice synthesizer and a set of LEDs across the lips (activated through a vumeter) have been employed to give more expressivity to the robot. Also, a set of sensors allows surrounding people and objects to be detected by the robot: an infrared movement detector perform medium-distance detection (0-4m), while four infrared reflexive sensors perform short-distance detection (0-50 cm) to detect objects or people in 4 different directions to the head.

The robot (see figure 1) has been equipped with a display to perform multimedia presentations and to accomplish guided tours in the museum.

Figure 1. The interactive robot museum.

The developed software for the head control (configuration of the different moods and vocal emissions) runs on a Pentiun III PC, under a Linux operating system. On the one hand the system has a reactive behaviour so that different moods are displayed depending on different situations. On the other hand, the system has a planner that performs different sequential tasks.

2. THE HARDWARE INTERFACE

2.1 The SCADAB (Servo Control And Data Acquisition Board)

The movement control has been achieved through two controlling boards (our own design) based on the PIC16F876 microcontroller. The main characteristicss of these boards are:
- Pulse With Modulation (PWM) control of up to 8 servomotors in position and speed with configurable speed ramps.
- 8 digital inputs.
- 5 analog inputs.
- RS232C serial connection.
- Configurable for the cascade connection of several boards.

The communication between the onboard personal computer (PC) and the controlling boards (SCADAB) follows the master-slave paradigm, where the master (PC) sends different commands to the specified slave (SCADAB). The slaves are arranged in a cascade configuration, in a USART bus. Up to 256 different slaves are possible (which gives the possibility of controlling 2048 servos, and reading 2048 digital signals and 1280 analogous signals). Two SCADA boards have been employed in this application. One board performs the control of the servos corresponding to the movement of eyebrows, mouth, lips commissure, eyelids and neck, as well as the reading of inputs from the presence sensors. The other board is used for the lateral movement of the eyes.

The communication between the PC and SCADAB is accomplished through an RS232 serial command communication. The different commands have the same structure: slave direction, command code (1 byte) and a CRC control byte.

Each servo shows different response to the PWM control signal because movement depends on servo size, torque, mechanical restrictions, frictions, etc. This situation can be managed by storing all the kinematic parameters in the internal EPROMs of the microcontrollers. This allows the kinematic parameters to be loaded when the board is turned on. Furthermore the head adopt a predefined face expression when the system is initialised, according to a predefined initial servo-angles set.

Figure 2. SCADA board with different servos.

An easy-to-use graphical interface has been also developed, in order to easily adjust all these parameters and store them in the microcontroller EPROMs, and in order to verify the sensor readings.

2.2 Other devices
The system is equipped with some other devices: Two microcameras have been introduced into the eyeballs, so that what the robot is looking at can be displayed.
- Two color microcameras located into the eyeballs. These cameras are employed for displaying what the robot is looking at, and can be also employed for machine vision. The cameras provide 330 000 pixel resolution, PAL images, and have a 65° field of view. They are small sized cameras (17x17x20'5 mm).

- Servos, to accomplish all the movements. Three different servo types have been used:
  - Futaba S3801: 14Kgr-cm torque, 59.2 x 28.8 x 49.8 mm size servos for neck movements.
  - Futaba S3001: 3,8 Kgr-cm torque, 40.4 19.8 x 36 mm size servo for mouth and eyebrows movement.
  - Futaba S3101: low torque, 40.4 x 19.8 x 36.0 mm size servo for eyes and eyelids movements.

3. THE SOFTWARE INTERFACE AND CONFIGURATION

Several functional modules have been implemented, in order for the robot to develop the different behaviours. The communication between these modules has been accomplished through TCA (Task Control Architecture) [8]. TCA is a process communication architecture oriented to robot programming. Each module is entrusted with a behavioural characteristic. Every module has some primary configuration parameters, which are read (at start-up) from a configuration file. Furthermore, the modules are equipped with a graphic user interface for checking whether the module is working correctly, as well as for programming new behaviours and face expressions. Figure 3 shows the structure and interconnection of the different modules.

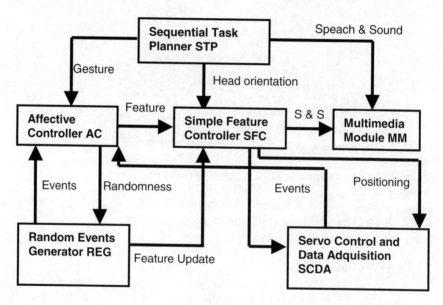

Figure 3. Control system structure. Communication between different modules is performed trough TCA (Task Control Architecture).

The following modules are employed:

SCDA. *Servo Control and Data Acquisition.* This module performs direct control of the hardware, sending commands to the SCADAB. In supervision mode, permits to perform different tasks:
- RS232 port control, to use in the communication.
- Servo kinematic and dynamic parameters remind.
- EPROM recording.
- Parameters load / Save (in a file), to keep different configurations

SFC. *Simple Feature Controller.* This module accomplishes the configuration of the different features that give expressiveness to the head. This is done by sending commands to the SSG and the SCDA modules. Each mood (sleepy, glad, happy, etc.) defines a set of simple movement features, with some degree of randomness defined in a configuration file. In this way, for example, "glad" mood does not match exactly with either a concrete face gesture, or a predefined speech sentence. In this way, face expressions are not repeated through time, which is one of the most surprising characteristics of the robot. Figure 4 shows different robot face expressions: neutral, happy and sad.

Figure 4. Different facial expressions.

Medium values and amount of randomness can be configured for the following features:
- Pan and tilt head position.
- Mouth aperture.
- Smile.
- Eyebrowns.
- Eyelids.
- Eyes rotation.

Furthermore, a file with different sentences can be specified for a mood. These sentences will be played by the speech generator in a reactive way, whenever there is not a situational plan specified by the STP.

REG. *Random Event Generator.* This module generates random events depending on a set of adjustable parameters. These parameters define the minimal and maximum repetition time for each type of event. Each event is associated to an elemental feature: eg. blinking, mouth opening, eye fixing, etc. When the REG is in supervision mode it is possible to configure many "states of activity", which corresponds to different vital rhythms of the head.

There is a correspondence between these activity states and the update rate for each feature in the SFC, that is to say, it is possible to configure the repetition frequency and the randomness for each feature. Thanks to this, the head updates feature in a non-foreseen instant.

AC. Affective Controller. This module defines the set of moods though a transitions graph. The transition state is accomplished probabilistically when a new event is produced. This module sends the feature configuration to be accomplished for the current state to the SCF and REG which defines the activity level. The AC module verifies whether the arriving events are present in the transition graph or not, for the current state. If the event is present, the system performs randomly a transition state to some of the successor states.

The transition graph can be configured in the supervision mode, in the following way: current state, events, successor states, probability of successor states.

The events are TCA messages that are sent to the AC. There are different kinds of events:

- Temporal event. Happens randomly and repeatedly in the time. It is generated by the REG module.
- Sensor detection. It is possible to configure a mask to indicate which sensor should be activated in order to produce an specific event. These events are sent by the SFC module.
- External event. Any event that is produced by an extern module.

MM Multimedia Module. This module is entrusted with all multimedia aspects: sounds, speech synthesis, pictures, video play, cameras, and other messages displayed on the screen. Each function can be activated by sending the corresponding TCA message to this module. The most important functions are the speech synthesis (which has to be synchronised with the mouth), and the display images from the eyes.

Speech synthesis is accomplished through the software provided by the University of Edinburgh, FESTIVAL (Festival Speech System) [10] and the image display is accomplished by SDL library. Other sounds are encoded and played as .wav files.

STP Sequential Task Planner: This module accomplishes the planning of tasks, issuing orders to the AC module for gesture change, to the SFC module for head reorientation, and to the SSG module for speech and sound generation. The plan is easily configured through a file containing a sequence of commands, such as "show a video", "say something", "play a wave file", "generate an event", etc. This way, sequential presentations (such as guided tours) can be performed by the robot, without losing its unforeseen and non-repetitive behaviour. This confers a great versatility to the robot. For example, it can perform a guided tour, showing videos and saying messages according to the museum contents, while behaves autonomously smiling, asking people to take off, or simply relating jokes.

Acknowledgment
This work is supported by the Comisión de Ciencia y Tecnología project 1FD97-1580, and the Junta de Castilla y León "Programa de apoyo a proyectos de investigación 2002"

References

1. Breazeal C. and Scassellati, How to Build Robots that Make Friends and Influence People, *In Proc. of IROS-99* Kyonju, Korea

2. Cañamero, L. Fredslund, J. 2001, "I Show You How I Like You. Can You Read it in My Face?." *IEEE Transactions On Systems, Man And Cybernetics. Part A: Systems and Humans, Vol. 31, Nº 5, September 2001. pp. 454-459*

3. Cassell J. Bickmore, Vilhjlmsson H., and Yan H. More than Just a Pretty Face: Affordances of Embodiment, *In Proc. of 2000 International Conference on Intelligent user Interfaces*, New Orleans, Lousiana.

4. Paiva A. Ed., Affective Interactions: Toward a New Generation of Computer Interfaces. New York: Springer Verlag, 2001, Lectures notes in Computer Science 1814.

5. Kirsch, The Affective Tigger: A study on the construction of an emotionally reactive toy, M.S. dissertation, Prog. Media Arts Scie., Mass Inst. Technol., Cambridge, 1999.

6. Rickel J., Gratch J., Hill S., Marsella, and Swartout. Steve Goes to Bosnia: Towards a New Generation of Virtual Humans for Interactive Experiences. In Proc. 2001 AAAI Spring Symposium on Artificial Intelligence and Interactive Entertainment, Tech. Report FS-00-04. Stanford University, CA.

7. Scheeff M.. Experiences with Sparky, a social robot, *in Proc. Workshop Interactive Robot entertaiment (WIRE)*, Pittsburgh, PA. 2000

8. Task Control Architecture. Department of Computer Sciences. Carnegie Mellon University.
http://www-2.cs.cmu.edu/afs/cs/project/TCA/release/tca.orig.html

9. Thrun, S., "Spontaneous short term interaction with mobile robots in public places", in *Proc. IEEE Int. Conference on Robotics and Automation, Detroit, MI. pp 10-15.*

10. The Festival Speech Synthesis System. The Centre for Speech Technology Research. University of Edinburgh. http://www.cstr.ed.ac.uk/projects/festival/

11. Trappl R. and Petta, P. Eds. Creating Personalities for Synthetic Actors: *Toward Autonomous Personality Agents.* New York: Springer Verlag, 1997.

Gesture Recognition for Commanding Robots with the Aid of Mechatronic Data-glove and Hidden Markov Model

K. P. Liu, S. K. Tso and B.L. Luk
Department of Manufacturing Engineering and Engineering Management, City University of Hong Kong
Hong Kong SAR, CHINA.

Abstract
This paper describes the design of a gesture-recognition system for commanding robot toys with the aid of a low-cost mechatronic data-glove and the hidden Markov model (HMM) technique. The mechatronic data-glove consists of a pair of orthogonal 2-D acceleration sensors that can measure acceleration in the x-y-z directions. Since the gesture is recorded in the form of noisy acceleration data, a wavelet-filtering technique is applied to smooth the data, and the velocity is calculated by integrating the smoothed acceleration data. The velocity profile is then transformed by the short-time discrete Fourier transform (STDFT) so that the time-domain profile is represented by a sequence of frequency spectrum vectors, which are more suitable for shape comparison. After the spectrum vector units are quantized to a finite number of symbols called observation sequence, it can be modeled and represented by HMM. Then, the gesture comparison and recognition is done by evaluating the observation sequence by all HMMs used to represent all the selected prototype gesture.

Keywords: Robotics, gesture recognition, hand-motion type, mechatonic data-glove, hidden Markov model.

1. INTRODUCTION

Due to the cost involved in purchasing and maintaining a robotic system, robots are traditionally considered mainly for industrial applications, such as car assembly. With the successful introduction of Sony's robotic dog, Aibo, many toy manufacturers have realised that there is a significant market demand for high-tech intelligent robot toys. More and more companies are interested in converting some of the state-of-the-art robotic technologies into real products. One such venture, initiated by a company in Hong Kong, is to develop a low-cost intelligent robotic arm for the toy market. In order to make the robot more fun to play with and responsive to the human intention, effective communication between the user and the robot becomes an important issue. Although speech is commonly used for transmitting human instructions, a more efficient and reliable way of sending human commands to a robot is based on gesture recognition [1-3]. The research issues of the gesture recognition involved are: (a) sensor technology to capture gesture and (b) algorithms to interpret the data. Regarding sensor technology, capturing of the dynamic features of human gesture requires a high sampling rate, and it is an important criterion for further data analysis. Obviously, traditional CCD camera technology with a maximum frame rate of 25 (or 30) frames per second is not fast enough. Generally speaking, 3-D feature extraction at a minimum of 60 frames per second [1] is required for gesture recognition purpose. Therefore, some high-speed capturing devices are designed for this application over the past few years. Two typical types of the devices are commonly applied. They are: artificial retina and data-glove. The artificial retina [2] is basically a CMOS vision chip [1] with signal processing being implemented on the same CMOS chip that allows a processing rate as high as a thousand frames per second. This processing rate improvement is due to the effective reduction of the bottleneck of data transfer in conventional vision systems, where the video signal is a time-multiplexed signal of pixel data by using scanning circuits.

In contrast, the CMOS chip can implement each pixel data by distributed processing elements associated with the pixel so that there is no such a bottleneck of pixel scanning by a frame grabber. However, the main disadvantage of this design is still low resolution with the present VLSI technology, and it is not capable of fine image analysis. The data-glove device was developed when the virtual reality system was commercially launched a decade ago. The sensor element for 6-D spatial measurement is conventionally based on an electro-magnetic device (e.g. Fastrak by Polhemus [4]). When applying this traditional data-glove to capture hand motion for gesture recognition, there are disadvantages of low sampling rate, poor accuracy and high cost. In order to provide a low-cost, high sampling-rate device to capture hand-motion gesture, a specially designed data-glove based on a low-cost accelerometer sensor is built locally in the laboratory to serve this application [5]. The sensor used is ADXL202 (Analog Device) and the sampling rate can be up to a few hundred samples per second. Typically, it is tuned to run at 60Hz for experimental evaluation, and applied to record motion gesture.

Some algorithms reported to interpret motion-gesture data are template matching [6], statistical matching [7], neural network [8], and hidden Markov model (HMM) approach [9]. The paper describes the HMM technique as applied to gesture recognition, based on a low-cost gesture capturing system. In Section 4, experimental results are presented to illustrate the method proposed.

2. DESIGN OF A LOW-COST GESTURE CAPTURING SYSTEM

The hand-motion capturing system consists of a host computer, an 8-bit MCU board, and a hand-glove as shown in Figure 1. The accelerometer chips on the hand-glove converts motion information to electrical signals. The MCU board processes the electrical signals, transforming them to 8-bit data. The host computer implements the data analysis algorithm for gesture recognition.

2.1 Principle of recording motion information by accelerometer

The accelerometer chip (ADXL202) is a dual-axis acceleration measurement device built on a single monolithic IC. For each axis of measurement, an output circuit converts the analog signal to a duty-cycle modulated digital signal that is ready for micro-controller TTL input. The accelerometer is capable of measuring both positive and negative accelerations up to, effectively, a maximum level of ±4g. The micro sensor is suspended on polysilicon springs on the surface of the wafer. Deflection of the structure is measured using a differential capacitor that consists of two independent plates and central plates attached to the moving mass. The fixed plates are driven by two square waves, which are 180° out of phase. Acceleration will deflect the central plates and unbalance the differential capacitor, resulting in two output square waves whose amplitudes are proportional to the acceleration in the two directions. The acceleration direction is recognised by the phase difference of the two output square waves. As one sensor provides two-directional information, a pair of them are applied to record 3-D hand motion, and they are orthogonally mounted on a hand-glove, as shown in Figure 2, where two signals are common with the Y-dimension so that either one of the signals is selected to give the information in this direction.

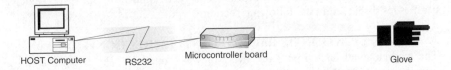

Fig. 1 Motion-detection glove structure

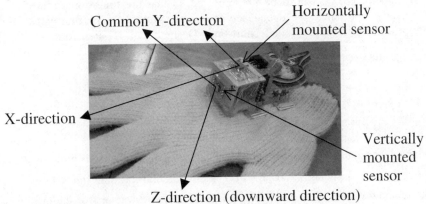

Fig. 2 Hand-glove with two accelerometer sensors mounted

Since the hand-motion is recorded in the form of noisy acceleration, the signal is first digitally filtered so that a more accurate velocity profile generated by integration can be obtained. The digital filter applied is based on the wavelet-type Daubechies filter [10], discussed in the next subsection.

2.2 Daubechies filter technique

Each acceleration signal is recorded in the form of a time series. A window of length 4, with positive Daubechies distribution, is applied to the time series. The dot product of the window and the time series segment is calculated as the 'average' value of the segment. A second window of similar type, but with alternating sign and revised in the Daubechies distribution, is applied to the same time segment. The corresponding dot product is regarded as the 'detail' value of the segment. Both windows are applied and moved along the whole time series. The resultant 'average' and 'detail' data series are called the Daubechies wavelet transformation of the original time series. A simple threshold comparison is applied to the 'detail' values so that all values below the threshold setting are floored to zero. Then an inverse process of the above wavelet transformation (called inverse wavelet transformation) is applied to the 'average' and the modified 'detail' values so that the original time series is recovered with unimportant noise removed. The advantage of this filtering technique over the traditional digital filter is a shorter computational time.

Since it is intended that the gesture information is based on the velocity profiles, an integration process is next applied to the filtered acceleration data. The gesture recognition by the HMM process is then applied to the 3-D velocity profiles, as discussed in the following section.

3. APPLICATION OF HMM TO GESTURE RECOGNITION

The mathematical background of HMM may be found in [9]. It is basically a probability approach to model or represent a gesture by an HMM parameter λ. Before applying the HMM to recognise a gesture, the gesture input in the form of a 3-D velocity profile has to be preprocessed so that the time domain profile is eventually represented by a sequence of discrete symbols. The first part of this pre-processing stage is called short-time Discrete Fourier transform (*STDFT*) modified from *STFT* [11]. The single-dimension velocity profile of the gesture is first processed increment by increment as shown in (1) below:

$$STDFT_x^{W_h}(i, f_r) = \sum_{i'=0}^{N-1} \left[x_{i+i'-\frac{N}{2}} \cdot W_h\left(i'-\frac{N}{2}\right) \right] e^{-j2\pi\varpi\frac{i'}{N}} \qquad (1)$$

where r represents the dimension of f, and the velocity profile x_i is multiplied by a moving 'analysis window' $W_h\left(i'-\frac{N}{2}\right)$, centered around the time index i, and N is the window width. This gives in fact a local spectrum vector f of the profile x_i around time index i. The process applies to X, Y, and Z dimension independently, and then the resulting three sequences of spectrum vectors are combined in cascade to form a single sequence of spectrum vectors with higher dimension. The frequency spectrum reflects the shape and amplitude of a short-time portion of the profile. In the second part of preprocessing stage, the frequency spectra are quantized to a limited number of spectrum-vector units. This part is processed differently for modeling and for evaluating the velocity profile. In the case of modeling gesture: the lists of spectrum vectors, transformed from the velocity profile of all possible prototype gestures, are quantized into a finite number of spectrum-vector units. As the quantization is multi-dimensional, it is called vector quantization (VQ). The algorithm chosen is the LBG algorithm [12]. The steps are summarized below:

1. Initialization: Set the number of partitions $K = 1$, and find the centroid of all spectrum vectors in the partition.
2. Splitting: Split K into $2K$ partitions.
3. Classification: Accept the kth partition C_k of each spectrum vector, v depending on the specified condition; i.e.

$$v \in C_k \text{ iff } d(v, \bar{v}_k) \le d(v, \bar{v}_{k'}) \text{ for all } k \ne k' \tag{2}$$

where $\bar{v}_k$ is the centroid vector of C_k and d is a distortion measure to be defined as a general norm.

4. Centroid updating: Recalculate the centroid of each accepted partition.
5. Termination: Steps 2 to 4 are repeated until the decrease in the overall distortion, at each iteration process, relative to the value at the previous process, is below a selected threshold. The number of partitions is increased to a value that meets the required level.

After termination, we will have a number of centroids, $\{\bar{v}_k\}$, of all the partitions. These centroids are in fact the spectrum vector units that represent spectrum vectors transformed from all possible short-time portions of the velocity profile. In the case of performing evaluation, the frequency spectra for an unknown gesture are mapped to the prototype spectrum vectors $\{\bar{v}_k\}$. The mapping is based on the minimum-distortion principle, with the distortion measure given by (3) below.

$$d(v, \bar{v}_k) = \|v, \bar{v}_k\| = \sum_{r=1}^{R} (v_r - \bar{v}_{k,r})^2 \tag{3}$$

where R is the total spectrum vector dimension. After completion of the mapping, the velocity profile is converted to a list of spectrum vectors $\{\bar{v}_k\}$. In the language of HMM, $\{\bar{v}_k\}$ is written as $\{O_k\}$, called the set of observation symbols, which will be sent through the tuned HMM for evaluating the likelihood index which is given by the conditional probability of getting $\{O_k\}$ given the HMM representing a certain gesture.

4. EXPERIMENTAL RESULTS AND DISCUSSION

To demonstrate the application of gesture recognition for commanding robots with the aid of the mechatronic data-glove and HMM, five prototype gestures are developed; they are (1) COME, (2) GO, (3) STOP, (4) LEFT and (5) RIGHT, which are shown in Figure 3. The recorded acceleration profile of a typical STOP gesture is shown in Figure 4a. After having applied the Daubechies filter process on the raw acceleration data, the resulting 3-D velocity profile, generated by integration process on the filtered acceleration data, is shown in Figure 4b. The *STDFT* process by (1) is applied to each dimension of the 3-D velocity profile. Three sequences of spectrum vectors are generated and then combined to form a single sequence of spectrum vectors with higher dimension, as shown in Figure 5. The spectrum vector units are then quantized into a number of discrete symbols according to (3). The outlook of the symbol listing is shown in Table 1.

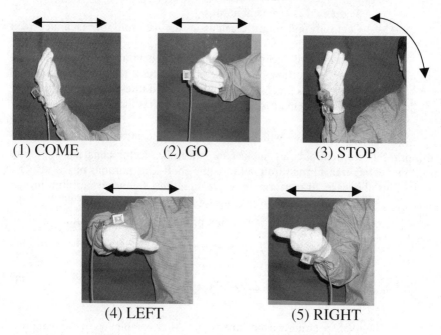

Fig. 3. Five prototype gestures with indications of swinging directions

Table 1 Outlook of the observation symbol sequence representing the sequence of spectrum vectors

Portion index	1	2	3	4	• • •	76	77	78
Symbol index	50	45	52	48	• • •	79	83	96

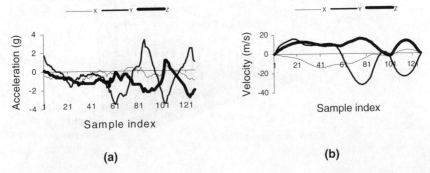

Fig. 4. Recorded acceleration and generated velocity profiles of a typical STOP gesture

Since the five prototype gestures have been modeled with the described treatment, they can thus be represented by five sequences of observation symbols. For each gesture, the exercise is repeated five times to improve the quality of the prototype. By the principle of trajectory selection reported in [13], the best exercise is selected to represent the prototype. Since the human cannot repeat exactly the trace of a certain motion, the profile shape may shift somewhat along the time axis even for the same gesture. The time-warping process [14] is applied to adjust the time scale to let a sequence of observation symbols from an unknown gesture map to a prototype one.

As a dynamic and probability-based time-warping process, HMM is applied to adjust this time scale. The details of the HMM application can be found in [9]. To put it simply, the observation sequence of each prototype gesture is represented by its respective HMM parameter λ_i, where $i = 1$ to 5, corresponding to the five prototype gestures. A test gesture is preprocessed by the same treatment as the prototype, and the output observation sequence O_t is evaluated by each λ_i by calculating the conditional probability $P(O_t | \lambda_i)$. The probability values obtained experimentally are $(0.16\ \ 0.18\ \ 0.43\ \ 0.12\ \ 0.11)$ after normalization. The test gesture is hence recognized as the third prototype, which is the STOP gesture, because λ_3 is distinctly highest. By using other test gestures for recognizing all the five prototypes, with each type repeated twenty times, the results show that the average successful recognition rate is 95.6%.

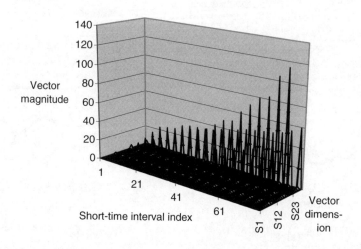

Fig. 5. Generated short-time frequency spectrum vectors by *STDFT*

5. CONCLUSION

A scheme of gesture recognition for commanding robots with the use of the mechatronic data-glove and HMM is reported. As the recognition accuracy is better than 95%, it can be applied to commanding robots by human hand gesture. Although the present prototype gestures are only COME, GO, STOP, LEFT and RIGHT, it is not difficult to extend the method to cover more gestures depending on the requirement. As the intended practical application is for robot toys, the cost is an important concern. The component cost of this simple mechatronic data-glove built in the laboratory is below a hundred US dollars.

Acknowledgement

The work is supported by an RGC Central Allocation grant in Hong Kong SAR on service robots awarded jointly to Chinese University of Hong Kong and City University of Hong Kong.

References

[1]   H. Miura, H. Ishiwata, Y. Iida, Y. Matunaga, S. Numazaki, A. Morisita, N. Umeki, M. Doi, ' A 100Frame/s CMOS Active Pixel Sensor for 3D-Gesture Recognition System,' *Proc. IEEE Int. Conf. on Solid-State Circuits, Digest of Technical Papers*, 1999, pp. 206-207.

[2] Y. Miyake, W. T. Freeman, J. Ohta, K. Tanaka, and K. Kyuma, 'A Gesture Controlled Human Interface Using an Artificial Retina Chip,' *Proc. IEEE Lasers and Electro-Optics Society Annual Meeting*, 1996, Vol. 1, pp. 292-293.

[3] H. Kage, K. Tanaka, and K. Kyuma, '3-D Human Motion Sensing by Artificial Retina Chips,' Proc. Third IEEE Int. Conf. On Automatic Face and Gesture Recognition, 1998, pp. 522-527.

[4] *Frastrak User's Manual*, Revision F, Polhemus Inc., Colchester, Vermont, U.S.A.

[5] A. Choy, J. Chung, K. P. Liu, and B. L. Luk, 'Development of a Mechatronic Chess-playing System,' Technical Report, Centre for Intelligent Design, Automation and Manufacturing (CIDAM), City University of Hong Kong, January 2002.

[6] J. S. Lipscomb, 'A Trainable Gesture Recognizer,' *Pattern Recognition*, Vol. 24, No. 9, 1991, pp. 895-907.

[7] D. H. Rubine, 'The Automatic Recognition of Gesture,' *PhD Dissertation*, Computer Science Department, Carnegie Mellon University, December 1991.

[8] S. S. Fels, and G. E. Hinton, 'Glove-talk: a Neural Network Interface between a Data-glove, and a Speech Synthesizer,' *IEEE* Trans. on Neural Networks, Vol. 4, No. 1, 1993.

[9] J. Yang, 'Hidden Markov Model for Human Performance Modeling,' *PhD Dissertation*, the University of Akron, August 1994.

[10] G. Strang, and T. Nguyen, *Wavelets and Filter Banks*, Wellesley-Cambridge Press, 1996, MA.

[11] F. Hlawatsch, and G.F. Boundeaux-Bartels, 'Linear and Quadratic Time-frequency Signal Representations,' *IEEE SP Magazine*, Vol. 9, No. 2, April 1992, pp. 21-67.

[12] Y. Linde, A. Buzo, and R.M. Gray, 'An Algorithm for Vector Quantizer Design,' *IEEE Trans. on Communication*, Vol. COM-28, 1980, pp. 84-95.

[13] Tso, S. K., and Liu, K. P. [1997]. 'Demonstrated Trajectory Selection by Hidden Markov Model,' *Proc. IEEE Int. Conf. on Robotics and Automation*, Albuquerque, New Mexico, USA, April 1997, pp. 2713-2718.

[14] X.D. Huang, Y. Ariki, and M.A. Jack, *Hidden Markov Models for Speech Recognition*, Edinburgh University Press, 1990.

A Mechatronic System for Non-Invasive Treatment of the Breast Tissue

Sunita Chauhan,
Division of Mechatronics and Design,
School of Mechanical and Production Engineering,
Nanyang Technological University,
Singapore 639 798

Abstract
Present methods of breast cancer treatment are either conventional/radical mastectomy or breast conservation procedure called lumpectomy. Lumpectomy procedures are limited to early growth treatment. This paper deals with a non-invasive means, using High Intensity Focused Ultrasound (HIFU) as the surgical modality, which can be applied to ablate early or moderate stage tumours in situ. The design and development of a mechatronic system operating partially in a water tank (as coupling medium) to guide the end-effector consisting of an assembly of multiple HIFU transducers, through a pre-determined trajectory is described. A PC based controller and treatment-planning module governs various sub-sections of the system and deploys a trajectory within a safe constrained work envelope under the surgeon's control. It is feasible to fragment the procedure and apply on an outpatient basis, thereby reducing the risks of either a further and fast growth of an early stage cancer or disfigurability and cosmetic concerns.

Keywords: *Mechatronics, non-invasive focal ultrasound surgery, treatment planning.*

1. INTRODUCTION

In the past approximately two decades, computer assisted or integrated imaging and surgery has dramatically improved the ways conventional open surgery procedures were performed. Computer integrated medical interventions and applications of robotic technologies in surgery have lead towards minimal invasive and safer procedures. These improvements include, but not limited to, smaller treatment times, less postoperative complications and accurate prognosis. Robotic/mechatronic assistance yields higher accuracy, precision, reliability and repeatability in manipulating surgical instruments in desired locations.

Remote ablation of deep-seated abnormalities by various modalities, such as lasers or the use of focused high intensity ultrasound, can provide completely non-invasive procedures if the energy in the beam is carefully targeted. High Intensity Focused Ultrasound (HIFU), alternatively known as Focal Ultrasound Surgery

(FUS), have shown promising clinical evidence, particularly in the field of urology and oncology [3,5,6,9].

1.1 Breast Cancer Treatment: Present-state-of the art

Breast cancer is reported to be the second leading cause of cancer death for all women after lung cancer, and also the leading cause of death in women between the ages of 40 and 55. Current surgical treatment for breast cancer is largely invasive. These include mastectomy or radical mastectomy and lymph node removal, wherein the complete tissue/organ is surgically removed. Breast conservation procedures, such as lumpectomy, involve removal of the affected tumour mass and its margin area from the breast. This process can also result in disruption of the breast contour (as a direct result of tissue loss), and scarring of the intervening normal tissue in the process during access to target site.

Hence, for optimum cosmesis for small or moderate sized lumps/tumour, non-invasive, trackless procedures such as tissue ablation using HIFU promise to achieve adequate local control whilst preserving the breast and thus provide better clinical acceptability.

1.2 Non-invasive HIFU Approach

Tissue ablation by HIFU is primarily due to conversion of mechanical energy of an ultrasound wave into heat energy at its focal point. A temperature range of 60-80°C is achieved and the thermal effect could lead to immediate coagulative necrosis within the focal zone, sparing overlying structures. Conventionally, FUS employs large aperture transducers (typically 100mm) for generating the desired intensities in the lesion sites. These comprise either single spherical disc type probes or multiple array configurations [1,3,4]. In both types, the secondary energy lobes can be high enough, leading to the formation of hot spots in off-focal regions, thus affecting intervening normal tissue.

One approach to minimise the occurrence of undesirable hot-spots, using low power multiple probes, was investigated by the author in a previous study [7,8]. In this technique, individual low intensity beams are not lethal to the healthy tissue and provide more flexibility to position the ultrasound probes over tissue surface. The spatial configurations of these probes, with respect to each other and with respect to the target tissue, were studied by developing 3-D numerical models and verified in *in vitro* and *ex vivo* tests. The computer simulations help in treatment planning for dosage levels and optimum range of inter-probe distances and angles of orientation (defining size and shape and intensity in the superimposed foci).

A multi-transducer HIFU manipulator system and its jig control sub-system for the treatment of breast tumours/cancers, are described in this paper.

2. DETAILED SYSTEM DESIGN

The dimension of the focal region produced in a single exposure, called a *lesion*, by the multi-transducer assembly, may not be able to cover the entire volume of the desired target region. It is necessary to mechanically scan the superimposed foci mutually over the entire tissue of interest. For scanning the con-focal region over the target area, a mechanical manipulator is developed for precise and accurate trajectory deployment in a given spatial configuration. The treatment is to be carried out by means of a suitable exposure duration of the HIFU beam at one spot, following a scanning motion of the probes and subsequent exposure, thus covering the entire volume of the lump in a three-dimensional manner.

The HIFU surgical system developed for breast cancer ablation comprises of various sub-sections, as described in the following text:

2.1 Energy Delivery Module

HIFU transducers, in the present work, have 25-mm active aperture, focal depth of 64–80 mm and operate at 2 MHz frequency. The focal dimensions of a single transducer of these specifications, in water, are measured as shown in figure 1.

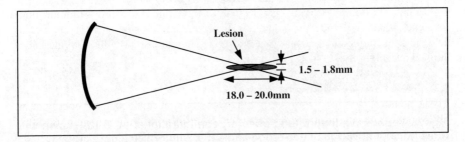

Fig. 1. Dimensions of the focal region

The spatial arrangement of the transducers is selected such that the constructive interference of the sound beams in the focal region leads to high intensity focal spot required for tissue ablation, whilst suppressing the presence of undesirable hot-spots in the regions outside the focal area (owing to destructive interference in those regions). The preferred spatial configurations (position and orientation with respect to each other and to the target) of the participating probes and the resulting acoustic field have been studied by developing numerical models and computer simulations [7].

Based upon the simulations and specific tissue anatomy (to infer accessibility to the target tissue), the transducers are arranged in a given spatial configuration in a jig fixture. This jig will serve as an end-effector for the manipulation system. A jig assembly for a typical arrangement of 90^0 with vertical, changeable inter-probe angles and provision of a diagnostic probe at the central shaft is shown in figure 2 (a & b). The jig is designed with collars of equal diameter to secure the transducers, such that the composite focal point is always achieved in the central axis. Individual probes are locked via grub-screws on the body of probe holders. A turn dial is provided to accommodate desired number of arms and each of the arms can be adjusted at a given tilt angle. The orientation of HIFU probes with respect to each other in spatial configuration can be pre-planned, such that the superimposed region is of desired dimension and intensity levels and is targeted in the diagnosed lesion.

Figure 2 (c) delineates a motorized end-effector configuration to achieve better flexibility and accuracy of angular orientations. A single large aperture probe is shown. It consists of a pinion and an arc (with gear connecting to a stem), at the end of which a probe holder is attached. The pinion is connected to gear-head shaft through direct coupling. A PC based programmed controller and planning routine provides stimulus to the probes through appropriate power amplification depending upon a selected protocol.

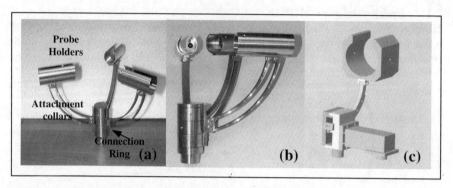

Fig. 2. Various Jig-fixture (end-effector) configurations: (a) & (b) shown with three holders at various inter-probe angles; (c) motorized single probe

2.2 Mechanical Manipulation Sub-system

The superimposed foci, as obtained in the jig end-effector arrangement, are required to be scanned mechanically to ablate the entire area of interest in three dimensions. In the manipulation of the transducers, it is critical to synchronize their movement to maintain the superimposed foci, and deploy the desired energy in the target region. In the present work, a 3-axis (x, z and θ) mechatronic manipulator, with a cylindrical work envelope, was designed and developed to guide the transducers through a desired scanning trajectory accurately and precisely.

Various desired movements and reach of individual axis of the manipulation system during breast surgery are calculated based upon the anthropomorphic data of an average female, as shown in figure 3. The surgical system, in its present configuration, can have extended applications for other soft tissue, such as kidney, liver, prostate, etc. as accessible through abdominal acoustic window. A PC based 4-axis controller has been used for trajectory programming and controlling various axis of the manipulator module and its registration with the target. The transducers, prefixed in the jig assembly, ensure that focusing is maintained during the entire scanning process.

The manipulation system partially works in a degassed water tank, so as to provide a couplant medium for effective transfer of ultrasound energy into the tissue, as shown in figure 4. The part of the manipulator submerged in water is shown in the inbox. The jig module is mounted on the central shaft and can be maneuvered in a desired region within the tank.

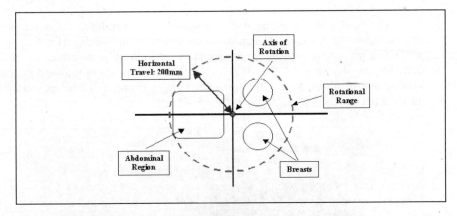

Fig. 3. Manipulation work envelope calculation

There are circular openings to allow the affected breast tissue to descend through a tissue immobilizing cup held in position and wrapped in a latex membrane into the water tank, while the patient lies in a prone position. In this manner, full access to the breast tissue is available for the robotic treatment to be performed underneath the operating table, as shown in figure 4.

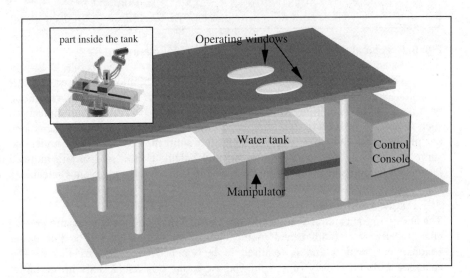

Fig. 4. Manipulation module and system set-up

2.3 Surgical Planning and User interface

During clinical set-up, treatment posture of the patient, as well as placement of the surgical equipment, must be carefully planned to ensure an efficient and accurate performance. A pre-surgical planning based upon numerical modelling of tissue-modality interaction and safety sub-system has been designed and integrated in the system [8].

For on-line planning, the HIFU transducer system seeks its reference target spot by accessing real-time data from the diagnostic scan module. The diagnostic probe is a part of the transducer jig assembly, whereby it is kept focused at the same region of interest as the HIFU transducers. Both diagnostic and surgical systems are orientated so as to be referenced with a common 'home' position and thus to each other.

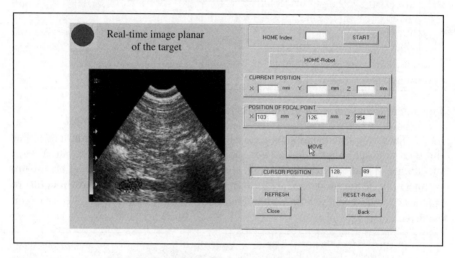

Fig. 5. Graphical user interface for on-line surgical planning

The graphical user interface, as shown in figure 5, allows the user to select and plan target points on subsequent on-line images to ablate the desired tissue with a soft switch control. All the stringent treatment parameters for sonication, such as power, duration of exposure, dead time between subsequent exposures etc. are planned and entered at this stage and the entire process is guided under the surgeon's control. The system automatically calculates the size of the induced lesion based upon a given HIFU probe configuration and imaging planes selected.

3. DISCUSSIONS

The ability to focus ultrasound beams in a defined region in order to ensure precise energy delivery is fundamental to the clinical application of HIFU. For larger lesions, this focal region is required to be scanned over the target area. The accuracy and consistency of scanning and creating adjacent lesions in the target can be improved by mechatronic means.

The visualisation of breast lesions with ultrasounds is well recognised in clinical practice. Non-invasive surgery using HIFU modality may provide an alternative solution at an early stage, when the size of the tumour is still small and has not spread beyond the original site. With this procedure, both risks and complication arising from open surgery can be avoided. Furthermore, patients undergo less trauma, can be treated on an outpatient basis and there are no ionising radiation risks. In extended cases of cancer stage, HIFU can be employed as an adjutant modality to control further tumour growth and invasion.

Various factors require attention in order to attain high therapeutic gain factors in biological tissues. The anatomy and functionality of the tissue under consideration should be properly studied for planning accessibility window and identifying potential targets to assess the severity of the risks. Tissue/organ location and geometry in respect to its surrounding areas in the human body will also affect the design of applicator systems specifications. The optimum specifications of the applicators (and thus the beam shape) can be evaluated by modelling the ultrasonic field in front of the transducers and considering the absorption characteristics of intervening tissues. The single focused transducers, however, are not very flexible in creating a desired spatial distribution of ultrasound energy within the treatment field.

In the present design of the multi-probe manipulator system, various components are immersed in water during treatment routine. For instance, the x-axis actuator, carrying the jig, is fully submerged in water without any cover, and driven by an encapsulated motor directly coupled to the actuator. Before this concept was approved, among various other alternatives such as water filled bellows for individual probes, a number of experiments were carried out to verify its feasibility. The other concern of exposing the drives to water is formation of bubbles due to air trapped in the openings in the drive. To eliminate this problem, the drive is run several times in the water to get rid of any air that might be trapped before degassing the water.

Hardware and software safety routines are augmented in the system so as to ensure that any movement to the subsequent points directed by a given protocol and the energy delivery system are not 'switched on' simultaneously at any time. During exposure routines, the manipulator is 'powered off' in order to ensure the referencing as well as to reduce interference, if any.

Means for detection of minor patient movements during the operation and to minimise patient movement are being studied. For excessive movement of the patient, or, wrongly planned trajectory, emergency shut down has been implemented in the system, which can then be restarted after targets are reconfirmed and registered again.

4. CONCLUSIONS

A surgical prototype system for breast cancer ablation using HIFU modality has been designed and tested for an overall manipulation accuracy within 2 mm. *Ex-vivo* tests have been performed in various fat and muscle tissue types such as beef, pork adipose tissue, lamb, etc. Further improvements and extension of the present system to treat other tissues through clear abdominal acoustic window are being studied.

Acknowledgement
The author would like to thankfully acknowledge the support from the Ministry of Education, Singapore, and the Agency for Science, Technology and Research, Singapore, for jointly funding the project. Thanks are also due to the staff and students at the Robotics Research Centre, NTU, who worked on various developmental parts of this project.

References
1. G. ter Haar, I. Rivens, L. Chen and S. Riddler: High intensity focused ultrasound for the treatment of rat tumors, Physics in Medicine and Biology, 1991, pp 1495-1501.
2. R. Yang , D.R. Reilly F.J. Rescorla, R.R. Faught, N.T. Sanghvi, F.J. Fry, T.D. Franklin, L. Lumeng and J.L. Grosfeld: High intensity focused ultrasound in the treatment of experimental liver cancer, Arch Surg, 1991, pp 126: 1002.
3. K. Chartier, D. Chopin and G. Vallancien: The effects of focused extracorporeal pyrotherapy on human bladder tumor cells, Urol., 1993, 149: pp 643-647.
4. K. Hynynen, A. Chung T. Fjield, M. Buchanan, D. Daum, V. Colucci, P. Lopath and F.A. Jolesz: Feasibility of using ultrasound phased arrays for MRI monitored non-invasive surgery, IEEE Trans. Ultrason., Ferroelect., Freq. Contr., 1996, 43, 6: pp 1043-1052.
5. A. Gelet , J.Y. Chapelon, R. Bouvier, O. Rouviere, Y. Lasne, D. Lyonnet, J.M. Dubernard: Transrectal high-intensity focused ultrasound: minimally invasive therapy of localized prostate cancer. J Endourol, 2000, 14, pp 519-524.
6. S. Madersbacher ,G. Schatl, B. Djavan, T. Stulnig and M. Marberger: Long-term outcome of transrectal high-intensity focused ultrasound therapy for benign prostatic hyperplasia. Euro Urol, 2000, pp 37: 687.
7. S. Chauhan, M.J.S. Lowe, B.L. Davies: A multiple focused probe approach for HIFU based surgery, Journal of Ultrasonics (Medicine and Biology), UK 2001, Vol.39, pp 33-44.
8. S. Chauhan: Field modelling for multiple focused ultrasound transducers. Proc M2VIP, 8th IEEE International Conference on Mechatronics and Machine Vision in Practice, Hong Kong, 27-29 Aug 2001.
9. K.U Köhrmann, M.S Michel, J. Gaa, E. Marlinghaus and P. Alken: High intensity focused ultrasound as noninvasive therapy for multilocal renal cell carcinoma: case study and review of the literature, The Journal Of Urology 2002;167, pp 2397-2403.

Author Index

Al Janobi, Abdulrahman, King Saud University, Saudi Arabia, 213
Bečanović, Vlatko, Fraunhofer Institute for Autonomous Intelligent Systems, Germany, 13
Billingsley, John, University of Southern Queensland, Australia, 221, 253
Bin, Li, Northwestern Polytechnical University, China, 295
Bradbeer, Robin, City University of Hong Kong, Hong Kong, China, 295
Brett, P N, University of Aston, UK, 205
Casanova, Eduardo Zalama, University of Valladolid, Spain, 105, 263, 341
Chauhan, Sunita, Nanyang Technological University, Singapore, 359
Chen, S Y, City University of Hong Kong, Hong Kong, China, 3
Chen, G, City University of Hong Kong, Hong, Kong, China, 287
Chew, Meng-Sang, Lehgih University, USA, 177
Chin, Kong Suh, University Science, Malaysia, 115
Chongstitvatana, Prabhas, Chulalongkorn University, Thailand, 169
Claesson, I, Blekinge Institute of Technology, Sweden, 97
Cowie, A, Massey University, New Zealand, 245
De Silva, C W, University of British Columbia, Canada, 161
Dillmann, Rüdiger, University of Karlsruhe, Germany, 63
Fernández, José Llamas, C.A.R.T.I.F., Parque Tecnológico de Boecillo, Spain, 105
Fongsamootr, Thongchai, Chiang Mai University, Thailand, 177
Fries, Thomas, University of Ulm, Germany, 153
Fung, Carmen K M, Chinese University of Hong Kong, Hong Kong, China, 125
Gaunekar, Ajit S, ASM Technology, Singapore, 55
García-Bermejo, Jaime Gómez, University of Valladolid, Spain, 105, 263, 341
Garg, Devendra P, Duke University, USA, 87
Gaungneng, Wang, ASM Technology, Singapore, 55
Giesler, Björn, University of Karlsruhe, Germany, 63
González, José R Perán, University of Valladolid, Spain, 263, 341
Håkansson, L, Blekinge Institute of Technology, Sweden, 97
Harris, Harry, University of Southern Queensland, Australia, 253
Ho, J H, Nanyang Technological University, Singapore, 305
Hui, Chen Xiong, ASM Technology, Singapore, 55
Indiveri, Giovanni, University of Lecce, Italy, 13
Jiang, X, National University of Singapore, Singapore, 187
Jin, Tae-Seok, Pusan National University, South Korea, 271
Kaiser, Jörg, University of Ulm, Germany, 153
Khor, Weng Li, University Science, Malaysia, 229
Ko, Jae-Pyung, Pusan National University, South Korea, 271
Kobialka, Hans-Ulrich, Fraunhofer Institute for Autonomous Intelligent Systems, Germany, 13
Kouzani, A Z, Deakin University, Australia, 333

Kumar, Manish, Duke University, USA, 87
Lai, King W C, The Chinese University of Hong Kong, Hong Kong, China, 197
Lau, M W S, Nanyang Technological University, Singapore, 305
Law, Eric, City University of Hong Kong, Hong, Kong, China, 295
Lee, Jang-Myung, Pusan National University, South Korea, 271
Lerones, Pedro Martín, C.A.R.T.I.F., Parque Tecnológico de Boecillo, Spain, 105
Li, Y F, City University of Hong Kong, Hong Kong, China, 3
Li, Wen J, The Chinese University of Hong Kong, Hong Kong, China, 125, 197
Li, Quing, Nanyang Technological University, Singapore, 145
Lim, Chee Peng, University Science, Malaysia, 229
Liu, K P, City University of Hong Kong, Hong Kong, China, 349
Low, E, Nanyang Technological University, Singapore, 305
Lu, R S, City University of Hong Kong, Hong Kong, China, 3
Luk, B L, City University of Hong Kong, Hong Kong, China, 287, 349
Mandava, Rajeswari, University Science, Malaysia, 115
McCarthy, Stuart G, University of Southern Queensland, Australia, 253
Mills, James K, University of Toronto, Canada, 23
Ming, Luo Xiao, ASM Technology, Singapore, 55
Modi, V J, University of British Columbia, Canada, 161
Ng, Chi-Kin, The University of Hong Kong, Hong Kong, China, 323
Olsson, S, Blekinge Institute of Technology, Sweden, 97
Orth, Alexandre, Laboratory of Machine Tools and Production Engineering, Aachen, Germany, 33
Pang, Grantham, The University of Hong Kong, Hong Kong, China, 77, 323
Pettersson, L, Blekinge Institute of Technology, Sweden, 97
Pfeifer, Tilo, Laboratory of Machine Tools and Production Engineering, Aachen, Germany, 33
Phythian, Marc, University of Southern Queensland, Australia, 237
Plöger, Paul, Fraunhofer Institute for Autonomous Intelligent Systems, Germany, 13
Quijada, Salvador Dominguez, C.A.R.T.I.F., Parque Tecnológico de Boecillo, Spain, 263, 341
Ratnam, Mani Maran, University Science, Malaysia, 115, 229
Roloff, Mario L, Federal University of Santa Catarina, Brazil, 33
Sack, Dominic, Laboratory of Machine Tools and Production Engineering, Aachen, Germany, 33
Salb, Tobias, University of Karlsruhe, Germany, 63
Seet, G, Nanyang Technological University, Singapore, 305
Shao, Guifeng, Seinan Gakuin University, Japan, 315
Sholanov, Korgan S, The Kazakh National Technical University, Kazakhstan, 135
Sooraksa, P, King Mongkut Institute of Technology, Thailand, 287
Stemmer, Marcelo R, Federal University of Santa Catarina, Brazil, 33
Stocker, Alan, University/ETH Zurich, Switzerland, 13
Sun, Dong, City University of Hong Kong, Hong Kong, China, 23
Suwannik, Worasait, Chulalongkorn University, Thailand, 169
Takaue, Royichi, University of East Asia, Japan, 315

Tamaki, Akikazu, University of East Asia, Japan, 315
Tan, K K, National University of Singapore, Singapore, 187
Tongpadungrod, P, King Mongkut Institute of Technology, Thailand, 205
Tsang, P W M, City University of Hong Kong, Hong Kong, China, 41
Tsang, W H, Thomson Multimedia (Ltd), Hong Kong, China, 41
Tso, Shiu-Kit, City University of Hong Kong, Hong Kong, China, 23, 287, 349
Widdowson, Gary P, ASM Technology, Singapore, 55
Wong, K H, University of British Columbia, Canada, 161
Wongratanaphisan, Theeraphong, Chiang Mai University, Thailand, 177
Wu, Fang Xiang, University of Saskatchewan, Canada, 145
Xi, Ning, Michigan State University, USA, 125
Xu, W L, Massey University, New Zealand, 245
Yang, Xuezhi, The University of Hong Kong, Hong Kong, China, 77
Yao, Fenghui, University of East Asia, Japan, 315
Yeung, L F, City University of Hong Kong, Hong, Kong, China, 295
Yung, Nelson, The University of Hong Kong, Hong Kong, China, 77
Zhongguo, Gu, Northwestern Polytechnical University, China, 295
Zhu, Jian, City University of Hong Kong, Hong Kong, China, 23

Subject Index

3D calibration, 108
3D modelling, 156

A
acoustic, 227
active vibration control, 97
active-X control, 223
actuators, 163
affine invariant shape matching, 41
agent architecture, 309
agriculture, 237
analog VLSI, 13
Augmented Reality (AR), 63
automated people counting, 323
autonomous robot, 237

B
back error propagation, 89
back-illuminated image, 231
Basic Functional Group (BFG), 135, 138
beacon, 264
behaviour, 344
boring bar, 99
brake disk, 105

C
calligraphy dictionary, 316
calligraphy robot, 315
camera calibration, 89
CAN-bus, 290
centrifugal force, 197, 200
climbing robot, 23
cluster isolation, 327
colour features, 215
communication protocols, 264
compliant motion, 116
computerised vision system, 55
Conic matching, 68
connection graph, 157
control performance index, 149
control, 258

D
date fruits, 213
Decomposed Affine Transform (DAT), 44
defect classification, 81
Denavit-Hartenberg transformation, 139
design for control, 145
dimensional characterisation, 109
discriminant analysis, 234
discriminate, 206, 208, 209, 210
distributive, 205, 206, 210

E
end-effector tool, 118
evolutionary robotic, 170

F
fabric defect inspection, 77
face, 341
facial expression, 333
Fast Fourier Transform (FFT), 299
feedback filtered X LMS algorithm, 99
fitness function, 173
Flexible Manufacturing System (FMS), 116
Flexible Structure Mounted Manipulators (FSMM), 177
flexure bearing, 55
Floating Point Genetic Algorithm (FPGA), 44
focal ultrasound surgery, 360
focussing mechanism, 55
force/torque sensor, 90
force-guided robot, 119
friction, 188
front-illuminated image, 231
fuzzy control, 292

fuzzy logic, 88, 246

G
generic programming, 170
genetic algorithm based contour matching, 43
gesture recognition, 350
GPS navigation, 241
graphical operator interface, 155
gravity compensation, 180

H
hand glove, 351
hardware, 290
harvester, 253
head tracking, 63
hidden Markov model, 350
homography, 5
hue, 224
human robot interaction, 333
hybrid system, 306

I
image cropping, 231
inspection, 213
Inverse Fast Fourier Transform (IFFT), 299

K
Kalman algorithm, 242
kinematic parameters, 141

L
landmark recognition, 68
laser, 264
least squares method, 189
linear frequency modulation (LFM), 297
linear motor, 191
line-to-point method, 3
localisation, 263

M
machine vision system, 35
Macro-micro Manipulator (MMM), 178

mass redistribution, 148
mass transfer, 249
matching broken edge contours, 46-48
meat chiller, 247
mechatronic manipulation system, 362
memoized function, 169-173
micro-assembly, 198
micro-biological systems, 132
micro-camera, 342
Micro-engineered Mechanical Structures (MEMS), 125
micro-manipulators, 126
micro-mirror, 129, 199
minimum error, 78
mobile robot navigation, 281
mobile robot, 287
model template, 327
morphing, 337
moving speed of robot hand, 320
multi-carrier modulation, 295, 298
Multi-layer Perceptron (MLP) network, 233
multimode, 306
multimodule deployable manipulator, 162
multi-sensor integration, 272
Multi-user MEMS Processes (MUMPs), 126, 197-202
multivariable control, 245

N
neural network, 78, 205, 208, 216, 232
neuromorphic, 18
notch-locked assembly, 116

O
omni-directional mirror, 17
optical flow, 17
output link, 137

P
parallel robot, 145
pattern projection, 4

PD control, 148
perspective projection, 325
PID, 292
piezo-actuator, 99
pixel, 224
point-to-point method, 3
polysilicon hinges, 199
Polyvinylidence Fluoride (PVDF), 125
pose recognition, 68
position estimation, 265
pressure control of writing brush, 319
probe-tip, 125, 130
pulse width modulation, 342
quadtree principal components analysis, 336

R
reactive agent, 307
recall rate, 173
recognition, 205, 206, 210, 334
refractometer, 255
relay, 189
remote control, 158
remote operated vehicle (ROV), 295
RGB, 215, 230-232
robot controller, 165
robot design, 164
robot localization, 279
robot model, 166
rotation angle of writing brush, 320

S
segmentation, 324
sensor feedback, 158
sensor fusion transformation, 272
sensor fusion, 88
sensor, 255
sensor-actuator, 20
servo, 343
servo model, 188
short-time discrete Fourier transform, 352

Single-planimetric Multimobile Manipulator (SMM), 135
six-mobile manipulator, 136
space and time filter, 277-278
Space and Time Sensor Fusion (STSF), 271
spectrum vector, 352
speed-up rate, 173
surface inspection, 110
surgical planning and user interface, 364
synchronisation, 297
synchronous imaging, 60
system safety discussions, 365

T
tactile, 205, 206, 208, 210
temperature control, 248
tissue ablation, 361
tool wear parameters, 34
topper, 254
turning operation, 98

U
ultrasonic communications, 295
underwater communications, 295
underwater vehicle, 295
URV, 305

V
ventilation inspection, 109
vibration avoidance, 179
vibration control, 180, 183
vision, 243
visual sensing technology, 24
voice coil motor, 55

W
wavelet, 78
wear detection, 37
wear measurement, 38
wear type and classification, 38
webcam, 221
winner-takes-all, 14